CRITICAL INFRASTRUCTURE STUDIES AND
DIGITAL HUMANITIES

DEBATES IN THE DIGITAL HUMANITIES

Matthew K. Gold and Lauren F. Klein, Series Editors

CRITICAL INFRASTRUCTURE STUDIES AND DIGITAL HUMANITIES

Alan Liu, Urszula Pawlicka-Deger, and James Smithies
EDITORS

DEBATES IN THE DIGITAL HUMANITIES

 University of Minnesota Press
Minneapolis
London

Copyright 2026 by the Regents of the University of Minnesota

All rights reserved. No part of this publication may be reproduced, stored in a retrieval system, utilized for purposes of training artificial intelligence technologies, or transmitted in any form or by any means, electronic, mechanical, photocopying, recording, or otherwise, without the prior written permission of the publisher.

Published by the University of Minnesota Press
111 Third Avenue South, Suite 290
Minneapolis, MN 55401-2520
http://www.upress.umn.edu

Available as a Manifold edition at dhdebates.gc.cuny.edu

ISBN 978-1-5179-1607-7 (hc)
ISBN 978-1-5179-1608-4 (pb)
ISBN 978-1-4529-7488-0 (Manifold)

A Cataloging-in-Publication record for this book is available from the Library of Congress.

Printed in the United States of America on acid-free paper

The University of Minnesota is an equal-opportunity educator and employer.

35 34 33 32 31 30 29 28 27 26   10 9 8 7 6 5 4 3 2 1

Contents

INTRODUCTION
"Object of Study": Digital Humanities and Critical Infrastructure Studies | *Alan Liu, Urszula Pawlicka-Deger, and James Smithies*   vii

PART I
Critical Infrastructure Studies (and Digital Humanities)

    1   Interfaces for the Anthropocene | *Anne Beaulieu*   3

    2   Replatforming | *Susan Brown*   21

    3   Networking the Nation: Settler Colonialism as an Analytic in Critical Infrastructure Studies | *Sarah Montoya*   44

    4   Manifesting Connection: Digital Humanities for the Critical Study of Logistics | *Matthew Hockenberry*   58

    5   Critical Studies of Tech Stacks: What Can Technologies Tell Us About a Lab Culture? | *Urszula Pawlicka-Deger, Arianna Ciula, and Miguel Vieira*   78

    6   Shadow Libraries and Pirate Infrastructures | *Martin Paul Eve*   96

PART II
Digital Humanities (and Critical Infrastructure Studies)

    7   Digital Humanities and the Energetics of Big Data | *Javier Cha and Ian M. Miller*   113

    8   Alternative Infrastructures for Digital Equity: Community-Based Internet Access | *Alex Wermer-Colan, Grant Wythoff, Allan Gomez, and Devren Washington*   132

    9   Understanding Multilingualism in Digital Humanities Infrastructures | *Paul Spence*   151

    10   What's Missing: Studying Digital Humanities and Critical Infrastructure in India | *Maya Dodd and Sharika Parmar*   166

11 Connecting Digital Systems by Whom and for Whom? Taking Stock of the Digital Humanities Infrastructures in China | *Lik Hang Tsui and Jing Chen* — 179

12 Reproducibility and Contestation in Humanities Digital Infrastructure | *Deb Verhoeven, Mike Jones, Toby Burrows, and Ann Borda* — 199

13 Scrounging | *Darren Wershler* — 216

PART III
(Re)envisioning Digital Humanities Infrastructure

14 Resisting BYOI (Bring Your Own Infrastructure) in Digital Humanities Learning Spaces | *Kush Patel, Ashley Caranto Morford, and Arun Jacob (Pedagogy of the Digitally Oppressed Collective)* — 233

15 Making Infrastructure Writable | *Lucie Kolb* — 253

16 Online Feminist Publishing and Content Creation as Feminist Infrastructure in India | *Puthiya Purayil Sneha and Saumyaa Naidu* — 261

17 Digital Humanities from Below: Speculating on Solidarity Infrastructure | *Matthew N. Hannah and Miriam Posner* — 271

18 Imagining a Future of Multimedia E-books | *Sylvia K. Miller* — 278

19 Subjective Functions: How Should Humanistic Research Be Quantified? | *Kyle Booten* — 290

APPENDIX: INFRASTRUCTURE MANIFESTS | *Alan Liu, Urszula Pawlicka-Deger, and James Smithies, Editors* — 301

CONTRIBUTORS — 307

Introduction

"Object of Study"
Digital Humanities and Critical Infrastructure Studies

ALAN LIU, URSZULA PAWLICKA-DEGER, AND JAMES SMITHIES

Critical Infrastructure Studies

Since the late 1990s and early 2000s, infrastructure has become a key interest among humanities and social science scholars well beyond the engineering, planning, and economics fields traditionally concerned with systems of material, built things. Ethnographers of organizational technology and others associated with science-technology studies (STS) led the way by studying "knowledge infrastructures," especially digital ones.[1] Also relevant was the history of technology (HOT) field adjacent to STS, whose research in such areas as large technical systems (LTS) bears on infrastructure studies.[2] Meanwhile, "thing theorists" explored the philosophy, social force, and aesthetics of "things."[3] Then others working on architecture, cities, the environment, various resources (e.g., oil, water, rare earth metals), waste and garbage, transportation and logistics, media (especially "media infrastructures"), feminism, race, ethnicity, postcolonialism, literature and art, and other areas extended "infrastructuralism," as it has also been called, in many directions, including historical infrastructures at the beginning of technological and industrial modernity.[4] Infrastructure has become a capacious, polyvalent framework for addressing the materials, technologies, agents, and actions that make, and also unmake and remake, societies.[5] Beyond academic publications, infrastructure studies also has a vibrant, public-facing presence in writing and photography from many disciplinary perspectives in such online journals as *Places, Scenario, e-flux, Mediapolis, Platform,* and *Society+Space.*

The study of infrastructure thus joins a lineage of previous intellectual paradigms, each of which looked at the world afresh from the viewpoint of a specific object of study, in *object*'s epistemic sense as a theme or topic.[6] Since the nineteenth century, this methodological lineage (eventually more a meshwork than a single line of descent) has taken up objects of study that include the following:

- *Ideas* (in traditions of intellectual history such as Hegelian philosophy and German *Geisteswissenschaften,* Anglo-American history of ideas, and French *histoire des mentalités*)
- *Mind* (from psychoanalytic theory to recent cognitive science)
- *Language* (in structuralist and poststructuralist approaches to linguistics, anthropology, psychoanalysis, literature, and other fields)
- *Design* (as professionalized and theorized after the Bauhaus era in architecture, industrial design, graphic arts, user interface design, and other design fields)
- *Institution* (in organizational and neoinstitutional studies)
- *Ideology and culture* (in cultural studies from the 1960s on)
- *Identity* (in race, ethnicity, gender, postcolonial, and other approaches to social groups and nationalities)
- *Environment* (in environmental studies)
- *Media* (in media and digital new media studies)

Each paradigm directed inquiry in novel directions, advancing new questions and new ways to ask those questions. Each also unpredictably inflected the others. After the linguistic turn of the 1970s, for example, *ideas* looked surprisingly different when seen as *language,* which in turn was transformed when seen from the viewpoint of *mind* (e.g., in Lacanian theory), *design* (e.g., in postmodern architecture), *identity* (e.g., in gendered or racial forms), *media* (e.g., in "post-truth" social media memes), and so on.[7]

As a recently minted object of humanistic and social science study, *infrastructure* is distinctive among the other paradigms that it joins because it surfaces the literal *objects* that, in Martin Heidegger's terms, "enframes" any epistemic object of study as a "standing-reserve" of technologized things "ordered to stand by, to be immediately at hand, indeed to stand there just so that it may be on call for a further ordering."[8] In infrastructure studies, all objects of study are constituted in their primary phenomena, as well as in the secondary instruments used to observe them, by materials, structures, and tools that are built, created, developed, supported, improved, maintained, protected, expanded, strengthened, enabled, implemented, installed, supplied, deployed (some of *infrastructure*'s most common verb collocates)[9] for the purpose, in Heidegger's vocabulary, of being ordered (like pressing an "execute" button) to operate *as* infrastructure. But for infrastructure studies, more existentially, all objects of study and their instruments of study are in an ontological sense part of the order of *being* itself. Thus, for instance, *ideas* and *language,* among the paradigms mentioned here, are shaped qua objects of study through technologies that are existentially part of what it means to be a human being today—whether in the form of human bodies using "verbomotor" techniques (Ong, 66) or in that of extensions of such corporeal expression as writing or print, modern analog information technologies (like the *Zettelkasten* index-card system that Niklas Luhmann famously used),[10] or today's digital information technologies. Infrastructures not only do, but are the core existential action of human being (with all its collocates of building, creating, developing, supporting, and so on among other verb collocates).

Indeed, to expand the scope of this line of thought, infrastructure studies is properly understood as addressing not just the infrastructure behind any epistemic object but, reciprocally, the epistemic in any infrastructure. From this point of view, which accords with actor-network theory (ANT) in STS, it is not just authors but assemblages of bodies-pens-printing presses-index cards-the internet and other entities, including, recently, artificial intelligence (AI) large language models (LLMs), that "know" ideas, mind, language, identity, or anything else.

One other principal characteristic of infrastructure studies is important to mention because it sets the stage for recent developments. Literal objects are never "just" literal, in the sense of being inertly or passively received as a given.[11] Instead, objects are composed as infrastructure through dynamic processes of ontological, epistemological, and social *relations*.[12] While one of the best-known axioms in Susan Leigh Star's "The Ethnography of Infrastructure" (1999), a founding work of infrastructure studies, is that infrastructure "becomes visible upon breakdown," meaning that it is "invisible" (taken for granted) until "the server is down, the bridge washes out, [and] there is a power blackout" (382),[13] the other crucial axiom is that "infrastructure is a fundamentally relational concept" (380), meaning that infrastructure toggles between invisibility and visibility depending on how one person is positioned relative to it versus another.[14] As Star observes, "the normally invisible quality of working infrastructure becomes visible when it breaks" (382) or, equally important, when it comes into focus from a relationally different social viewpoint as a "topic, or difficulty":

> For a railroad engineer, the rails are not infrastructure but topic. For the person in a wheelchair, the stairs and doorjamb in front of a building are not seamless subtenders of use, but barriers. . . . One person's infrastructure is another's topic, or difficulty. As Star and Ruhleder (1996) put it, infrastructure is a fundamentally relational concept, becoming real infrastructure in relation to organized practices. . . . So, within a given cultural context, the cook considers the water system as working infrastructure integral to making dinner. For the city planner or the plumber, it is a variable in a complex planning process or a target for repair (380).[15]

For ordinary consumers or for cooks, therefore, the plumbing and its water source are givens. The system just works. But infrastructure suddenly comes into focus as a "topic, or difficulty" (a perspective shift that Geoffrey Bowker calls "infrastructural inversion") when it stops working—and never more so than when seen from a different social standpoint.[16] Thus, for instance, a city planner might see a broken municipal water system as an opportunity for a contract bid, while marginalized communities afflicted with contaminated water systems or water scarcity—important topics in recent infrastructure studies[17]—must see it as an existential threat.

A powerful idea will eventually arise from fusing the brokenness and relationality axioms of infrastructure studies. Recognizing this idea for what it is requires seeing it in sharp relief against an easily missed, historically specific premise of infrastructure: infrastructure as a *concept* is a modern idea.[18] Of course, built structures of all kinds ranging from small to immense, and simple to complex, certainly existed in prior ages around the world. Massive or intricate architectural, defensive, and transportation structures from prehistoric through classical and medieval times are cases in point. But "infrastructure" as an idea and as a literal word rose into prominence only after modern (especially twentieth-century) military, industrial, technological, organizational, and governmental systems normalized the notion that all things, especially engineered things and the populations that they are supposed to socially engineer, are structured as *systems* through planned and standards-based development, support, improvement, maintenance, protection, expansion, implementation, supply, and other means (repeating here in the nominative previously cited verb collocates of *infrastructure*).[19] This is why, as before, we speak today not just of water but of *water systems* designed around technical standards, regulatory codes, and administrative procedures embedding infrastructure in systems and, seen in overview, enveloping systems of systems (e.g., not just plumbing systems but overall city planning).[20] Borrowed from a late nineteenth-century French term, according to the *Oxford English Dictionary (OED)*, the word "infrastructure" emerged in English c.1927 to designate "substructures" and "foundations" (initially military ones) that are "subordinate parts of an undertaking" (i.e., systems).[21] Infrastructure thus correlates conceptually with modernity, understood as an ipso facto system from as early as the system thinkers and planners of the French Revolution or, later, of Marxian thought (according to which *base of production* is now indelibly tied to the idea of infrastructure) through the heyday of twentieth-century social systems theorists such as Talcott Parsons, Jürgen Habermas, Niklas Luhmann, and, for large technical systems theory, Thomas P. Hughes.[22]

Against that background, the powerful, contrasting idea that today emerges from fusing the brokenness and relationality axioms in infrastructure studies may be put as follows: infrastructure is not just material systems functioning in social systems. It also embodies the relational differences in any social system that are profoundly dysfunctional and antisystemic—like *Kristallnacht* fractures in a pane of glass never meant by underlying or overlording social forces to be anything other than broken. In other words, infrastructure studies fuses the concepts of brokenness and social relationality to depict infrastructure as *systematically antisystemic.* Infrastructure is a system that works for some, but for that very reason, is broken for others. More cruelly, as in the case of city benches made with protrusions to prevent unhoused poor people from sleeping on them, it is *designed* to be broken for others.[23] Systemically antisystemic infrastructure is an extreme example of what Foucault called a *dispositif* (usually translated as "apparatus"): "a thoroughly heterogeneous ensemble . . . of discourses, institutions, architectural forms, regulatory decisions, laws, administrative measures, scientific statements, philosophical, moral and philanthropic propositions" (Foucault, *Power/Knowledge,* 194). In the context of infrastructure studies, heterogeneous ultimately means—beyond temporary, makeshift patchworks of ill-sorted and, in the last instance, conflicting apparatuses—*broken.* For Foucault, after all, *hetero*-anything—e.g., "heterotopia" (Foucault, "Of Other Spaces")—signals the break or fracture in any supposed unity or identity. Detecting the *hetero* in the homogeneous or universal is Foucault's indictment of any fantasy that humanity can ever be systemically self-identical.

As infrastructure studies evolved alongside other humanities and social science approaches, "heterogeneous" next became "intersectional," which has now become an important concern of infrastructure studies. At its inception in the 1990s and 2000s, infrastructure studies was not yet in dialogue with intersectionality theory as introduced by Kimberlé Crenshaw in her 1989 article depicting differences between, and among, racial, ethnic, and gender groups—an article that itself was infrastructuralist at the key moment when, to define intersectionalism, it imagined in a strikingly detailed metaphor a "basement" with people "stacked" one atop another at different levels of disadvantage beneath a ceiling "hatch" (Crenshaw, 151–52). But the very fact that infrastructure studies from its onset emphasized socially relational differences—including the class differences implicit in Star's example of infrastructure's city planners versus plumbers—prepared for many of today's works of intersectional infrastructuralism focused, for example, on feminism, race, ethnicity, postcolonialism, disability, and other areas related to social justice (see these and other relevant topics in Cistudies.org, "Bibliography"). Three recent examples of infrastructure studies can serve as a synecdoche: Nikhil Anand's *Hydraulic City* and "The Banality of Infrastructure" (discussing water infrastructures that harmed marginalized citizens in Mumbai and Flint); Adrienne Brown's *The Black Skyscraper: Architecture and the Perception of Race* (arguing "not only that race proved crucial to this architecture's inception, but that the skyscraper also impeded the perception of race," 2); and Ara Wilson's "The Infrastructure of Intimacy" (discussing "intimacy" as "an analytical term in studies of gender, sexuality, kinship, or social relations" [247] in connection with the intimate infrastructures of public restrooms and mobile phones). The following statement in Wilson's richly nuanced essay could be generalized to all such contemporary infrastructure studies simply by varying the adjective in its phrase "*intimate* relations" to extend to any mode of social, and thus necessarily also differential, relations: "As I hope the examples of toilets and phones have shown, infrastructure offers a useful category for illuminating how intimate relations are shaped by, and shape, materializations of power: it offers a vehicle for translating (operationalizing) broader theories of power, system, materiality, space, ideology, and discourse into observations of concrete situations" (263).

The emphasis on relational difference in contemporary infrastructure studies sets the frame for *Critical Infrastructure Studies and Digital Humanities*. As indicated in the title of this new book in the Debates in the Digital Humanities series, that frame is *critical* infrastructure studies. After infrastructure studies expanded beyond its initial STS orientation to an ensemble of disciplines and approaches, the ensemble as a whole increased in dialogic, intersectional, and therefore also critical potential. Any one infrastructural approach can now more readily debate the standpoints and emphases of another. Critical infrastructure studies today thus focuses on "topics, or difficulties" (in Star's apt phrase) reflecting contesting views of *what* infrastructure is, *where* it is sited, *how* it is created and maintained, *why* it exists (and breaks down), *whom* it is for (or threatens), and *if* it is possible to envision better infrastructures—ones that are more just, caring, and sustainable.[24]

Recent infrastructural politics around the world have also raised the stakes terrifically. In North America, for example, U.S. President Donald Trump's border wall, President Joe Biden's Infrastructure Investment and Jobs Act, and the Keystone XL oil pipeline between Canada and the United States all incited controversy. In China, the immense Three Gorges Dam project and the recent overbuilding of housing infrastructure did the same. And in multinational or international regional areas such as Europe (where the Nord Stream 2 Baltic Sea gas pipeline from Russia became a vital concern), Africa, Asia, Europe, and Latin America (the scenes of the Chinese Belt and Road and Digital Silk Road initiatives), or maritime zones (such as in the case of China's Maritime Silk Road initiative or Google's Firmina open subsea cable between the eastern United States and Argentina), infrastructural politics scaled up to geopolitical stakes.[25] Critical infrastructure studies focuses on these and many other infrastructures created in the continual clash—material, conceptual, social, and digital—between power and resistance, constraint and freedom, privilege and want, globalism and localism, design and mess, functionality and breakdown, engineering and repair, standardization and anomaly, technologically closed and open, and many other variances.

Intriguingly, critical infrastructure studies may also advance today's discussion of the powers and limitations of critique itself. The gauntlet was recently thrown down on humanities and social science critique in a movement that may be called the "critique of critique," as forcefully articulated by such thinkers as Rita Felski in *The Limits of Critique* and, a key influence on Felski, Bruno Latour in writings like "Why Has Critique Run out of Steam?" and "An Attempt at a 'Compositionist Manifesto.'" As noted by Felski and others, the critique of critique actually originated as far back as Paul Ricouer's *Freud and Philosophy* in 1970, with its meme about the "hermeneutics of suspicion," or Stephen Best and Sharon Marcus's landmark call for "surface reading" in 2009 and subsequent calls by others for noncritical modes of thought and affect in academic "reading." What is of note in the context of infrastructure studies is that critique-of-critique's arguments against critical *close* reading, reading *under* the surface, reading *behind* manifest content, and so on often rely on what are fundamentally spatial and material paradigms expressed through modifiers and prepositions such as those italicized here. In critique-of-critique, such paradigms can precipitate into full-blown mise-en-scènes of infrastructure.

While Felski's primal scene in her incisive chapter "Digging Down and Standing Back" in *The Limits of Critique* is archaeological ("digging down" imagines an archaeological dig in which critics excavate buried meaning), the scene of Latour's critique of critique is clearly infrastructural. In a Piranesi-like infrastructural reverie, for example, Latour in "An Attempt at a 'Compositionist Manifesto'" describes critique as a misdirected attempt to penetrate the "wall of appearances" of "composited" things (e.g., any technology) to an illusory foundation of reality. By contrast, he characterizes his own "compositionism" as acknowledging that reality is always constructed from chains of prior mediating agents and actions. The exact language in which he makes this argument stages what he terms the "wall of appearances" as a fully infrastructural version of Plato's cave, complete with the tools needed for its construction or destruction (including Latour's version of Heidegger's "ready-to-hand" "hammer" in *Being and Time*):

> It is really a mundane question of having the right tools for the right job. With a hammer (or a sledge hammer) in hand you can do a lot of things: break down walls, destroy idols, ridicule prejudices, but you cannot repair, take care, assemble, reassemble, stitch together. It is no more possible to compose with the paraphernalia of critique than it is to cook with a seesaw. Its limitations are greater still, for the hammer of critique can only prevail if, behind the slowly dismantled wall of appearances, is finally revealed the netherworld of reality. But when there is nothing real to be seen behind this destroyed wall, critique suddenly looks like another call to nihilism. What is the use of poking holes in delusions, if nothing more true is revealed beneath? (Latour, "An Attempt at a 'Compositionist Manifesto,'" 475)

In its metaphors of walls and destroyed walls, hammers and the holes that they make—and even in its prepositions ("behind," "beneath")—this passage may be paired with a film like Michelangelo Antonioni's *Red Desert* (1964), in which infrastructure is not just the mise-en-scène but the main character.

Latour's theory of compositionism (related to ANT) represents one of the major branches of STS. For that reason, remarking on it here closes the circuit opened by our introduction's initial mention of the "knowledge infrastructures" branch of STS. Knowledge infrastructures are certainly "compositions" in Latour's sense. But today, the acknowledgement that they are constructs all the way down (analogous to mythical "turtles all the way down") just opens whole other circuits—other discussions in the ongoing contemporary reconsideration of the powers and limitations of critique. It is thus important to recognize, for example, that Latour's call to "repair, take care, assemble, reassemble, stitch together" overlaps in its vocabulary with such recent areas of infrastructure studies as the "repair and care" movement (e.g., Jackson; Russell and Vinsel). Critical infrastructure studies is an experiment not just in *critiquing-to-break* (to show the antisystemic brokenness of the modern world of infrastructure) but also in *critiquing-to-build-and-repair*—that is, to *make, remake,* and *make whole* again. Critical infrastructure studies—to which the digital humanities (DH), the specific subject of this book, contributes—can be reparative in helping make critique itself whole again.

Digital Humanities and Critical Infrastructure Studies

As reprised in the discussion thus far, infrastructure studies started in the late 1990s and early 2000s, when STS scholars began studying digital "knowledge infrastructures," while in parallel historians of technology worked on "large technical systems." Of course, that was also the period when the spread of personal computers in the workplace and home converged with the popularization of the internet to prompt a Cambrian explosion of networked digital media, platforms, and services. From the 2000s on, therefore, the STS infrastructure studies group increasingly discussed *cyberinfrastructure*—the term that the National Science Foundation's Atkins report of 2003 introduced to refer to "computation, data, information, and networks" for the sciences (Atkins et al., 4); and that the *Our Cultural Commonwealth* report by the American Council of Learned Societies (ACLS) Commission on Cyberinfrastructure for the Humanities and Social Sciences of 2006 extended to the "organized use of networks and computation" in the humanities and social sciences (ACLS, 1).[26] While the word "cyberinfrastructure" began to fall out of general use after 2007,[27] the digital and networked research knowledge infrastructures that it announced expanded across the digital sciences (e.g., in so-called *in silico* medicine or data-sharing astronomy),[28] digital social sciences (e.g., in research on the internet's impact on politics),[29] digital new media studies (e.g., in network critique),[30] and—the particular domain of this book—in the digital humanities.

Critical Infrastructure Studies and Digital Humanities focuses on the longstanding but now-refreshed relation between DH and infrastructure studies. Even when after the 2000s, infrastructure studies widened to encompass many material, social, economic, epistemic, aesthetic, and other infrastructural topics, its original topic of knowledge infrastructures (and digital ones in particular) continued to be prominent. Attention to the digital was no doubt boosted by the sheer novelty at the time of digital and networked applications—a novelty that also led dialectically to pushback in such offshoots of STS infrastructure studies as the "repair and care" movement against the very notion of innovation in Silicon Valley's "move fast and break things" style.[31] But the outsize importance of the digital also owed much to how digital sensing, surveillance, control, communication, storage, analysis, and simulation technologies began to permeate the urban, transportation, and other systems paradigmatic of modern infrastructure so thoroughly that they seemed to become part of the very weave, or World Wide Web, of "postindustrialism," "late capitalism," and "late modernity."[32] If modern infrastructure was a "pillar" of modern systems (to cite an infrastructural metaphor that was central to the neoinstitutionalist approach to organization theory in the 1980s and 1990s), then digital infrastructure is the pillar (now called a "stack") of late modernity as a *digital* system.[33] Digital systems—whether as add-ons, extensions, or full-on "digital twins"[34]—became the peak of what James Beniger called the "control revolution," Manuel Castells updated as the "network society," and James Smithies further updated as "the digital modern." Exemplary are "smart cities," one of the topics of recent infrastructure studies (e.g., Karnoven et al.; Mattern). *Smart cities* refers to city infrastructures that are so inlaid with digital sensor and control systems that they are typically visualized in "cyberpunk" iconography as a ghostly (or "ghost in the shell") wire-frame, virtual architecture superimposed over urban architecture.[35]

Against the backdrop of such digital infrastructural studies, *Critical Infrastructure Studies and Digital Humanities* depicts DH both as a continuance of legacy infrastructure studies and as part of the renewal of that legacy in *critical* infrastructure studies.

Continuance is an important emphasis in this book because infrastructural thought and practice were central to DH from the first. Under its original name of *humanities computing*, DH was infrastructure studies *avant la lettre*.[36] This was true even if at the time—for example, soon after the start of the Text Encoding Initiative (TEI) in 1987 or of the first annual joint conference of the Association for Literary and Linguistic Computing & Association for Computers and the Humanities (ALLC-ACH) in 1989[37]—there were few known connections in either direction between the humanities computing scholarly community and the STS circle that explicitly thematized the study of infrastructure. It was true because of DH's long-standing hands-on thinking, as it might be called, about corpus construction, text-encoding and metadata standards, text-analysis tools, minimal computing, maker or builder studies, media archaeology, and the creation of DH labs, centers, and programs. The photo of a hard drive's damaged, naked platter on the cover of Matthew G. Kirschenbaum's influential media-archaeological book *Mechanisms* (2007) is perhaps perfectly emblematic of DH's interest in infrastructure. So is the recent surge of writings about DH labs, which directly acknowledges the originating infrastructuralism of STS with its studies of labs, medical facilities, and other research environments.[38]

But the *renewal* of infrastructuralism is just as important an emphasis in this book. DH is an important contributor to the current evolution of infrastructure studies as *critical* infrastructure studies. Ironically, this is the case despite the fact that DH was first seen as critically naive in comparison to its sibling field, "new media studies." "Where is cultural criticism in the digital humanities?" it was asked.[39] Typically, after all, new media studies faced outward from its base in academe to criticize digital society at large, critiquing digital and media technical infrastructures as the means toward that end. Exemplary are such branches of new media studies as social-media platform studies (e.g., Helmond, Nieborg, and van der Vlist; Lovink) and media infrastructure studies (e.g., Parks and Starosielski), where incisive infrastructural critique aimed ultimately to critique digital society's overlords (e.g., Facebook or Google). By contrast, DH seemed at first not to have developed ways to merge its critique of academic technologies with cultural critique.[40] DH appeared to be academically introverted, focusing on such scholarly infrastructural technical problems previously mentioned as corpus construction, text-encoding and metadata standards, text-analysis tools, and creating labs and centers.

Yet the critical potential of DH was definitely there, charging up in the academy as if in a battery.[41] If DH focused on technologies of digital scholarship, that also meant that it could critique and change those technologies (and the academic practices that they reflected) in a hands-on way that new media studies in its early phases, commenting at a hands-off distance from the Silicon Valley empires and other regimes that it observed in fascinated horror, typically could not.[42] From the beginning, for example, DH's critical potential was evident in the passionate commitment of many digital humanists to developing scholarly platforms for collaboration and open access that circumvented, or simply ignored, the offerings of both the overlord FANG companies (an acronym originally encompassing Facebook, Amazon, Netflix, and Google; since expanded to include firms like Apple) and the so-called EdTech industry.[43]

With regard to collaboration infrastructures, for instance, a notable case of do it yourself (DIY) in DH is the series of collaborative and open-comment forums and publishing platforms that originated in whole or part from the CUNY Academic Commons and ultimately resulted in Knowledge Commons (formerly named Humanities Commons) and the Manifold online publishing system (the digital platform of the Debates in Digital Humanities series).[44] Another notable case is the Zotero bibliography platform from the Roy Rosenzweig Center for History and New Media, whose features include collaborative group libraries. Yet another example is the Mukurtu content management system (CMS), which was originally developed by the scholars Kimberly Christen and Craig Dietrich in collaboration with members of the Indigenous Warumungu community in Australia. Mukurtu is a unique platform that adapted the Drupal CMS for the purpose of sharing Indigenous cultural heritage according to "cultural protocols" based on "traditional knowledge" conventions permitting selective access to materials depending on a group's understanding of who has culturally appropriate access (e.g., the public or only a member of the group, a woman, a man, or others).[45] Mukurtu is a case study in alternatives to mainstream Western notions of collaboration and openness on the internet, which typically assume the dominance of a blend of liberal, libertarian, capitalist, or what Richard Barbrook and Andy Cameron term "Californian ideology" cultural protocols.

And with regard to open access infrastructures, breaking down the paywalls of closed publication empires has from the first been DH's version of storming a university administration building in the 1970s. Just as during the French Revolution it was the ethos of *power to the people* that razed the national records in the Bastille and led to a new National Archives along with a whole new theory of archives (*respect des fonds*), so in DH it was *knowledge to the people* that led to the creation of open digital archives hardly recognizable as archives at all to professional archivists[46]—for example, archives practicing what, in an insightful article actually discussing the French Revolution, Jefferson Bailey (the Internet Archive's director of archiving and data services) calls "*disrespect des fonds*." (Chapter 6 in this book, by Martin Paul Eve, discusses what may be the ultimate example of *disrespect des fonds*: "shadow libraries."[47]) One important difference from the French Revolution, however, was that DH sought less to raze the Bastilles of academic knowledge infrastructure—research libraries—than to collaborate with library and information science's own progressive initiatives for open access, fair use (in U.S. copyright law), the extension of fair use to nonconsumptive data analytics, and so on.[48]

DH focused on critiquing knowledge infrastructures in its own academic backyard. But the ethos of those critiques cannot be truly understood unless seen as directed *through* the academy toward the larger society for which the academy serves both as tutor—a role that is so far from being only symbolic that it directly entrains hundreds of millions of students each year[49]—and, along with such other institutions as journalism and nongovernmental organizations (NGOs)—as guardian, watchdog, or, at minimum, *alternative* mode of work and governance. In other words, DH joins other academic fields, especially in the humanities and social sciences, in pursuing one of the most fundamental aspirations of the academy: to conduct scholarship in a manner that models how scholars believe society at large should ideally function to resolve difficult issues of fairness and truth. In this sense, the academy itself is for DH the "medium" for thinking about society; and digital academic infrastructures are thus DH's equivalent of "new media studies." After all, what philosophers of science call "epistemic virtues" (values for the conduct of scholarly knowledge, such as rationality, rigor, thoroughness, openness, and richness) are always also social virtues, especially when instituted in shared governance and other organizational forms of collaboration or openness, even if these themselves require vigilance, as in critical university studies.[50] When digital humanists code or critique in the spirit of collaboration or openness, they are trying to change the Heideggerian "hammer" of digital technologies and its EdTech extensions into an infrastructure that is true not just to the institution but to the *spirit* of the humanities as it contributes to society. What would a humane, nuanced, supple, or empathetic hammer be? Such are questions for the digital humanities.

In short, DH's role is to help shape digitally smart but also ethical academic infrastructures that can advance normative scholarly work while, through such work, critically transferring "the best," and therefore not all, values and practices in both directions between higher education and today's other powerful institutions in society: business (including big tech), law, medicine, government, media, and creative industries, among others. This is DH's charge in helping higher education fulfill its general critical mission of being relationally different from other major social institutions, and not just a pale copy of them. DH develops and studies infrastructure as a standing reserve not only of instrumentalized and ordered technology designed for some master plan, but of technology in the service of different plans—which is to say, ultimately, of difference as itself a design principle. By calling on such digital-technology and DH principles for working across differences as Findable, Accessible, Interoperable, Reusable (FAIR); Collective Benefit, Authority to Control, Responsibility, and Ethics (CARE Principles for Indigenous Data Governance); collaboration; public humanities; global humanities; feminism; postcolonialism; multilingualism; and so on,[51] DH practices what artificial intelligence (AI) computer scientist and cultural critic Philip Agre influentially calls "critical technical practice."

(Infra)structure of this Book

Critical Infrastructure Studies and Digital Humanities focuses on DH in relation to current critical infrastructure studies. In doing so, it builds on and accompanies other writings in DH, beginning with such relatively early works addressing DH infrastructure as Sheila Anderson's "What Are Research Infrastructures?" (2013). Volumes of the Debates in the Digital Humanities series itself are important precedents, including *Making Things and Drawing Boundaries* (ed. Jentry Sayers, 2017), *The Digital Black Atlantic* (ed. Roopika Risam and Kelly Baker Josephs, 2021), and *People, Practice, Power* (ed. Anne McGrail, Angel David Nieves, and Siobhan Senier, 2021).[52] These and other books in the Debates in the Digital Humanities series, along with their original calls for papers, addressed such topics as "digital humanities and its institutions and infrastructures" (McGrail, Nieves, and Senier, "Introduction"); DH's "uneven distributions of [infrastructural] resources—on national, institutional, organizational, and cultural levels" (Gold and Klein); and makers' or builders' "conceptual matter" of "humanities scholarship as built, assembled, or constructed" (Sayers, "Introduction" and "CFP: Making Things and Drawing Boundaries"). Patrik Svensson's long-standing focus on DH infrastructure is also notable, including his institutional work (e.g., directing Humlab at Umeå University), event organization (e.g., "Humane Infrastructures" at the University of California, Los Angeles [UCLA] in 2020 ["Curated Events"]), and scholarly writings (e.g., "The Humanistiscope: Exploring the Situatedness of Humanities Infrastructure" and *Humane Infrastructures*).[53]

Critical Infrastructure Studies and Digital Humanities also follows in the wake of many DH conferences and workshops on infrastructure studies in recent years, including the following:[54]

Introduction [xix

- Infrastructure | Space | Media, Umeå University (2012)
- The Frontiers of DH: Humanities Systems Infrastructure Workshop, University of Canterbury, New Zealand (2015)
- DH Infrastructure Symposium at UCLA (annual workshops started in 2015)
- Creating Feminist Infrastructure in the Digital Humanities Panel, DH 2016 (2016)
- Interrogating Infrastructure Symposium, King's College London (2016)
- Romanticism and Critical Infrastructures Studies Seminar, North American Association for the Study of Romanticism (NASSR) conference (2018)
- Critical infrastructure Workshop, King's Digital Lab, King's College London (2018)
- Critical Infrastructure Studies Special Session, Modern Language Association convention, New York City (2018)
- Workshop on Humanistic Infrastructure, Stockholm (2018)
- Radical Transparency Infrastructure Design Workshop, King's College London (2018)
- Public Humanities Infrastructure: A Post-Harvey Introduction to Critical Infrastructure Studies Panel, Rice University (2019)
- Humane Infrastructures, UCLA (2020)
- Digital Humanities & Critical Infrastructure Studies Workshops coorganized by King's Digital Lab, King's College Department of Digital Humanities, and CIstudies.org ("Infrastructural Interventions"; "Interrogating Global Traces of Infrastructure") (2021)

More generally, *Critical Infrastructure Studies and Digital Humanities* builds on and helps consolidate work on such topics of DH critical infrastructure studies (some previously mentioned) as the following:

- *Cultural heritage infrastructures* (see, e.g., Benardou et al.)
- *DH and media labs* (e.g., Oiva and Pawlicka-Deger; Wershler, Emerson, and Parikka; and Pawlicka-Deger and Thomson)
- *Environmental digital humanities* (e.g., Baillot et al.; Miya, Rossier, and Rockwell; Ryan, Hearn, and Arthur; and Digital Humanities Climate Coalition)
- *Feminist infrastructures* (e.g., Brown et al.; McPherson; and D'Ignazio and Klein)
- *Labor and infrastructure* (e.g., Graban et al.)
- *Maker and builder studies* (e.g., Sayers) and *experimental media archaeology* (e.g., Fickers and van den Oever, *Doing Experimental Media Archaeology: Theory* and "Experimental Media Archaeology"; and Heijden and Kolkowski)

- *Minimal computing*, or "computing done under some set of significant constraints of hardware, software, education, network capacity, power, or other factors" (Minimal Computing)
- *Open access* (e.g., Eve; and Eve and Gray)
- *Postcolonial and transnational digital infrastructures* (e.g., Risam and Gairola; and Smithies, Flohr, Bala'awi et al.)
- *Sustainable research infrastructures* (e.g., Edmond and Morselli; Gourley and Viterbo; Smithies, Westling, Sichani et al. and Tucker)

To extend and innovate on such works and topics, *Critical Infrastructure Studies and Digital Humanities* is organized into three sections with different emphases, as described next. Of course, as Bowker and Star demonstrated in *Sorting Things Out: Classification and Its Consequences* (an influential work of the STS and ethnography circles of infrastructure studies), classifying chapters or anything else involves unresolvable conflicts in understanding and priorities. It is useful to array chapters according to *some* scheme of difference. But overlapping themes and arguments among the chapters mean that it is disputable whether some chapters belong in one section versus another in the scheme. As editors of a volume in the DH field, indeed, we wondered whether computational modeling might assist as what would amount to a fourth editor, providing a different opinion about how the volume should be ordered. Indeed, we experimented with machine learning for this purpose, using topic modeling, K-means and dendrogram clustering, and other kinds of querying for similarity to explore lexical and thematic chapter groupings that might be a different way of "sorting things out."[55] For example, topic modeling showed that chapters 11 (Tsui and Chen), 2 (Brown), 5 (Pawlicka-Deger, Ciula, and Vieira), and 18 (Miller) are closely associated with a topic whose most frequent words are *digital, project, platforms, platform, technology, software, institutional, press, services,* and *based*—a similarity that suggests that these chapters could be the nucleus for a section in an alternative scheme for this book (whereas in the organizing scheme that we settled on, they are distributed across three sections). Or, again, K-means analysis (see Figure I.1) showed that the following chapters form distinct clusters:

- Chapters 3 (Montoya), 18 (Miller), and 19 (Booten)
- Chapters 1 (Beaulieu), 6 (Eve), and 14 (Patel, Caranto Morford, and Jacob)
- Chapters 4 (Hockenberry) and 7 (Cha and Miller)
- Chapters 2 (Brown), 8 (Wermer-Colan, Wythoff, Gomez, and Washington), and 12 (Verhoeven, Jones, Burrows, and Borda)

These groupings could also be the nuclei around which to accrete sections in an alternative version of this book. It would be intriguing to follow up with a close reading of the chapters involved to seek the significance of their lexical and thematic

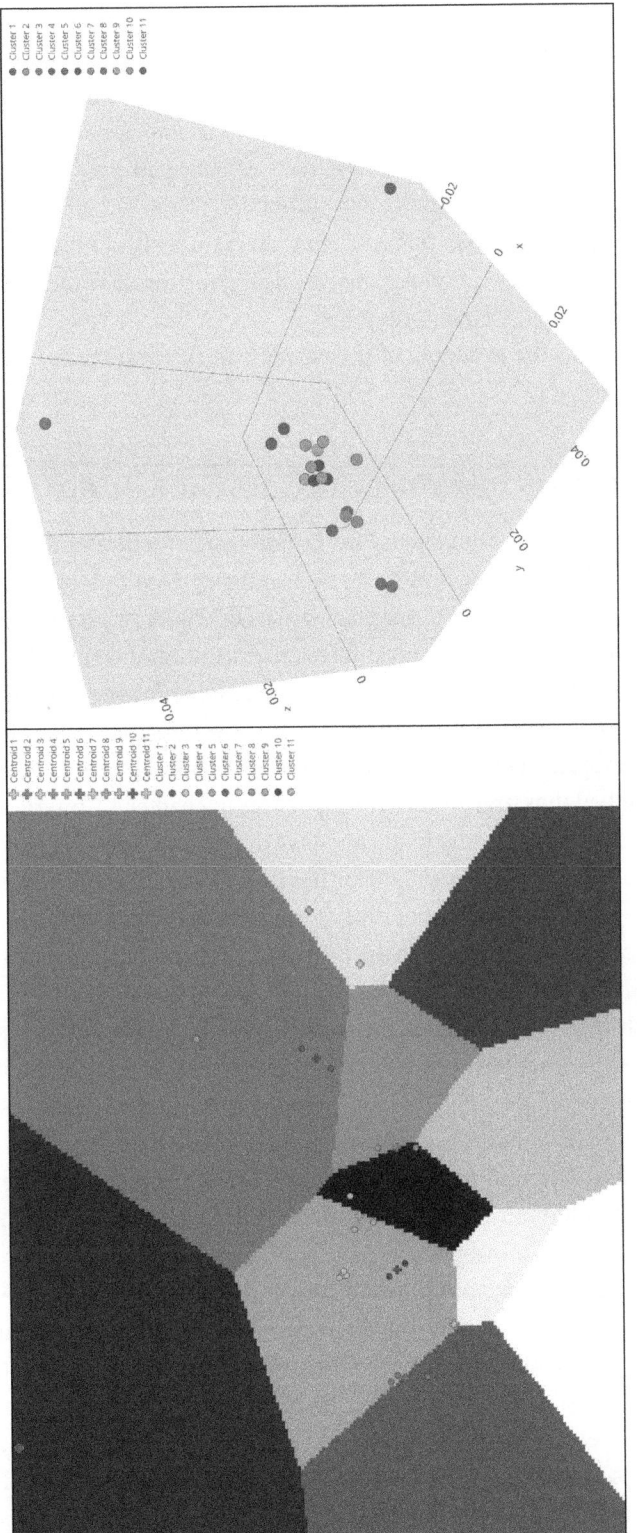

Figure I.1. K-means cluster analysis of lexical similarity of chapters in this book, visualized in Voronoi space (left) and 3D space (right). Nodes represent chapters clustered nearer or farther from other clusters of chapters based on statistical analysis of their words. Created using Lexos; Lexos v4.0 © 2019 Wheaton Lexomics.

similarities that the computer noticed but that we as human editors did not, or that we unconsciously suppressed to bring forward other themes.

In the end, however, we did not pursue the machine learning exercise further because we had higher confidence in our judgment as domain experts in the DH field working with a small document set that we had read deeply (down to the level of "track changes" in word-processing programs). However, as scholarly books increasingly move to online platforms, there is no reason why it might not be good practice in the future to use computational modeling to provide readers with selectable playlists of different chapter orderings.

The following is the grouping of the chapters in this volume as settled on by its human editors.

CRITICAL INFRASTRUCTURE STUDIES (AND DIGITAL HUMANITIES)

The first section of this book explores how critical infrastructure studies provides a wider context for DH—that is, how issues such as the environment, decolonization, logistics, digital platforms, and proprietary versus open-access knowledge, among other themes in contemporary critical infrastructure studies, appear from, or are inflected by, the viewpoint of DH (and, in the case of some chapters, also new media studies). As the parenthetical "(and Digital Humanities)" in this section's title represents, DH here appears nested in the larger frame of critical infrastructure studies.

Anne Beaulieu thus draws on DH perspectives in chapter 1, "Interfaces for the Anthropocene," to reflect on the implications of digital interfaces in the Anthropocene age. She argues that the tendency for interfaces to fall into the background of our awareness poses risks for responding to environmental change.[56] In reflecting on the interfaces of such projects as the Svalbard Integrated Arctic Earth Observing System (SIOS), she "aims to find effective ways to question and stop reproducing an architecture of universalism and frictionless circulation of knowledge, an architecture that furthermore hides the uneven distribution of violence and profit." Beaulieu's goal is to use DH to "propose a route to better interfaces through an examination of how different epistemic values might be inscribed in interfaces, enabling (but not determining) different interactions." Importantly, this extends to decentering the human subject in interface design and privileging nonhuman actants affected by the Anthropocene.

In chapter 2, "Replatforming," Susan Brown draws on digital humanities initiatives such as Project Orlando, the Canadian Writing Research Collaboratory (CWRC), the Linked Infrastructure for Networked Cultural Scholarship (LINCS), and others to contribute to the critical discussion of "platforms." Drawing on insights from the extensive work that she and collaborators have done on scholarly digital platforms, she discusses the unexpectedly complex ethics of "re-platforming" cultural data by republishing it on a different platform, and the sustainable and curative potential—curatorial in an expanded sense—for "replatforming" (without a hyphen). For her, the prefix "re" in the latter "replatforming" should align in its ethos with that of the same prefix in such watchwords of careful and caring digital infrastructural work as "repair" and "rebuilding." Recognizing all the responsibilities of platforming, she concludes, opens "a path to better situated and more ethical knowledge production and circulation."

Sarah Montoya's chapter 3, "Networking the Nation: Settler Colonialism as an Analytic in Critical Infrastructure Studies," contributes to postcolonial, anticolonial, and decolonial approaches in recent DH. She explores how the assumptions and biases of settler colonialism—from the time of the telegraph to that of electronic and digital colonialism—inform (in both generic and specific senses) the infrastructure of U.S. information and communications technology. Historicizing technical infrastructure in this way, Montoya argues, allows us to "confront and dismantle narratives of settler technological supremacy and consider anticolonial and decolonial approaches in critical infrastructure studies." Unlike those who during the early era of the internet and cyberlibertarianism relied on the trope of the digital "frontier," she suggests, infrastructure designers need to resist "the framing of cyberspace-as-frontier space" and imagine new forms of decolonial infrastructure—ones that reimagine the world created by settler colonialism. Montoya ends by asking: "What if, instead of a worlding practice that is predicated on death to preserve settler subjectivity, we ask ourselves how we can create a world that promises protection and life to those currently rendered most vulnerable?"

Another kind of historical and contemporary worlding—that of supply chains and trade networks—is Matthew Hockenberry's topic in chapter 4, "Manifesting Connection: Digital Humanities for the Critical Study of Logistics." He draws on DH and other projects for visualizing supply chains and trade networks (including the Manifest tool, descended in part from his earlier Sourcemap) to think about modern "supply chain capitalism" and its impact on the world. Positioning his argument and practice as part of the emerging field of "critical study of logistics," he argues for digital and mapping models "centered around the act of 'manifesting connection'—not only of logistical networks, but of the external points of contact that situate these networks in broader patterns of sociality, while also outlining the ethical dimensions at stake and the future challenges that will emerge."

Urszula Pawlicka-Deger, Arianna Ciula, and Miguel Vieira's chapter 5, "Critical Studies of Tech Stacks: What Can Technologies Tell Us About a Lab Culture?," then also foregrounds digital visualization in making its argument. To provide an ethnographic observation (inspired in part by STS research) of the complex "technology stack" at the King's Digital Lab (KDL) at King's College London, they create an interactive network visualization to explore a detailed dataset of the industry sectors, developers, functions, and intellectual property status (proprietary or open source) associated with the lab's ensemble of tools. The result is a case study for understanding DH knowledge infrastructure in its complex weave of conflicting needs and constraints. Studying the play of forces between institutional and technological actors at King's College London—where *play* here means something like the flexibility in movement among machine parts that accommodates inexact tolerances (as when we say, "There should be some play in that shaft coupling")—Pawlicka-Deger, Ciula, and Vieira come to broad conclusions about the balance to be maintained in university knowledge infrastructures between proprietary and open technologies. "By observing the combination of everyday technological and human practices at KDL," they say, "we can gain insights into what we call KDL's *open-source pragmatism* culture."

At the close of part I, chapter 6, "Shadow Libraries and Pirate Infrastructures," by Martin Paul Eve, then follows by focusing on knowledge infrastructures like Sci-Hub, Library Genesis, Memory of the World, UbuWeb, Monoskop, and AAARG that break the fragile balance of the proprietary and open observed in Pawlicka-Deger, Ciula, and Vieira's chapter 5. Shadow libraries, to continue with the *play* trope, play with knowledge infrastructures not in the manner of the loose play of one machine part coupling to another within tolerances, but more radically of *détournement* in situationism (Debord) or *tactics* in "tactical media" (see, e.g., Garcia and Lovink; and Raley). They subversively throw all the parts off track to bring into question the very alignment of the tracks, which is to say the system. Some of the shadow libraries studied by Eve are playful, in the sense of having an avant-garde sensibility. But others like Sci-Hub are deadly serious, in the sense that they legitimate themselves on the grounds of making medical and other research open as a life-and-death matter. "How broken is our current infrastructure of scholarly communications?" Eve asks in his first sentence, alluding to Star's axiom that infrastructure is invisible until it breaks. Ultimately, he shows how profoundly undecidable the whole idea of brokenness is in critical infrastructure studies. Are paywalled, proprietary knowledge infrastructures broken? Or, from the perspective of the paywalled systems—and of systems in general—is it the shadow libraries transgressing paywalls that are the breakage? Eve's reflections on shadow libraries from a DH perspective are a pitch-perfect example of this book's focus in its first part on "critical infrastructure studies (and digital humanities)."

DIGITAL HUMANITIES (AND CRITICAL INFRASTRUCTURE STUDIES)

The book's second part reverses the emphasis of the first to nest the wider area of critical infrastructure studies parenthetically in discussions focused on DH. The chapters here look primarily at infrastructural thought and practice in DH itself, suggesting how attention to critical infrastructure studies can assist in developing more equitable, responsible, sustainable, and in many other senses "better" DH infrastructure. Such infrastructure, for example, engages with environmental issues, as in chapter 7, by Javier Cha and Ian M. Miller, "Digital Humanities and the Energetics of Big Data"; democratizes internet access in the United States, as in chapter 8, by Alex Wermer-Colan, Grant Wythoff, Allan Gomez, and Devren Washington,

"Alternative Infrastructures for Digital Equity: Community-Based Internet Access"; fosters multilingualism, as in chapter 9, by Paul Spence, "Understanding Multilingualism in Digital Humanities Infrastructures"; widens humanities scholarship and public humanities in India and China, as in Chapter 10, by Maya Dodd and Sharika Parmar, "What's Missing: Studying Digital Humanities and Critical Infrastructure in India," and chapter 11, by Lik Hang Tsui and Jing Chen, "Connecting Digital Systems by Whom and for Whom? Taking Stock of the Digital Humanities Infrastructures in China"; embraces contestation as an ideal of knowledge production, as in chapter 12, by Deb Verhoeven, Mike Jones, Toby Burrows, and Ann Borda, "Reproducibility and Contestation in Humanities Digital Infrastructure"; and cannily navigates university constraints on infrastructure, as in chapter 13, by Darren Wershler, "Scrounging." Attention to critical infrastructure studies can lead to innovative approaches that not only make DH infrastructure more robust but enhance its relevance and responsiveness to local and global challenges.

In chapter 7, Cha and Miller address the significant environmental impact of big data, which is increasingly crucial to digital humanities and other computational work. For instance, they explore the energy consumption of digital storage and DH's ecological footprint. Drawing on examples from East Asia and the history of media, they suggest that attention to historical, technical, and regional nuances can lead to more sustainable DH practices. Their chapter thus contributes to the growing number of studies of the environmental impact of the digital humanities (e.g., Baillot et al.; Miya et al.; and Ryan et al.) and shows how digital humanists can create nuanced, context-specific strategies to mitigate big data's environmental impact.

Wermer-Colan, Wythoff, Gomez, and Washington, focus on community technology projects as an alternative model for public internet access in chapter 8. By detailing their work on Philly Community Wireless (PCW), they showcase how community-driven mesh networks support equitable, sustainable internet access. The infrastructural approach that they advocate helps bridge the digital divide and serves as a platform for community empowerment and activism. Their chapter also highlights the potential of DH to support such initiatives. Working on a community technology project, they write, "has encouraged the digital humanists within PCW to rethink some of the canonical concepts in their field." "We argue that more digital humanists should join community technology projects and organizations, especially those neighboring their places of employment and residence, not just to see how their theories hold up to the rigor of praxis, but also to ensure that their research and academic institutions help empower the local communities upon which they depend."

In chapter 9, Spence surveys recent advances and continuing challenges in creating a "sociotechnical infrastructure" that can support the recent DH emphasis on multilingualism for "geocultural and linguistic diversity." Looking at projects "driven by both language activists and language technology researchers," he observes that "different models exist for approaching multilingual challenges in DH research infrastructures," including ones focused on "access to language resources," "literacy and ideation," "translation," and "tactical response." Approaching the topic in a way representative of critical infrastructure studies more broadly, Spence argues that such solutions help "redress geolinguistic imbalance in the [DH] field from numerous perspectives such as decolonial/postcolonial studies, modern languages, biocultural diversity and Indigenous pedagogies."

In chapter 10, Dodd and Parmar ask from a DH perspective who is really benefiting from massive investments in commercial and government network services in India. "The story of Indian DH," they observe, "is an extension of public humanities in India." There is thus an opportunity and a pressing need for Indian DH to contribute to the development of national infrastructure in India rather than to be siloed within academic boundaries. This is especially the case in India, where the gap between the nation's many languages and the English-centric internet may widen further because of the emergence of LLMs that are trained almost exclusively on English content. Dodd and Parmar suggest that addressing issues related to AI and LLMs in India will need to occur at a scale beyond what DH can accomplish in its institutional silos.

Next, Tsui and Chen in chapter 11 contribute a broad survey and commentary on digital and scholarly infrastructures in China from a DH viewpoint. Looking at trends, policies, and specific projects in China—and focusing on both China's national and higher education initiatives—they offer "a critical analysis" of research infrastructures "that can link various resources and systems not just for digital scholarship generally, but for the study of China specifically." They contextualize the growth of Chinese digital scholarly infrastructure in relation to developments such as China's New Liberal Arts in higher education and conclude on "future challenges for the Chinese DH community."

Verhoeven, Jones, Burrows, and Borda draw in chapter 12 on their familiarity with DH knowledge infrastructures created in, or associated with, Australia in order to make an innovative, contrarian argument about the aim of such systems. Typically, they point out, knowledge infrastructure systems try to minimize contestation in favor of "infrastructure focused on reproducibility and consensus." But they argue instead that contestation should be a principal aim as "a form of collaborative knowledge production." Drawing on the examples of the Mukurtu content management system (previously mentioned in this introduction) and the Humanities Networked Infrastructure (HuNI), they advocate for "contestability and contestation" that can "bear on digital research and information infrastructure" to create "spaces in which existing meanings can be challenged, and new or alternative meanings and truths constructed . . . to explore new processes of validation in the humanities and sciences alike . . . embracing complexity, multiplicity, and contestation."

Rounding up part II, Wershler in chapter 13 delves into "scrounging" as a vital yet undertheorized and underappreciated "cultural technique" supporting such fields as DH and new media studies that are engaged materially with infrastructure but typically starved for resources. Based not just on his own long experience as a scrounger but on interviews that he conducted with other scrounger-scholars who lead new media or DH programs, centers, labs, and projects, he portrays scrounging as a unique method of resource acquisition and redistribution, distinct from formal institutional processes for creating infrastructure, and even from other informal methods such as gleaning and poaching. Scrounging varies in definition and perception across disciplines and cultural contexts. But Wershler argues that it is an activity that is crucial for facilities in new media and DH that many university infrastructures, designed to pump big money into the sciences, do not adequately support. In a spirit somewhat like that of early digital (and, in this case, material) "hacking," Wershler memorably concludes: "If university infrastructure is going to function more efficiently, sometimes its guardians need to look the other way for a moment as a researcher scoops up a box of heterogeneous oddities from the floor of a back corridor and scurries away with it, scarcely believing their luck."

REENVISIONING DIGITAL HUMANITIES INFRASTRUCTURE

To complement the primarily analytical, theoretical, and critical work of the chapters in the book's first two sections, a final section includes briefer chapters that conclude the volume more speculatively, creatively, or playfully (thus segueing from the playful zest of Wershler making a virtue of scrounging). These final chapters include position statements, what-if scenarios or plans, and multimodal arguments (visualizations, infographics, other forms of graphics, and a Twine narrative). They argue for, and in some cases enact, critical infrastructure studies by imagining *con brio* how DH infrastructures could be different and better.

In chapter 14, "Resisting BYOI (Bring Your Own Infrastructure) in Digital Humanities Learning Spaces," Kush Patel, Ashley Caranto Morford, and Arun Jacob of the Pedagogy of the Digitally Oppressed Collective present a fictitious narrative set in higher education institutions in India, the U.S., and Canada during the Covid-19 pandemic. Their story unfolds in a Twine narrative of fictional emails between campus administrators, department chairs, faculty members, students, and others. As the pandemic unfolds and remote-working takes over, "responsibilization" (whereby neoliberal corporatization places pressure on individuals to use their own resources to keep things running) and differences in infrastructure across institutions and classes become more apparent. Ultimately, Patel, Caranto Morford, and Jacob argue in a powerful indictment, responsibilization—"a legacy and aspect of capitalist colonialism"—is "a supposed freedom that strips others of their agency and safety."[57]

In chapter 15, "Making Infrastructure Writable," Lucie Kolb is inspired by two experimental search interfaces for library materials—Feminist Search Tools and Infrastructural Manoeuvres—to imagine "the possibility of a 'writable library infrastructure.'" Such an infrastructure would be "a discursive and performative space exposing the ways that our actions and reflections are shaped by infrastructure and creating the possibility for a more emancipatory practice." It would be what she calls the interface as a "thinking infrastructure" able to create "awareness of the infrastructural work and procedures" behind library searching.

Puthiya Purayil Sneha and Saumyaa Naidu in chapter 16, "Online Feminist Publishing and Content Creation as Feminist Infrastructure in India," present an extensive, detailed infographic they created (based on their ongoing research and interviews) of digital infrastructure in India for feminist scholarship. Their goal is "to understand how the engagement with digital spaces has been both empowering and challenging for structurally marginalized communities, where often systemic forms of injustice perpetuate within modern and neoliberal frameworks." Informed by feminist, postcolonial, and global DH, they propose a framework for building a feminist internet in India as "networks of solidarity and care." Indian DH, they observe, has markedly shifted towards socially-engaged inquiries into access, ownership, and regulation of the internet. Ultimately, their vision of feminist infrastructures extends beyond the Indian context, offering a foundation for critical infrastructures rooted in values of openness, accessibility, multimodality, and inclusion.

Matthew N. Hannah and Miriam Posner in chapter 17, "Digital Humanities from Below: Speculating on Solidarity Infrastructure," draw on the inspiration of "history from below" in Marxist historiography to issue a strong call for a "digital humanities from below" attentive to labor issues. Focused on the U.S., their position statement envisions a DH "solidarity infrastructure" "offering a speculative infrastructural space that addresses the needs of information workers as a laboring class, redistributes resources from top to bottom, and employs an implicit commitment to openness." Solidarity infrastructure in DH, they thus "speculate" (with a ghostly, etymological echo we might hear of the "spectre ... haunting Europe" in Marx and Engels), can critique "late capitalism and the higher-education industrial complex."

In chapter 18, "Imagining a Future of Multimedia E-books," Sylvia K. Miller provides a bold infrastructural roadmap for future "routine" publishing of digital or online scholarly books in the humanities and social sciences that integrally include multimodal materials. Based on her extensive experience in scholarly publishing and her consultations with other publishing professionals, Miller offers an authoritative, and not science-fiction, plan for near-future scholarly publishing. In paired *now* versus *future* scenarios, she draws a contrast between what currently happens catch-as-catch-can to incorporate multimedia in books and what in the future could ideally happen. The result is a beginning-to-end blueprint for multimedia authoring, editing, and publishing.

Finally, Kyle Booten in chapter 19, "Subjective Functions: How Should Humanistic Research Be Quantified?" critiques how humanities work fits—which is to say, does not fit—into quantitative metrics for scholarship. His critique is in the form of a creative, speculative plan for a metrics able to gauge the humanities subjectively as much as objectively. Rejecting as inadequate for this purpose such bibliometrics as the h-index and i10-index (and also altmetrics for capturing references to academic work on social media and other online sites), he offers instead as a provocation several kinds of "subjective function" measures complete with visual badges and graphs. His bibliometrics "aspire to subjectivity rather than objectivity, encoding obviously opinionated, controversial, nonuniversal notions of what virtues scholars should manifest," and "attend not to impact but to the gesture, not to the ends but to the means." His "Persistence Score," therefore, measures researchers' "relentless attention to a text or question" over the course of their career. Other scores measure a scholar's tendency to discuss unusual combinations of authors and works, make contrary arguments, and marshal complex ideas and jargon into conceptual simplicity. In the playful spirit of this section of the book, Booten presents his subjective functions not as rigid new data points to be implemented (and gamed) but as thought experiments helping to conceive new alternatives.

Specific approaches and topics aside, *Critical Infrastructure Studies and Digital Humanities* also brings forward for attention three overarching emphases that are important for understanding how critical infrastructure studies bears alike on scholarship and contemporary society. These emphases are *global, social,* and *disciplinary diversity.* As also witnessed in other volumes of the Debates in the Digital Humanities series, these are priorities now widely represented in the digital humanities.

Global: This book includes chapters about, or from the viewpoint of, varied regions and nations around the world. These include the West and global North, where DH has had a relatively long presence. But also included are East Asia (e.g., South Korea in Cha and Miller); South Asia (e.g., India in Dodd and Parmar; and Sneha and Naidu); and Indigenous, First Nations, or Aboriginal peoples in North America (Montoya) and Australia (Verhoeven, Jones, Burrows, and Borda). Additionally, chapters like Hockenberry's on supply-chain logistics or Beaulieu's and Cha and Miller's on environmental concerns address the global in another world-spanning sense. In DH infrastructure studies as elsewhere, attending to the worldwide and (in material, social, and representational senses) world-*making* infrastructures responsible for globalization—knowledge infrastructures like the World Wide Web or ubiquitous concrete ones like shipping containers or, for that matter, literal concrete[58]—opens discursive and performative spaces for critical activities challenging dominant knowledge, social, political, economic, and cultural world systems.

Social: Equally important in *Critical Infrastructure Studies and Digital Humanities* is the sociocultural diversity of its contributors and topics. Authors in this volume include senior and early-career faculty, graduate students, other academic or library staff professionals, and designers. They also include contributors from national, gender, racial, and ethnic backgrounds historically underrepresented in areas of scholarship or practice related to the bricks, mortar, and (today) chips of infrastructure. Topics include ones centered on, or substantially involving, themes of sociocultural diversity and inclusion—e.g., decolonization, Indigenous peoples, underrepresented (and under-networked) communities, feminism, multilingualism, and labor justice.

Disciplinary: In addressing the influence of critical infrastructure studies and the digital humanities on each other, this book engages a wide variety of fields cognate to DH, including (just as some examples): book studies, East and South Asian studies, environmental studies, feminist studies, Indigenous studies, information science, labor studies, library studies, logistics, media studies, platform studies, postcolonial studies, history of technology, and science-technology studies. This disciplinary breadth suggests that critical infrastructure studies for DH can be a crossing point—a bridge, switching yard, or router (among other historical or recent infrastructural metaphors)—across the disciplines of the digital and the humanities. As represented by the chapters in this volume, both critical infrastructure studies and DH route new and unexpected connections between disciplines. And this is not even to mention chapters in this volume such as those by Eve on "Shadow Libraries and Pirate Infrastructures" and Wershler on "Scrounging" that address the destabilization of the very notions of disciplined and institutionally sanctioned knowledge infrastructures. Infrastructure and the digital humanities, Eve and Wershler show, are as much a matter of what Michel de Certeau influentially called street "tactics" as strategic design.

The "topics, or difficulties" of infrastructure, as Star called them, occur across a broad set of global, social, and disciplinary contexts that are foundational for critical infrastructure studies today. *Critical Infrastructure Studies and Digital Humanities* explores how the digital humanities can continue to lay, and renew, its cornerstone in that foundation.

Infrastructure Manifests

A distinctive feature of *Critical Infrastructure Studies and Digital Humanities* is that each chapter is accompanied by an "infrastructure manifest" (sometimes multiple ones for coauthored chapters), declaring the principal infrastructures behind that chapter's creation or, in a few cases, that of the digital platforms, services, or tools that are a chapter's topic. (Manifests can be accessed at https://dhdebates.gc.cuny.edu/projects/critical-infrastructure-studies-and-digital-humanities/resource-collection/infrastructure-manifests. See the appendix, "Infrastructure Manifests," for a fuller explanation.) Such infrastructures include unceded Indigenous land; natural resources; labor; and major platforms, networks, machines, tools, and institutional or other apparatuses. Akin in mission to the "datasheets for datasets" influentially proposed by Gebru et al. for documenting datasets, these manifests are offered to start DH along the path of evolving a shared protocol for declaring the infrastructures used in knowledge production. Of course, infrastructure manifests in this book do not offer definitive answers, as if there were only one set of infrastructures important for DH. They are less a mode of answering than of *asking* what the DH community and others represented in this volume think are key infrastructures to make "visible" (pace Star) as part of DH's "object of study."

NOTES

1. The STS circle of ethnographers and information scientists who inaugurated infrastructure studies beginning in the mid-1990s by studying organizational technologies, information science, and knowledge infrastructures include (alphabetically) Christine L. Borgman, Geoffrey C. Bowker, Paul N. Edwards, Steven J. Jackson, and Susan Leigh Star. For examples of works by these authors, see under each name in this chapter's bibliography. In the present volume, Urszula Pawlicka-Deger, Arianna Ciula, and Miguel Vieira's "Critical Studies of Tech Stacks: What Can Technologies Tell Us About a Lab Culture?" (chapter 5) draws in part on methods influenced by this STS circle, and STS more generally.

2. For key or representative examples of LTS, see Hughes, *Networks of Power* and *Rescuing Prometheus,* and Williams, "Cultural Origins and Environmental Implications of Large Technological Systems."

3. Jane Bennett and Bill Brown wrote influentially on thing theory starting in the early 2000s. Relevant also are Bruno Latour's writings roughly at the same time (1990s through 2000s) on the "Parliament of Things" (*We Have Never Been Modern,* 142–45) and *Dingpolitik* ("From Realpolitik to Dingpolitik"). Latour's STS work on things ("actor-network theory") provides a bridge between the work of the STS infrastructure-studies group (cited previously) and thing theory. It should be noted, however, that there have been surprisingly few linkages between infrastructure studies and another important scholarly context for thinking about objects: object-oriented ontology (OOO) philosophy. An excellent introduction to, and reflection on, OOO to read alongside infrastructure studies is Ian Bogost's *Alien Phenomenology.* For other works on technological things and objects relevant to the notion of infrastructure, see, for example, Verbeek, *What Things Do*; Kroes, *Technical Artefacts;* and Rosenberger, "Technological Multistability and the Trouble with the Things Themselves."

4. For a bibliography of works in infrastructure studies organized topically in these and other areas, see CIstudies.org, "Bibliography" and "Primer." (CIstudies.org is the website of the Critical Infrastructure Studies initiative.) Writings after the inaugural period of STS infrastructure studies that were influential in articulating infrastructuralism as a general approach include Brian Larkin's work in anthropology; a special issue of *Modern Fiction Studies* on infrastructuralism in literature (see the introductory article by Rubenstein, Robbin, and Beal); and Lisa Parks's and Nicole Starosielski's works on "media infrastructures" in media studies (e.g., Parks; Starosielski; and Parks and Starosielski). For an example of the study of historical infrastructures at the onset of the modern technological and industrial era, see Speitz.

5. Cf., Urszula Pawlicka-Deger's reflections on "the multiformity of infrastructure that makes infrastructure a powerful and engaging object with the embedded power to form and re-form social, cultural, and intellectual configurations" ("The Multiformity of Infrastructure").

6. An important recent example of the epistemic usage of the phrase *object of study* is John Guillory's discussion in *Professing Criticism* of the historical and current confusion

about the object of study of language and literary studies. Is the object of language and literary studies—as represented by the Modern Language Association (MLA), for example—language, literature, or society and culture? See the discussion in Guillory, 54 and passim on the "object of study" of modern language disciplines.

7. Richard Rorty's edited volume on *The Linguistic Turn* (1967) marked an early, influential use of the phrase "linguistic turn" to refer to the generalization of the paradigm of language. Jencks's *The Language of Post-modern Architecture* is an example of the impact of the linguistic turn on architectural design.

8. Heidegger, "The Question Concerning Technology," 17. Rereading Heidegger's essay today from the point of view of infrastructure studies brings into prominence its specifically infrastructural examples—for example, also on page 17, see the following observation about transportation infrastructure as a case of "standing-reserve": "Yet an airliner that stands on the runway is surely an object. Certainly. We can represent the machine so. But then it conceals itself as to what and how it is. Revealed, it stands on the taxi strip only as standing-reserve, inasmuch as it is ordered to ensure the possibility of transportation. For this it must be in its whole structure and in every one of its constituent parts, on call for duty, i.e., ready for takeoff."

9. The verbs in this sentence are among the most frequent collocates (among verbs) of *infrastructure* in the Corpus of Contemporary American English (COCA) (Davies)—a corpus-linguistics corpus from English-Corpora.org that "contains more than one billion words of text (25+ million words each year 1990–2019) from eight genres: spoken media, fiction, popular magazines, newspapers, academic texts, TV and movies subtitles, blogs, and other web pages." Specifically, the top verb collocates (in order of frequency beginning with most frequent) are the following: *build, support, provide, create, invest, rebuild, develop, spend, improve, require, destroy, maintain, repair, protect, establish, fund, upgrade, lack, fix, expand, finance, project, damage, strengthen, ensure, enable, target, house, modernize, dismantle, implement, restore, sustain, install, construct, enhance, supply, collapse, facilitate, accommodate, crumble, deploy*. . . . English-Corpora.org's other corpora (e.g., for British English) could be similarly explored for collocates.

10. For a media-archaeological study of index-card systems exemplifying the "analog humanities," see Kil, "Excavating Infrastructure in the Analog Humanities' Lab" and "From the Digital to the Analog Humanities."

11. For an important STS study of the changing history of ideas about "objects" and "objectivity," see Daston and Galison, *Objectivity*. For a recent media-archaeological and literary study of how complex the idea of "literalism" is, see Shoemaker.

12. The word "composed" in this sentence alludes to Bruno Latour's theory of "compositionism" (see, e.g., Latour, "On Technical Mediation" and "An Attempt at a 'Compositionist Manifesto'")—a linkage to STS approaches that could be expanded in a fuller discussion of the relation of infrastructure studies to STS.

13. Cf., Brown, "We begin to confront the thingness of objects when they stop working for us: when the drill breaks, when the car stalls, when the windows get filthy, when their flow within the circuits of production and distribution, consumption and exhibition, has been arrested, however momentarily" (4).

14. Cf., Rosenberger's comment that "One way to move forward here is to consider how the aspects of technology articulated . . . are addressed by accounts that conceive of it in terms of a kind of fundamental relationality. A number of theoretical perspectives can be understood to subscribe to a 'relational ontology,' including feminist new materialism, actor-network theory, embodied and extended cognition, postphenomenology, and, more broadly speaking, critical theory and American pragmatism, among others. . . ." (377). See also McGrail, Nieves, and Senier, "Introduction": "[infrastructure] is also profoundly relational. Seeing digital humanities infrastructure in this way—as a set of evolving relations and dependencies and not merely static resources—supports a critical digital humanities practice that acknowledges institutional constraints and engages in purposeful, reflexive action."

15. See also Star and Ruhleder's "Steps Toward an Ecology of Infrastructure," which Star's "Ethnography of Infrastructure" cites. For a more recent discussion of the relationality of infrastructure in the digital humanities, see Pawlicka-Deger, "Infrastructuring Digital Humanities."

16. Bowker, another founding member of the original infrastructure studies STS group, coined as early as 1994 his phrase "infrastructural inversion" to describe how taken-for-granted infrastructure in the background can suddenly come into focus in the foreground (*Science on the Run,* 10 and passim). The concept and term "infrastructural inversion" is also important in another founding work of infrastructural studies of the 1990s: Bowker and Star's *Sorting Things Out* (1999).

17. See CIstudies.org, "Bibliography," under the heading "Water." Nikhil Anand's writings on water, which make water infrastructures the basis for general theses about infrastructure, are exemplary.

18. On infrastructure and modernity, see Edwards, "Infrastructure and Modernity," esp. 186, 191.

19. While the infrastructure concept arose most prominently and explicitly in the twentieth century (the word "infrastructure" originated in the context of military built structures in 1927 and according to the *OED* and *Google Books, Ngram Viewer* climbed steeply in usage in a general sense after 1960), principal elements of the concept such as standardization clearly descended from a somewhat earlier technological and industrial modernity—for example, from the nineteenth-century invention of standard-parts manufacturing (based on the measurement of parts against systems of "standards"). See Liu, "Transcendental Data," 63–73, on standards-based industrial processes from the time of John Hall's rifle works at Harpers Ferry in the nineteenth century through twentieth-century Taylorism and ultimately postindustrialism.

20. For an example of how the standards, codes, permits, licenses, etc. together define a "water system," see California State Water Resources Control Board, "What Is a Public Water System?"

21. In full, the *OED*'s definition of infrastructure is as follows: "A collective term for the subordinate parts of an undertaking; substructure, foundation; spec. the permanent installations forming a basis for military operations, as airfields, naval bases, training establishments, etc." A paradigmatically conjoined usage of *infrastructure* and *system*

today is the title of the Infrastructure Systems program in Cornell University's College of Engineering, which defines its subject in this way: "Infrastructure systems involves the design, analysis, and management of infrastructure supporting human activities, including, for example, electric power, oil and gas, water and wastewater, communications, transportation, and the collections of buildings that make up urban and rural communities." The *OED*'s hierarchical notion of infrastructure as a system organized into "subordinate parts" is usefully expanded upon in the last sentence of Sheila Anderson's important article, "What Are Research Infrastructures?," which characterizes infrastructures nonhierarchically "as an ecosystem in which the component parts interact, shift and change in a constant process of engagement, adjustment, and readjustment" (21). Other important discussions in infrastructure studies of infrastructure as "systems" include the following. Star writes that "people commonly envision infrastructure as a *system of substrates*—railroad lines, pipes and plumbing, electrical power plants, and wires" ("Ethnography," 380, emphasis added). Brian Larkin observes in "The Politics and Poetics of Infrastructure" that "what distinguishes infrastructures from technologies is that they are objects that create the grounds on which other objects operate, and when they do so they operate as systems" (329). He adds: "Placing the system at the center of analysis decenters a focus on technology and offers a more synthetic perspective, bringing into our conception of machines all sorts of nontechnological elements. . . . the focus is on system building" (330). For a "systems analysis" of humanities and digital humanities infrastructures, see chapter 5 of Smithies's *Digital Humanities and the Digital Modern,* esp. pp. 118–123, on the historical, philosophical, and social science modes of systems-thinking that provide context for studying technological systems. Chapters in the present volume that turn crucially on the idea of "systems" and systematicity of one kind or another include those by Anne Beaulieu (chapter 1) and Matthew Hockenberry (chapter 4).

22. For a sweeping survey, and unsparing critique, of modern systems theory across many fields and applications, see Lilienfeld, "Systems Theory as an Ideology." Edwards's "Infrastructure and Modernity," which in part discusses large technical systems theory, fully embeds the concept of infrastructure within a systems view—as indicated by the essay's subtitle: "Force, Time, and Social Organization in the History of Sociotechnical Systems."

23. See R. Baker's "Broken Infrastructures, Invisible Subjects" for resources related to infrastructures designed to work against groups of people. An instance that Baker cites is infrastructure intended to discourage the presence of the unhoused or homeless. An intriguing artist's project protesting such infrastructure that Baker mentions is Sarah Ross's "Archisuits." *Archisuits* are wearable outfits designed "for specific architectural structures in Los Angeles" so that, for example, they fit over, and allow the unhoused to sleep on, benches with protrusions that otherwise make it impossible to lie flat on them (Ross).

24. Cf., Pawlicka-Deger, "Multiformity": "Infrastructure is made up of social and technical components—community and platforms—and their interrelations disclose divergences and disagreements between them. The quality of infrastructure is built by making connections, enabling participation, and facilitating practices. These values prompt the question of the social side of infrastructure that discloses hidden interrelations between

the system and society, such as *who* is connected, *who* is enabled to participate, and *whose* practices are facilitated. Infrastructure reveals a set of disturbing questions of inclusion/exclusion, connection/disconnection, and scaling up/down in the social configurations." See also Harvey, Jensen, and Morita (quoted in Pawlicka-Deger): "The fact that infrastructure is a divergent phenomenon does not need to lead to mutual indifference among differently invested actors. To the contrary, as we see it, one of the most exciting things about the current STS and anthropological interest in infrastructure is that it *draws into* unfolding conversations an increasingly varied array of infrastructural actors materials, offering an ever-expanding range of resources for thinking and acting."

25. Infrastructure projects like these on the national and geopolitical scale are examples of the intended visibility of infrastructure that Larkin argues is a notable exception to Star's rule that infrastructure is normally "invisible" (334–36). As Larkin puts it: "Invisibility is certainly one aspect of infrastructure, but it is only one and at the extreme edge of a range of visibilities that move from unseen to grand spectacles and everything in between" (336).

26. For the work of the STS infrastructure studies circle on cyberinfrastructure, see, for example, Bowker, Edwards, Jackson et al., "The Long Now of Cyberinfrastructure"; and Bowker, Baker, Miller, et al., "Toward Information Infrastructure Studies."

27. As shown in *Google Books Ngram Viewer*, *cyberinfrastructure* spiked sharply upward in usage during 2000–2007 before appearing to fall off in frequency.

28. For a definition and history of in silico medicine, see Insigneo Institute (University of Sheffield), "About in Silico Medicine." For the importance of data sharing in contemporary astronomy, see Pepe et al.

29. For a representative early instance of the digital social sciences, see the work of Bruce Bimber (e.g., "Information and Political Engagement in America"). For an overview on the digital social sciences in relation to the digital humanities, see Spiro.

30. Manovich's *The Language of New Media* is an early, influential work on new media studies. The Institute of Network Cultures in Amsterdam is an early example of new media studies' interest in network critique.

31. Exemplary of "repair and care" studies are works by Steven J. Jackson; and Andrew Russell and Lee Vinsel. "Move fast and break things" was the early internal motto of Facebook. With regard to the latter, Facebook founder Mark Zuckerberg noted a change in 2014 that bears interestingly on the topic of infrastructure: "We've changed our internal motto from 'Move fast and break things' to 'Move fast with stable infrastructure'" (Levy).

32. Works such as Jameson's *Postmodernism, or, The Cultural Logic of Late Capitalism* (1990) and Giddens's *Modernity and Self-Identity* (1991) use the terms *postmodern, postindustrial, late capitalist,* and *late modernity* to designate different but overlapping social, economic, and cultural or aesthetic features of the late twentieth (and now early twenty-first) century.

33. The infrastructural metaphor of *pillar* became well-known in neoinstitutionalist work on the sociology of organizations when W. Richard Scott in 1995 introduced his analytic of "the three pillars of institutions" (the "regulative," "normative," and "cultural-cognitive") that *make* institutions out of organizations (Scott, 47–71). Refocusing on

digital society mutates such metaphors as pillars into ones like *stacks* based on computer-programming and network architectures. An extreme version of the latter metaphor is Benjamin H. Bratton's universalizing theory of the world and society as what he calls (always with initial caps) The Stack.

34. Used to facilitate design, testing, and manufacture, "a digital twin is a digital representation of a physical object, person, or process, contextualized in a digital version of its environment" (McKinsey & Company, "What Is Digital-Twin Technology?").

35. A typical visualization of smart cities in this style is the image by metamorworks (Getty Images) used in Mondschein, Clark-Ginsberg, and Kuehn's post about smart cities on the Rand Blog. Such visualizations are the descendants of fictional cyberpunk depictions of urban spaces that originated in the mid-1980s in Ridley Scott's *Blade Runner* film and William Gibson's *Neuromancer* novel. Gibson's novel explicitly foreshadowed the smart-city imaginary in its now canonical metaphor of cyberspace as "city lights receding": "Cyberspace. . . . A graphic representation of data abstracted from the banks of every computer in the human system. Unthinkable complexity. Lines of light ranged in the nonspace of the mind, clusters and constellations of data. Like city lights, receding. . . ." (51; second ellipses are Gibson's). "Ghost in the shell" alludes to a Japanese cyberpunk story and its offshoots that spread internationally from manga to television, films, video games, and other media. The main character in the fiction franchise is a person whose mind is transplanted as a cyberbrain into a full-body cyborg prosthesis. See Wikipedia, "Ghost in the Shell," May 1, 2025, https://en.wikipedia.org/w/index.php?title=Ghost_in_the_Shell&oldid=1288199461.

36. Previously referred to as *humanities computing,* DH took its new name, *digital humanities,* in the 2000s. The rebranding is usually traced to the publication in 2004 of *A Companion to Digital Humanities,* edited by Susan Schreibman, Ray Siemens, and John Unsworth.

37. The first joint ALLC-ACH conference occurred in 1989 under the name "16th International ALLC Conference and 9th ICCH" and focused on the theme of "The Dynamic Text." ALLC-ACH conferences later became known as today's annual DH conferences with calendrically iterated names such as "DH 2024." (Thanks to respondents on the Humanist list, including Willard McCarty, Geoffrey Rockwell, John Bradley, Maurizio Lana, Max Kemman, and Manfred Thaller, for their assistance in confirming the title and date of the first joint ALLC-ACH conference. See Alan Liu, "Re: [Humanist] 37.367: name of the first ALLC/ACH conference," Humanist, January 4, 2024, 18:12:47, archived in "Humanist Archives: Jan. 5, 2024, 8:31 a.m. Humanist 37.369," https://dhhumanist.org/volume/37/369/.)

38. For the original work of the STS infrastructure studies circle on labs and related research, medical, and other facilities, see, for example, Star and Ruhleder (who studied genetics research facilities) and Bowker and Star's *Sorting Things Out* (especially the chapters on nurses in medical organizations). Recent writings in DH about labs include the special issue of *Digital Humanities Quarterly* on "Lab and Slack" (see the introduction by Oiva and Pawlicka-Deger); Pawlicka-Deger and Thomson's *Digital Humanities and Laboratories* (2023); and other work by Pawlicka-Deger and her colleagues (see Pawlicka-Deger

in this chapter's bibliography). See also Wershler, Emerson, and Parikka's work on labs in media archaeological and DH contexts: *The Lab Book* (2022). In the current volume, Pawlicka-Deger, Ciula, and Vieira, "Critical Studies of Tech Stacks: What Can Technologies Tell Us About a Lab Culture?" (chapter 5) is a new instance of scholarship about DH labs that makes KDL its case study. Additional analyses of KDL can be found in Smithies and Ciula, "Humans in the Loop: Epistemology & Method in King's Digital Lab" (2020); and Smithies, Ffrench, and Ciula, "*Droit de cité*: The Digital Lab as Digital Milieu" (2024).

39. See an early statement on this by Liu, "Where Is Cultural Criticism in the Digital Humanities?"

40. For examples of previous writings by the present authors themselves contributing to this perception of DH, see not just Liu's "Where Is Cultural Criticism in the Digital Humanities?" (cited in note 39), but Smithies, *The Digital Humanities and the Digital Modern* and "The Dark Side of DH"; and Smithies et al., "*Droit de cité:* The Digital Lab as Digital Milieu."

41. This argument about DH reflects a reconsideration and restatement by one of the editors of this volume, Alan Liu, of his essay of 2012: "Where Is Cultural Criticism in the Digital Humanities?"

42. Recent projects by media infrastructures scholars such as Nicole Starosielski contradict the simplification that new media studies approaches need to be at a distance or be hands-off from making an impact outside the academy. Starosielski's work, for example, increasingly involves working with and in relation to industry to influence technological development. See, for example, the Sustainable Subsea Networks research initiative that she coleads, which is an international "academic-industry partnership."

43. The digital humanities community diverged early from the "computers and composition" and education studies communities, which have been more aware of and engaged with EdTech. Thus, there have been relatively few direct engagements in the DH field itself (in its projects, events, and writings) with such associations concerned with EdTech as EDUCAUSE, the nonprofit for "technology, academic, industry, and campus leaders" dedicated to "the strategic use of technology and data to further the promise of higher education" (EDUCAUSE, "About EDUCAUSE"). (Some DH scholars, however, have participated in authoring informative writings about DH for EDUCAUSE; see, for instance, Dombrowski and Lippincott; Lippincott et al.) DH has also rarely engaged with the work of noted critics of EdTech, such as Audrey Watters (see, e.g., *The Monsters of Education Technology*).

44. Thanks to Matthew K. Gold for confirming the history by which the CUNY Academic Commons descended through Commons in a Box to MLA Commons and Humanities Commons (later renamed Knowledge Commons) (personal communication, December 27, 2019). On the origin of the CUNY Academic Commons, see Gold and Otte. Manifold originated in a collaboration between the CUNY Graduate Center's Digital Scholarship Lab and the University of Minnesota Press and Cast Iron Coding (Manifold, "A Brief History").

45. On Mukurtu and cultural and knowledge protocols, see Mukurtu CMS, "Our Mission"; and Christen, "Does Information Really Want to Be Free?" See also discussions of Mukurtu by Shepard, and by Machulak. In the present volume, see chapter 12 by Deb

Verhoeven, Mike Jones, Toby Burrows, and Ann Borda, "Reproducibility and Contestation in Humanities Digital Infrastructure," for an extended discussion of Mukurtu.

46. See Kate Theimer for the bemused reflections of an archivist on what can appear to be the complete disconnect between established notions of an "archive" and DH's "break-all-rules" notion of archives.

47. In general, Eve's chapter in this volume and his other work on open access (e.g., *Open Access and the Humanities*) show how open-access issues continue to be formative for DH.

48. For explanations of the nonconsumptive use of copyrighted materials for data analytics, see HathiTrust; and Samberg and Hennesy.

49. In 2024, "254 million students are enrolled in universities around the world—a number that has more than doubled in the last 20 years and is set to expand" (UNESCO). For example, 15 percent of the U.S. population were higher education students in 2024 (Hanson).

50. On epistemic virtues, see, for instance, Peels. For examples of "critical university studies," see the book series of that name from Johns Hopkins University Press (Williams and Newfield).

51. For the FAIR principles, see the report of the ALLEA E-Humanities Working Group (Harrower et al.). For the CARE principles, see Carroll, Garba, Figueroa-Rodríguez et al. Many of the other principles mentioned are represented in the present volume (see the description of chapters that follows).

52. The volume edited by McGrail, Nieves, and Senier focuses on what its "Introduction," which contains extensive discussion of infrastructure, calls the "human side of digital humanities . . . infrastructure" (e.g., on what one of the book's sections titles "Human Infrastructures: Labor Considerations and Communities of Practice").

53. Svensson's book, *Humane Infrastructures,* is now published but was still forthcoming from MIT Press at the time the present book was written and completed.

54. These are among the events currently known to the authors of this chapter, who participated in or organized some of them. The following are URLs for the events' websites or materials: https://patriksv.com/curatorship/#infrastructure; https://web.archive.org/web/20151116061420/http://dh.canterbury.ac.nz/blog/2015/11/10/the-frontiers-of-dh-humanities-systems-infrastructure/; https://humtech.ucla.edu/symposium/; https://dh2016.adho.org/abstracts/233;https://docs.google.com/document/d/1M4ofHwM7exM9qnVdIx4LZbLsi3s-AFoA124r0L-pvBA/edit?usp=sharing; https://cistudies.org/events/romanticism-and-critical-infrastructure-studies/; https://cistudies.org/events/romanticism-and-critical-infrastructure-studies/; https://cistudies.org/wp-content/uploads/critical-infrastructure-workshop-march-29-2018.pdf; https://criticalinfrastructure.hcommons.org/; https://patriksv.com/curatorship/#huminfra; https://cistudies.org/events/radical-transparency-infrastructure-design-workshop/; https://cistudies.org/events/public-humanities-infrastructure-a-post-harvey-introduction-to-critical-infrastructure-studies/; https://patriksv.com/curatorship/#humaneinfrastructures; https://cistudies.org/events/digital-humanities-critical-infrastructure-studies-workshop-series/; https://cistudies.org/events/digital-humanities-critical-infrastructure-studies-workshop-series/infrastructural-interventions/; https://cistudies.org/events/digital-humanities-critical-infrastructure-studies-workshop-series/interrogating-global-traces-of-infrastructure/.

55. Treating the chapters in this volume as whole documents (with a stopwords list applied), we conducted MALLET LDA topic modeling for 50 topics; and we used the Lexos tool (Kleinman, LeBlanc, and Zhang) to conduct K-means clustering visualized in two-dimensional (2D), three-dimensional (3D), and Voronoi space, and also to perform a dendrogram analysis. More extensive exploration would require iterative experimentation with different modeling parameters, "chunk" sizes of the documents, stopwords, etc. Word embedding analysis could also usefully be added as a method. Readers of this volume can explore an elementary topic model of the chapters by going to the digital resources for this chapter at https://dhdebates.gc.cuny.edu/projects/critical-infrastructure-studies-and-digital-humanities/resources?tag=introduction and downloading a .zip file. Ater extracting the compressed .zip file, opening the "all_topics.html" file in the "output_html" subfolder will show a web page with an interactive view of the topic model. Each of the 50 topics is represented by a clickable list of the most frequent words in a topic. Clicking on a topic's list will show a ranked list of the chapters most associated with that topic. Clicking on one of the chapters will then show other topics associated with that chapter. Readers can also use the Lexos suite of tools to explore this volume's chapters by going to this chapter's digital resources at https://dhdebates.gc.cuny.edu/projects/critical-infrastructure-studies-and-digital-humanities/resources?tag=introduction, downloading a .zip file with Lexos data preloaded, and after decompressing the file, inputting it into Lexos's "upload" screen at http://lexos.wheatoncollege.edu/upload. The workspace.lexos file contains the cleaned and stopworded text of this book's chapters, accompanied by Lexos settings and options. Loading it in Lexos reproduces the editor's workspace for exploring the chapters in Lexos.

56. In the present volume, see also the beginning of Lucie Kolb's chapter 15, "Making Infrastructure Writable," for a general discussion of interfaces.

57. The Twine narrative in Patel, Caranto Morford, and Jacob's chapter 14, "Resisting Bring Your Own Infrastructure in Digital Humanities Learning Spaces," can be downloaded as a web page or as a Twine file by searching for the word "Twine" among this book's digital resources at https://dhdebates.gc.cuny.edu/projects/critical-infrastructure-studies-and-digital-humanities/resources. A static text version is also included in that chapter.

58. See the CIstudies.org, "Bibliography," under "Transportation" (https://cistudies.org/critical-infrastructures-bibliography/ci-bibliography-transportation/) for examples of infrastructure studies works on shipping containers (e.g., Klatskin; Klose; Martin; and Schwarzer) and under "Materials" (https://cistudies.org/critical-infrastructures-bibliography/ci-bibliography-materials/) for works on concrete (e.g., Forty; Gandy; and Idorn).

BIBLIOGRAPHY

Agre, Philip E. "Toward a Critical Technical Practice: Lessons Learned in Trying to Reform AI." In *Bridging the Great Divide: Social Science, Technical Systems, and Cooperative Work,* edited by Geoffrey C. Bowker et al., 131–57. Psychology Press, 1998.

American Council of Learned Societies (ACLS). *Our Cultural Commonwealth: The Report of the American Council of Learned Societies Commission on Cyberinfrastructure for the Humanities and Social Sciences.* American Council of Learned Societies, 2006. https://www.acls.org/wp-content/uploads/2021/11/Our-Cultural-Commonwealth.pdf.

Anand, Nikhil. "The Banality of Infrastructure." In *Items: Insights from the Social Sciences*, Social Science Research Council, 2017. https://items.ssrc.org/the-banality-of-infrastructure/.

Anand, Nikhil. *Hydraulic City: Water and the Infrastructures of Citizenship in Mumbai*. Duke University Press, 2017.

Anderson, Sheila. "What Are Research Infrastructures?" *International Journal of Humanities and Arts Computing* 7, no. 1–2 (2013): 4–23. https://doi.org/10.3366/ijhac.2013.0078.

Antonioni, Michelangelo. *Red Desert*. Film Duemila, Francoriz Distribution, 1964. 120 minutes.

Atkins, Daniel E., et al. *Revolutionizing Science and Engineering Through Cyberinfrastructure: Report of the National Science Foundation Blue-Ribbon Advisory Panel on Cyberinfrastructure*. National Science Foundation, January 2003. https://web.archive.org/web/20121017100852/http://www.nsf.gov/od/oci/reports/atkins.pdf.

Bailey, Jefferson. "Disrespect des Fonds: Rethinking Arrangement and Description in Born-Digital Archives." *Archive Journal*, no. 3 (2013). https://web.archive.org/web/20170919162159/http://www.archivejournal.net/essays/disrespect-des-fonds-rethinking-arrangement-and-description-in-born-digital-archives/.

Baker, R. "Broken Infrastructures, Invisible Subjects: A 'Starter Kit' of Readings and Curios." Critical Infrastructure Studies: Science Fiction and Social Dystopia—A Dynamic Systems Look at Inhabiting Alien Worlds and Making Do with Broken Things (blog), December 10, 2018. https://subalternsf.wordpress.com/critical-infrastructures-of-sci-fi-biotech-and-the-subaltern-a-starter-kit-of-readings-and-curios/.

Baillot, Anne, James Baker, Madiha Zahrah Choksi, et al. "Digital Humanities and the Climate Crisis: A Manifesto." 2022. https://dhc-barnard.github.io/dhclimate/.

Barats, Christine, Valérie Schafer, and Andreas Fickers. "Fading Away . . . The Challenge of Sustainability in Digital Studies." *Digital Humanities Quarterly* 14, no. 3 (2020). https://www.digitalhumanities.org/dhq/vol/14/3/000484/000484.html.

Barbrook, Richard, and Andy Cameron. "The Californian Ideology." Alamut: Bastion of Peace and Information, August 1995. http://www.alamut.com/subj/ideologies/pessimism/califIdeo_I.html.

Beniger, James R. *The Control Revolution: Technological and Economic Origins of the Information Society*. Harvard University Press, 1989.

Bennett, Jane. "The Force of Things: Steps Toward an Ecology of Matter." *Political Theory* 32, no. 3 (2004): 347–72.

Benardou, Agiatis, Erik Champion, Costis Dallas, et al., eds. *Cultural Heritage Infrastructures in Digital Humanities*. Digital Research in the Arts and Humanities. Routledge, 2018.

Best, Stephen, and Sharon Marcus. "Surface Reading: An Introduction." *Representations* 108, no. 1 (2009): 1–21. https://doi.org/10.1525/rep.2009.108.1.1.

Bimber, Bruce. "Information and Political Engagement in America: The Search for Effects of Information Technology at the Individual Level." *Political Research Quarterly* 54, no. 1 (2001): 53–67. https://doi.org/10.2307/449207.

Bogost, Ian. *Alien Phenomenology, or, What It's Like to Be a Thing*. Posthumanities 20. University of Minnesota Press, 2012.

Borgman, Christine L., Andrea Scharnhorst, and Milena S. Golshan. "Digital Data Archives as Knowledge Infrastructures: Mediating Data Sharing and Reuse." *Journal of the Association for Information Science and Technology* 70, no. 8 (2019): 888–904. https://doi.org/10.1002/asi.24172.

Bowker, Geoffrey C. *Science on the Run: Information Management and Industrial Geophysics at Schlumberger, 1920–1940.* Inside Technology. MIT Press, 1994.

Bowker, Geoffrey C., Karen Baker, and Florence Millerand, et al. "Toward Information Infrastructure Studies: Ways of Knowing in a Networked Environment." In *International Handbook of Internet Research,* edited by Jeremy Hunsinger, Lisbeth Klastrup, and Matthew Allen, 97–117. Springer Netherlands, 2010. https://doi.org/10.1007/978-1-4020-9789-8_5.

Bowker, Geoffrey C., Paul N. Edwards, Steven J. Jackson, et al. "The Long Now of Cyberinfrastructure." In *World Wide Research: Reshaping the Sciences and Humanities,* edited by William H. Dutton and Paul W. Jeffreys, 40–44. MIT Press, 2010. https://direct.mit.edu/books/edited-volume/1940/chapter/4231318/The-Long-Now-of-Cyberinfrastructure.

Bowker, Geoffrey C., and Susan Leigh Star. *Sorting Things Out: Classification and Its Consequences.* Inside Technology. MIT Press, 2000.

Bratton, Benjamin H. *The Stack: On Software and Sovereignty.* Software Studies. MIT Press, 2015.

Brown, Adrienne. *The Black Skyscraper: Architecture and the Perception of Race.* Johns Hopkins University Press, 2017. https://muse.jhu.edu/book/56718.

Brown, Bill. "Thing Theory." *Critical Inquiry* 28, no. 1 (2001): 1–22.

Brown, Susan, Tanya Clement, Laura Mandell, Deb Verhoeven, et al. "Creating Feminist Infrastructure in the Digital Humanities." In *Digital Humanities 2016: Conference Abstracts,* 47–50. Jagiellonian University and Pedagogical University, 2016. https://dh2016.adho.org/abstracts/233.

California State Water Resources Control Board. "What Is a Public Water System?" n.d. https://www.waterboards.ca.gov/drinking_water/certlic/drinkingwater/documents/waterpartnerships/what_is_a_public_water_sys.pdf.

Carroll, Stephanie Russo, Ibrahim Garba, Oscar L. Figueroa-Rodríguez, et al. "The CARE Principles for Indigenous Data Governance." *Data Science Journal* 19, art. no. 43 (2020): 1–12. https://doi.org/10.5334/dsj-2020-043.

Castells, Manuel. *The Rise of the Network Society.* The Information Age: Economy, Society, and Culture, v. 1. Blackwell Publishers, 1996.

Certeau, Michel de. *The Practice of Everyday Life.* University of California Press, 2013.

Christen, Kimberly. "Does Information Really Want to Be Free? Indigenous Knowledge Systems and the Question of Openness." *International Journal of Communication* 6 (2012): 2870–93. https://ijoc.org/index.php/ijoc/article/view/1618.

CIstudies.org (Critical Infrastructure Studies). "Bibliography." 2022. https://cistudies.org/critical-infrastructures-bibliography/.

CIstudies.org (Critical Infrastructure Studies). Home page. 2022. https://cistudies.org/.

CIstudies.org (Critical Infrastructure Studies). "Primer." 2022. https://cistudies.org/critical-infrastructure-studies-primer/.

Cornell University, College of Engineering. "Infrastructure Systems," 2024. https://www.engineering.cornell.edu/infrastructure-systems.

Crenshaw, Kimberlé. "Demarginalizing the Intersection of Race and Sex: A Black Feminist Critique of Antidiscrimination Doctrine, Feminist Theory and Antiracist Policies." *University of Chicago Legal Forum* 1989, no. 1 (1989): 139–67.

Daston, Lorraine, and Peter Galison. *Objectivity.* Zone Books, 2010.

Davies, Mark. Corpus of Contemporary American English (COCA). 2008. English-Corpora.org. https://www.english-corpora.org/coca/.

Debord, Guy. *The Society of the Spectacle.* Zone Books, 1994.

Digital Humanities Climate Coalition. "The Digital Humanities Climate Coalition Toolkit," May 23, 2024. https://sas-dhrh.github.io/dhcc-toolkit/.

D'Ignazio, Catherine, and Lauren F. Klein. *Data Feminism.* Strong Ideas Series. MIT Press, 2020.

Dombrowski, Quinn, and Joan Lippincott. "Moving Ahead with Support for Digital Humanities." *EDUCAUSE Review,* March 12, 2018. https://er.educause.edu/articles/2018/3/moving-ahead-with-support-for-digital-humanities.

Edmond, Jennifer, and Francesca Morselli. "Sustainability of Digital Humanities Projects as a Publication and Documentation Challenge." *Journal of Documentation* 76, no. 5 (2020): 1019–31. https://doi.org/10.1108/JD-12-2019-0232.

EDUCAUSE. "About EDUCAUSE." 2024. https://www.educause.edu/about.

Edwards, Paul N. "Infrastructuration: On Habits, Norms and Routines as Elements of Infrastructure." In *Thinking Infrastructures,* edited by Martin Kornberger et al., 355–66. Emerald, 2019.

Edwards, Paul N. "Infrastructure and Modernity: Force, Time, and Social Organization in the History of Sociotechnical Systems." In *Modernity and Technology,* edited by Thomas J. Misa, Philip Brey, and Andrew Feenberg, 185–225. MIT Press, 2003.

Edwards, Paul N., Steven J. Jackson, Melissa K. Chalmers, et al. "Knowledge Infrastructures: Intellectual Frameworks and Research Challenges." Deep Blue, 2013. https://escholarship.org/uc/item/2mt6j2mh.

e-flux Journal. Home page. n.d. https://www.e-flux.com/journal/.

European Commission. "Right to Repair: Commission Introduces New Consumer Rights for Easy and Attractive Repairs." March 22, 2023. https://ec.europa.eu/commission/presscorner/home/en.

Eve, Martin Paul. *Open Access and the Humanities: Contexts, Controversies and the Future.* Cambridge University Press, 2014. https://doi.org/10.1017/CBO9781316161012.

Eve, Martin Paul, and Jonathan Gray, eds. *Reassembling Scholarly Communications: Histories, Infrastructures, and Global Politics of Open Access.* MIT Press, 2020.

Felski, Rita. *The Limits of Critique.* University of Chicago Press, 2015.

Fickers, Andreas, and Annie van den Oever. *Doing Experimental Media Archaeology: Theory.* De Gruyter Oldenbourg, 2022. https://www.degruyter.com/document/doi/10.1515/9783110799774/html?lang=en.

Fickers, Andreas, and Annie van den Oever. "Experimental Media Archaeology: A Plea for New Directions." In *Technē/Technology: Researching Cinema and Media Technologies—Their Development, Use, and Impact,* edited by Annie van den Oever, 272–78. Amsterdam University Press, 2014. https://mediarep.org/entities/bookpart/c98bb81a-528b-40e6-b3ed-609ab323b75f.

Forty, Adrian. *Concrete and Culture: A Material History.* Reaktion Books, 2012.

Foucault, Michel. "Of Other Spaces." Translated by Jay Miskowiec. *Diacritics* 16, no. 1 (1986): 22–27.

Foucault, Michel. *Power/Knowledge: Selected Interviews and Other Writings, 1972–1977.* Edited by Colin Gordon. Translated by Colin Gordon, Leo Marshall, John Mepham, and Kate Soper. Pantheon Books, 1980.

Gandy, Matthew. *Concrete and Clay: Reworking Nature in New York City.* Urban and Industrial Environments. MIT Press, 2003.

Garcia, David, and Geert Lovink. "The ABC of Tactical Media." Nettime-l, May 16, 1997. http://www.nettime.org/Lists-Archives/nettime-l-9705/msg00096.html.

Gebru, Timnit, Jamie Morgenstern, Briana Vecchione, et al. "Datasheets for Datasets." arXiv:1803.09010, March 23, 2018. http://arxiv.org/abs/1803.09010.

Gibson, William. *Neuromancer.* Ace Science Fiction. Ace Books, 1984.

Giddens, Anthony. *Modernity and Self-Identity: Self and Society in the Late Modern Age.* Stanford University Press, 1991.

Gold, Matthew K., and Lauren F. Klein. "CFP: Debates in the Digital Humanities 2021." Debates in the Digital Humanities, n.d. https://dhdebates.gc.cuny.edu/page/cfps-ddh2021.

Gold, Matthew, and George Otte. "The CUNY Academic Commons: Fostering Faculty Use of the Social Web." *On the Horizon* 19, no. 1 (January 1, 2011): 24–32. https://doi.org/10.1108/10748121111107681.

Gourley, Donald, and Paolo Battino Viterbo. "A Sustainable Repository Infrastructure for Digital Humanities: The DHO Experience." In *EuroMed 2010: Digital Heritage,* 473–81. Lecture Notes in Computer Science. Springer, 2010. https://doi.org/10.1007/978-3-642-16873-4_38.

Graban, Tarez Samra, Paul Marty, Allen Romano et al., ed. Invisible Work in Digital Humanities. Special issue of *Digital Humanities Quarterly* 13, no. 2 (2019). http://www.digitalhumanities.org/dhq/vol/13/2/index.html.

Guillory, John. *Professing Criticism: Essays on the Organization of Literary Study.* University of Chicago Press, 2022.

Hanson, Melanie. "College Enrollment & Student Demographic Statistics." Education Data Initiative, January 10, 2024. https://educationdata.org/college-enrollment-statistics.

Harrower, Natalie, Maciej Maryl, Timea Biro, et al. "Sustainable and FAIR Data Sharing in the Humanities: Recommendations of the ALLEA Working Group E-Humanities." ALLEA-All European Academies (2020). https://doi.org/10.7486/DRI.TQ582C863.

Harvey, Penelope, Casper Bruun Jensen, and Atsuro Morita, eds. "Introduction: Infrastructural Complications." In *Infrastructures and Social Complexity: A Companion,* edited by Penelope Harvey, Casper Bruun Jensen, and Atsuro Morita, 1–22. Culture, Economy and the Social. Routledge, 2017.

HathiTrust. "Non-consumptive Use Policy." 2017. https://www.hathitrust.org/the-collection/terms-conditions/non-consumptive-use-policy/.

Heidegger, Martin. "The Question Concerning Technology." In *The Question Concerning Technology, and Other Essays,* translated by William Levitt, 3–35. Harper & Row, 1977.

Heijden, Tim van der, and Aleksander Kolkowski. *Doing Experimental Media Archaeology: Practice.* De Gruyter Oldenbourg, 2023. https://www.degruyter.com/document/doi/10.1515/9783110799767/html?lang=en.

Helmond, Anne, David B. Nieborg, and Fernando N. van der Vlist. "Facebook's Evolution: Development of a Platform-as-Infrastructure." *Internet Histories* 3, no. 2 (2019): 123–46. https://doi.org/10.1080/24701475.2019.1593667.

Hughes, Thomas P. *Networks of Power: Electrification in Western Society, 1880–1930.* Johns Hopkins University Press, 1983.

Hughes, Thomas P. *Rescuing Prometheus: Four Monumental Projects That Changed the World.* Vintage, 2000.

Humanist Discussion Group (Archives), 37.369 ("Name of the First ALLC/ACH Conference"), January 5, 2024. https://dhhumanist.org/volume/37/369/.

Idorn, Gunnar M. *Concrete Progress: From Antiquity to Third Millenium.* Thomas Telford, 1997.

Insigneo Institute (University of Sheffield). "About in Silico Medicine," n.d. https://www.sheffield.ac.uk/insigneo/overview/silico-medicine.

Institute of Network Cultures. Home page. n.d. https://networkcultures.org/.

Jackson, Steven J. "Rethinking Repair." In *Media Technologies: Essays on Communication, Materiality, and Society,* edited by Tarleton Gillespie, Pablo J. Boczkowski, and Kirsten A. Foot, 221–40. MIT Press, 2014.

Jameson, Fredric. *Postmodernism, or, The Cultural Logic of Late Capitalism.* Post-Contemporary Interventions. Duke University Press, 1990.

Jencks, Charles. *The Language of Post-modern Architecture.* 6th ed. Rizzoli, 1991.

Karvonen, Andrew, Federico Cugurullo, and Federico Caprotti, eds. *Inside Smart Cities: Place, Politics and Urban Innovation.* Routledge, 2019.

Kil, Aleksandra. "Excavating Infrastructure in the Analog Humanities' Lab: An Analysis of Claude Lévi-Strauss's Laboratoire d'anthropologie Sociale." *Digital Humanities Quarterly* 14, no. 3 (2020). http://digitalhumanities.org/dhq/vol/14/3/000468/000468.html.

Kil, Aleksandra. "From the Digital to the Analog Humanities: Index Cards as a Humanistic Knowledge Apparatus." Ph.D. diss., University of Wrocław, 2021.

Kirschenbaum, Matthew G. *Mechanisms: New Media and the Forensic Imagination.* MIT Press, 2012.

Klatskin, Alex. "Trade as Form." Translated by Charles Marcrum II. *Scenario Journal,* no. 6 (2017). https://scenariojournal.com/article/trade-as-form/.

Kleinman, Scott, Mark D. LeBlanc, and Cheng Zhang. "Lexos." Lexomics Project, 2019. http://lexos.wheatoncollege.edu/.

Klose, Alexander. *The Container Principle: How a Box Changes the Way We Think*. Infrastructures. MIT Press, 2015.

Knowledge Commons (previously Humanities Commons). Home page. n.d. https://hcommons.org/.

Kroes, Peter. *Technical Artefacts: Creations of Mind and Matter: A Philosophy of Engineering Design*. Philosophy of Engineering and Technology 6. Springer Netherlands, 2012. https://doi.org/10.1007/978-94-007-3940-6.

Larkin, Brian. "The Politics and Poetics of Infrastructure." *Annual Review of Anthropology* 42 (2013): 327–43. https://doi.org/10.1146/annurev-anthro-092412-155522.

Latour, Bruno. "An Attempt at a 'Compositionist Manifesto.'" *New Literary History* 41, no. 3 (2010): 471–90. https://www.jstor.org/stable/40983881.

Latour, Bruno. "From Realpolitik to Dingpolitik, or How to Make Things Public." In *Making Things Public: Atmospheres of Democracy*, edited by Bruno Latour and Peter Weibel, 4–32. MIT Press; ZKM/Center for Art and Media Karlsruhe, 2005.

Latour, Bruno. "On Technical Mediation: Philosophy, Sociology, Genealogy." *Common Knowledge* 3, no. 2 (1994): 29–64. http://www.bruno-latour.fr/sites/default/files/54-TECHNIQUES-GB.pdf.

Latour, Bruno. *We Have Never Been Modern*. Translated by Catherine Porter. Harvard University Press, 1993.

Latour, Bruno. "Why Has Critique Run out of Steam? From Matters of Fact to Matters of Concern." *Critical Inquiry* 30, no. 2 (2004): 225–48. https://doi.org/10.1086/421123.

Levy, Steven. "Mark Zuckerberg on Facebook's Future, From Virtual Reality to Anonymity." *Wired*, April 30, 2014. https://www.wired.com/2014/04/zuckerberg-f8-interview/.

Lilienfeld, Robert. "Systems Theory as an Ideology." *Social Research* 42 (1975): 637–60.

Lippincott, Joan, Lisa Spiro, Annelie Rugg, et al. "7 Things You Should Know About Digital Humanities." EDUCAUSE Library, November 7, 2017. https://library.educause.edu/resources/2017/11/7-things-you-should-know-about-digital-humanities.

Liu, Alan. "Transcendental Data: Toward a Cultural History and Aesthetics of the New Encoded Discourse." *Critical Inquiry* 31, no. 1 (2004): 49–84. https://doi.org/10.1086/427302.

Liu, Alan. "Where Is Cultural Criticism in the Digital Humanities?" In *Debates in the Digital Humanities*, edited by Matthew K. Gold, 490–509. University of Minnesota Press, 2012. https://dhdebates.gc.cuny.edu/read/untitled-88c11800-9446-469b-a3be-3fdb36bfbd1e/section/896742e7-5218-42c5-89b0-0c3c75682a2f#ch29.

Lovink, Geert. *Stuck on the Platform: Reclaiming the Internet*. Making Public. Valiz, 2022.

Machulak, Erica. "Mukurtu: A Digital Platform That Does More than Manage Content." *Humanities: The Magazine of the National Endowment for the Humanitiess*, 41, no. 4 (2020). https://www.neh.gov/article/mukurtu-digital-platform-does-more-manage-content.

Manifold. "A Brief History of Manifold," 2024. https://manifoldapp.org/history.

Manifold. Home page. 2024. https://manifoldapp.org/.

Manovich, Lev. *The Language of New Media*. Leonardo. MIT Press, 2001.

Martin, Craig. "Shipping Container Mobilities, Seamless Compatibility, and the Global Surface of Logistical Integration." *Environment and Planning A: Economy and Space* 45, no. 5 (2013): 1021–36. https://doi.org/10.1068/a45171.

Mattern, Shannon. *A City Is Not a Computer: Other Urban Intelligences*. Places Books 2. Princeton University Press, 2021.

McGrail, Anne, Angel David Nieves, and Siobhan Senier. "Introduction." In *People, Practice, Power: Digital Humanities Outside the Center*, edited by Anne McGrail, Angel David Nieves, and Siobhan Senier. University of Minnesota Press, 2021. https://dhdebates.gc.cuny.edu/read/people-practice-power/section/90aa6320-4ab1-4a1d-8314-f9bf309c0d7f#intro.

McGrail, Anne, Angel David Nieves, and Siobhan Senier, eds. *People, Practice, Power: Digital Humanities Outside the Center*. University of Minnesota Press, 2021. https://dhdebates.gc.cuny.edu/read/people-practice-power/section/12936dca-2e8e-4b5d-b6ed-84cc5afda912.

McKinsey & Company. "What Is Digital-Twin Technology?" McKinsey Blog, July 12, 2023. https://www.mckinsey.com/featured-insights/mckinsey-explainers/what-is-digital-twin-technology.

McPherson, Tara. *Feminist in a Software Lab: Difference + Design*. MetaLABprojects. Harvard University Press, 2018.

Mediapolis. Home page. n.d. https://www.mediapolisjournal.com/.

Minimal Computing. "About." GO::DH, n.d., https://go-dh.github.io/mincomp/about/.

Miya, Chelsea, Oliver Rossier, and Geoffrey Rockwell, eds. *Right Research: Modelling Sustainable Research Practices in the Anthropocene*. Open Book Publishers, 2021. https://www.openbookpublishers.com/books/10.11647/obp.0213.

Mondschein, Jared, Aaron Clark-Ginsberg, and Andreas Kuehn. "Tech Alone Isn't Enough to Create a Successful Smart City." Rand Blog, February 10, 2021. https://www.rand.org/pubs/commentary/2021/02/tech-alone-isnt-enough-to-create-a-successful-smart.html.

Mukurtu CMS. Home page. Mukurtu CMS, n.d. https://mukurtu.org/.

Mukurtu CMS. "Our Mission." Mukurtu CMS, n.d. https://mukurtu.org/about/.

Oiva, Mila, and Urszula Pawlicka-Deger. "Lab and Slack. Situated Research Practices in Digital Humanities—Introduction to the DHQ Special Issue." *Digital Humanities Quarterly* 14, no. 3 (2020). https://www.digitalhumanities.org/dhq/vol/14/3/000485/000485.html.

Oiva, Mila, and Urszula Pawlicka-Deger, ed. Special issue on "Lab and Slack. Situated Research Practices in Digital Humanities." *Digital Humanities Quarterly* 14, no. 3 (2020). https://www.digitalhumanities.org/dhq/vol/14/3/index.html.

Ong, Walter J. *Orality and Literacy: The Technologizing of the Word*. Methuen, 1982.

Parks, Lisa. "Stuff You Can Kick: Toward a Theory of Media Infrastructures." In *Between Humanities and the Digital*, edited by Patrik Svensson and David Theo Goldberg, 355–73. MIT Press, 2015. https://direct.mit.edu/books/edited-volume/4494/chapter/192061/Stuff-You-Can-Kick-Toward-a-Theory-of-Media.

Parks, Lisa, and Nicole Starosielski, eds. *Signal Traffic: Critical Studies of Media Infrastructures*. The Geopolitics of Information. University of Illinois Press, 2015.

Pawlicka-Deger, Urszula. "Infrastructuring Digital Humanities: On Relational Infrastructure and Global Reconfiguration of the Field." *Digital Scholarship in the Humanities* 37, no. 2 (2022): 534–50. https://doi.org/10.1093/llc/fqab086.

Pawlicka-Deger, Urszula. "A Laboratory as the Infrastructure of Engagement: Epistemological Reflections." *Open Library of Humanities* 6, no. 2 (2020). https://doi.org/10.16995/olh.569.

Pawlicka-Deger, Urszula. "The Laboratory Turn: Exploring Discourses, Landscapes, and Models of Humanities Labs." *Digital Humanities Quarterly* 14, no. 3 (2020). http://www.digitalhumanities.org/dhq/vol/14/3/000466/000466.html.

Pawlicka-Deger, Urszula. "The Multiformity of Infrastructure." DH Infra (blog), March 29, 2021. https://dhinfra-org.github.io/197/the-multiformity-of-infrastructure/.

Pawlicka-Deger, Urszula, and Christopher Thomson, eds. *Digital Humanities and Laboratories: Perspectives on Knowledge, Infrastructure and Culture*. Digital Research in the Arts and Humanities. Routledge, 2023.

Peels, Rik. "Epistemic Values in the Humanities and in the Sciences." *History of Humanities* 3, no. 1 (2018): 89–111. https://doi.org/10.1086/696304.

Pepe, Alberto, Alyssa Goodman, August Muench, et al. "How Do Astronomers Share Data? Reliability and Persistence of Datasets Linked in AAS Publications and a Qualitative Study of Data Practices Among US Astronomers." *PLOS One* 9, no. 8 (August 28, 2014): e104798. https://doi.org/10.1371/journal.pone.0104798.

Places Journal. Home page. 2024. https://placesjournal.org/.

Platform. Home page. n.d. https://www.platformspace.net/home/platform-the-state-of-the-built-environment-an-editorial.

Raley, Rita. *Tactical Media*. Electronic Mediations 28. University of Minnesota Press, 2009.

Repair Association. Home page. Repair Association, n.d. https://www.repair.org.

Ricoeur, Paul. *Freud and Philosophy: An Essay on Interpretation*. Translated by Denis Savage. Yale University Press, 1970.

Risam, Roopika, and Kelly Baker Josephs, eds. *The Digital Black Atlantic*. Debates in the Digital Humanities. University of Minnesota Press, 2021. https://dhdebates.gc.cuny.edu/read/the-digital-black-atlantic/section/612d809f-3f10-4dad-9de8-58375d4d7e95.

Risam, Roopika, and Rahul K. Gairola, eds. *South Asian Digital Humanities: Postcolonial Mediations Across Technology's Cultural Canon*. Routledge, 2021.

Rorty, Richard, ed. *The Linguistic Turn: Recent Essays in Philosophical Method*. University of Chicago Press, 1967.

Rosenberger, Robert. "Technological Multistability and the Trouble with the Things Themselves." In *The Oxford Handbook of Philosophy of Technology*, edited by Shannon Vallor, 374–92. Oxford University Press, 2022. https://doi.org/10.1093/oxfordhb/9780190851187.013.42.

Ross, Sarah. "2005–2006, Archisuits." Sarah Ross (blog), n.d. https://insecurespaces.net/archisuits-2005-2006/.

Roy Rosenzweig Center for History and New Media. Home page. 2023. https://rrchnm.org/.

Rubenstein, Michael, Bruce Robbins, and Sophia Beal. "Infrastructuralism: An Introduction." *Modern Fiction Studies* 61, no. 4 (2015): 575–86. https://doi.org/10.1353/mfs.2015.0049.

Russell, Andrew, and Lee Vinsel. "Hail the Maintainers." Edited by Sam Haselby *Aeon*, April 7, 2016. https://aeon.co/essays/innovation-is-overvalued-maintenance-often-matters-more.

Ryan, John, Lydia Hearn, and Paul Longley Arthur. "The Digital Environmental Humanities (DEH) in the Anthropocene: Challenges and Opportunities in an Era of Ecological Precarity." *Digital Humanities Quarterly* 17, no. 3 (2023). https://www.digitalhumanities.org/dhq/vol/17/3/000714/000714.html.

Samberg, Rachael G., and Cody Hennesy. "Law and Literacy in Non-Consumptive Text Mining: Guiding Researchers Through the Landscape of Computational Text Analysis." In *Copyright Conversations: Rights Literacy in a Digital World*, edited by Sara R. Benson, 289–315. Association of College and Research Libraries, 2019. https://escholarship.org/uc/item/55j0h74g.

Sayers, Jentery. "CFP: Making Things and Drawing Boundaries." Debates in the Digital Humanities, n.d. https://dhdebates.gc.cuny.edu/page/cfps-mtdb.

Sayers, Jentery. "Introduction: 'I Don't Know All the Circuitry.'" In *Making Things and Drawing Boundaries*, edited by Jentery Sayers. Debates in the Digital Humanities. University of Minnesota Press, 2017. https://dhdebates.gc.cuny.edu/read/untitled-aa1769f2-6c55-485a-81af-ea82cce86966/section/7d8fca82-c6ca-480f-bf17-1df4a2cdb577#intro.

Sayers, Jentery, ed. *Making Things and Drawing Boundaries: Experiments in the Digital Humanities.* Debates in the Digital Humanities. University of Minnesota Press, 2017. https://dhdebates.gc.cuny.edu/projects/making-things-and-drawing-boundaries.

Scenario Journal. Home page. 2024. https://scenariojournal.com/.

Schreibman, Susan, Ray Siemens, and John Unsworth, eds. *A Companion to Digital Humanities.* Blackwell, 2004. https://companions.digitalhumanities.org/DH/.

Schwarzer, Mitchell. "The Emergence of Container Urbanism." *Places Journal*, February 2013. https://doi.org/10.22269/130212.

Scott, Ridley. *Blade Runner.* Science fiction. Warner Bros., 1982. 117 minutes.

Scott, W. Richard. *Institutions and Organizations: Ideas and Interests.* 3rd ed. SAGE, 2008.

Shepard, Michael. "Review of Mukurtu Content Management System." *Language Documentation & Conservation* 8 (2014): 315–25. https://scholarspace.manoa.hawaii.edu/items/7865dfd4-9dd8-4735-af99-3787829b82af.

Shoemaker, Tyler. "Literalism: Reading Machines Reading." PhD diss., University of California, Santa Barbara, 2020.

Smithies, James, "The Dark Side of DH." In *The Bloomsbury Handbook to the Digital Humanities,* edited by James O'Sullivan, 111–23. Bloomsbury Academic, 2022.

Smithies, James. *The Digital Humanities and the Digital Modern.* Palgrave Macmillan, 2017.

Smithies, James, and Arianna Ciula. "Humans in the Loop: Epistemology & Method in King's Digital Lab." In *Routledge International Handbook of Research Methods in Digital*

Humanities, edited by Kristen Schuster and Stuart Dunn, 155–72. Taylor & Francis, 2021.

Smithies, James, Patrick ffrench, and Arianna Ciula. "*Droit de cité*: The Digital Lab as Digital Milieu." In *Digital Humanities Laboratories: Perspectives on Knowledge, Infrastructure and Culture,* edited by Urszula Pawlicka-Deger and Christopher Thomson, 52–66. Routledge, 2024.

Smithies, James, Pascal Flohr, Fadi Bala'awi, et al. "MaDiH (مديح): A Transnational Approach to Building Digital Cultural Heritage Capacity." *ACM Journal on Computing and Cultural Heritage* 15, no. 4(2022): 1–14. https://doi.org/10.1145/3513261.

Smithies, James, Carina Westling, Anna-Maria Sichani, et al. "Managing 100 Digital Humanities Projects: Digital Scholarship & Archiving in King's Digital Lab." *Digital Humanities Quarterly* 13, no. 1 (2019). https://www.digitalhumanities.org/dhq/vol/13/1/000411/000411.html.

Society+Space [journal]. Home page. n.d. https://www.societyandspace.org/journal.

Speitz, Michele. "The Infrastructural Sublime and Imperial Landscape Aesthetics: Robert Southey, Poet Laureate, and Thomas Telford, Father of Civil Engineering." *European Romantic Review* 32, no. 1 (2021): 41–63. https://doi.org/10.1080/10509585.2020.1865163.

Spiro, Lisa. "Defining Digital Social Sciences." *dh+lib,* April 9, 2014. https://acrl.ala.org/dh/2014/04/09/defining-digital-social-sciences/.

Star, Susan Leigh. "The Ethnography of Infrastructure." *American Behavioral Scientist* 43, no. 3 (1999): 377–91. https://doi.org/10.1177/00027649921955326.

Star, Susan Leigh, and Karen Ruhleder. "Steps Toward an Ecology of Infrastructure: Design and Access for Large Information Spaces." *Information Systems Research* 7, no. 1 (1996): 111–34. https://www.jstor.org/stable/23010792.

Starosielski, Nicole. *The Undersea Network.* Sign, Storage, Transmission. Duke University Press, 2015.

Sustainable Subsea Networks. "About." SustainSubseaNetwork, n.d. https://www.sustainablesubseanetworks.com/about.

Svensson, Patrik. "Curated Events." Patrik Svensson (blog), n.d. https://patriksv.com/curatorship/.

Svensson, Patrik. *Humane Infrastructures.* MIT Press, 2025.

Svensson, Patrik. "The Humanistiscope: Exploring the Situatedness of Humanities Infrastructure." In *Between Humanities and the Digital,* edited by Patrik Svensson and David Theo Goldberg. Cambridge London, 337–53. MIT Press, 2015.

Theimer, Kate. "Archives in Context and as Context." *Journal of Digital Humanities* 1, no. 2 (2012). http://journalofdigitalhumanities.org/1-2/archives-in-context-and-as-context-by-kate-theimer/.

Tsesmelis, Theodore, I. Hasan, M. Cristani, et al. "An Integrated Light Management System with Real-Time Light Measurement and Human Perception." *Lighting Research & Technology* 53, no. 1 (January 2021): 74–88. https://doi.org/10.1177/1477153520947464.

Tucker, Johanna. "Facing the Challenge of Digital Sustainability as Humanities Researchers." *Journal of the British Academy* 10 (2022): 93–120. https://doi.org/10.5871/jba/010.093.

UNESCO. "What You Need to Know About Higher Education." July 4, 2024. https://www.unesco.org/en/higher-education/need-know.

Verbeek, Peter-Paul. *What Things Do: Philosophical Reflections on Technology, Agency, and Design.* Pennsylvania State University Press, 2005.

Watters, Audrey. *The Monsters of Education Technology.* Audrey Watters, 2014. https://monsters.hackeducation.com/.

Wershler, Darren, Lori Emerson, and Jussi Parikka. *The Lab Book: Situated Practices in Media Studies.* University of Minnesota Press, 2022. https://doi.org/10.5749/9781452958408.

Williams, Jeffrey J. and Christopher Newfield, eds. "Critical University Studies" (book series) Johns Hopkins University Press, 2024. https://press.jhu.edu/books/series/critical-university-studies.

Williams, Rosalind. "Cultural Origins and Environmental Implications of Large Technological Systems." *Science in Context* 6, no. 2 (1993): 377–403. https://doi.org/10.1017/S0269889700001459.

Wilson, Ara. "The Infrastructure of Intimacy." *Signs* 41, no. 2 (2016): 247–80.

Zotero. Home page. n.d. https://www.zotero.org/.

PART I

CRITICAL INFRASTRUCTURE STUDIES (AND DIGITAL HUMANITIES)

PART I][Chapter 1

Interfaces for the Anthropocene

ANNE BEAULIEU

Infrastructures are prime sites for the work of modernization and globalization, and they have been amply critiqued as conduits and enablers of liberalization, globalization, and colonialism (Aouragh et al.; Jensen and Morita). These attributes coincide with carbon-intensive economic structures. Quite frequently, infrastructures are entwined with systems of reification, abstraction, and standardization that have led to widespread destructive exploitation and environmental pollution. The predominant values that infrastructures support are efficiency and removal of specificity in favor of the generic and universal, much in line with the aspirations of the dominant disciplines and sectors from which they originate: engineering, transportation, and information and communication technologies. When extended to research, knowledge infrastructures have served to increase the circulation of data, monitoring, and the creation and management of global objects (Beaulieu, "Organising Knowledge"; Eren and Beaulieu). It is important to accept the multifaceted effect that infrastructures have on global communities, however. It is all too easy to focus on infrastructures as instruments of exploitation and conduits of unsustainable practices because these are the forms that tend to loom large in the imaginaries of Western scholars. There is a danger in overidentifying infrastructures with instruments of destruction since such unitary treatment can contribute to the loss of the ability to imagine that things might be otherwise, with dire consequences (Graeber and Wengrow). It is crucial to note how infrastructures have also been the focus of calls for novel engagement for scholars (Liu, "Toward Critical Infrastructure Studies") and for the creation of urgently needed knowledge (Pawlicka-Deger), and that infrastructures, such as those described later in this chapter, can convey other values.[1]

In this chapter, I explore how knowledge infrastructures can help address the crises of the Anthropocene.[2] I use the phrase "knowledge for the Anthropocene" to indicate a project that improves knowledge production as a process of creating new relations—ones in which humans are neither invisible nor standing outside nature but are accountably, reflexively present. Specifically, I delve into how recent research in digital humanities can contribute to conceptualizing and realizing better infrastructures—better in the sense that they further enable us to care (Bellacasa) by bringing issues of interdependence, responsibility, and discernment to the forefront, and in the sense that they garner sufficient trust and reliability to support action.

I focus on one aspect of infrastructures (namely, the interface) to contribute meaningfully to the vast issue of how to reinvent them. This focus makes tractable the question of what kind of knowledge we need for the Anthropocene by showing the importance of articulating and reflecting on assumptions about the source, user, and use of knowledge that is built into interfaces. It also turns our attention toward the significance of interactions and how they are shaped by the interfaces available to us. Interfaces can be designed, tackled, built, tinkered with, and/or demonstrated—and as such, they form sites of intervention toward better knowledge practices.

What Is an Interface?

An interface can be defined in more or less specific terms. Our contemporary, everyday understanding tends to associate it with the digital such that an interface is "a mediating structure that supports behaviors and tasks" and that "disciplines, constrains, and determines what can be done in a digital environment" (Drucker, *Graphesis*, 138–39). To develop knowledge for the Anthropocene, however, it is useful to step back and see how interfaces are a key feature of infrastructures in general, and more specifically of the subcategory of infrastructures that science and technology studies (STS) scholars have defined as "knowledge infrastructures."

Therefore, while the digital is currently a predominant object of concern when studying contemporary knowledge infrastructures, I extend this definition to other settings, defining an interface as a space of interaction between a system and an agent. While interfaces are often assumed to act as *information spaces* that show, they are also *action spaces* that enable doing. While this definition shares the inclusive approach to a diversity of interfaces also embraced by writers such as Alexander Galloway, it remains grounded in the analysis of design and the materiality of interfaces rather than focusing on an "effect." In addition, while an interface is often treated as a site that enables the connection of entities that have different powers—typically, humans and computers (Cramer and Fuller)—my approach is informed by STS. "Agent" should be understood here as a category open to humans, more-than-humans, and technologies, as well as hybrid forms of these. "Interaction" should also be understood as a layered, performative concept. In both cases, "interface situates us in a mediating relation to information, communication, and experience. It also provides a platform for interpretative work in knowledge production" (Drucker, *Visualization*, 92). To understand how interfaces matter for knowledge, therefore, we must consider how what we are shown (and/or what we can do) in this space is shaped by material, symbolic, and time-bound elements, as well as how interfaces are rendered meaningful through actual use by skilled agents.

Interfaces are linked to wide-ranging technologies that vary from mundane to high-tech, and analog to digital—think of a volume dial on a transistor radio or a programmable setting on an electron microscope. The complexity of interfaces further increases when interfaces are connected to infrastructures since they constitute the space of action that links agents to entire suites of technologies. In the context of knowledge infrastructures, as defined in STS (Beaulieu, *Revealing Relations*; Edwards), interfaces rely on multiple layers of conventions and protocols that articulate the relations among the components of an infrastructure and shape what can be shown or done. This layering makes such interfaces potentially powerful, but it also constrains the range of possibilities of interaction since interfaces must remain aligned with the infrastructural components.

A focus on interfaces is especially helpful to start thinking differently about infrastructures because we associate interfaces with possibility, potential, and dynamic realization. We come to interfaces with the expectation that they are a site for activity. In contrast, infrastructures tend to fall into the background, to be experienced as neutral enablers, and we are able to see them as performative only through infrastructural inversions that require deep analysis (Bowker and Star; Brown et al.). While there are very different takes on what it means to interact with an interface, there is an important contrast between how we experience interfaces and how we tend to experience infrastructures—as invisible, stable, monolithic, or just there. We feel much more engaged by interfaces and experience our encounters with them as active and normative, whether judged as user-friendly or not. We are also increasingly becoming aware that interfaces read us, gathering data about our search terms, patterns, and life rhythms (Beaulieu and Leonelli). This contrast is one of the elements that makes interfaces productive in rethinking knowledge for the Anthropocene. Interfaces could be the thin end of the wedge, helping to clarify and modulate how infrastructures function as a "generative matrix" that constitutes routines, subjects, and relations (Povinelli, 150; Jensen and Morita). This is what makes interfaces especially potent sites to start rethinking and redoing infrastructures.

In what follows, I sketch out how we can study and develop interfaces as specific encounters between systems and agents that are situated and rely on specific conditions and skills. Rethinking interfaces is part of a project that aims to find effective ways to question and stop reproducing an architecture of universalism and frictionless circulation of knowledge, an architecture that furthermore hides the uneven distribution of violence and profit. Starting from interfaces, we can articulate knowledge infrastructures that help us construct liveable futures. Concretely, interfaces that show rather than elide the connection of knowledge to its context and process of production—such as those discussed later in this chapter around the SVALUR project—form essential contributions.

Approaches to Interfaces

A vast body of work focuses on understanding and designing interfaces. While several other fields are also concerned with interfaces, by contrasting human-computer interaction (HCI) and digital humanities, we can better situate dominant paradigms of interface design, implemented in the great majority of digital environments, and show how digital humanities can be a source of inspiration for the potential of interfaces.

At the root of HCI is the goal-oriented approach of helping humans get things done, whether in a military or industrial context (Harrison, Sengers, and Tatar), and more recently in the context of ubiquitous computing in a wealth of everyday, mundane settings. Despite this range of settings, the elements of requirements elicitation and evaluation based mainly on usability and efficiency have remained core concerns in HCI. These serve the aim of enabling people to work effectively, using systems designed around their needs. A core assumption underlying the elicitation of requirements and evaluation according to usability and efficiency involves the definition and stabilization of users and technologies as separate actions. The project of creating interfaces, therefore, remains one of optimization, of seeking the best fit between human and machine, of enabling efficient information exchange, and, with ubiquitous computing, of supporting situated action. These approaches yield interfaces as discrete technologies. Importantly, this engineering approach aligns knowledge infrastructures with infrastructures elaborated for mobility, communication, or energy needs.

In contrast, approaches to interfaces informed by digital humanities have revealed the interface as a site of creation, at times focused on novel relations and actions (Hookway; Cramer and Fuller) or on the effects of the interfaces and digital technologies in terms of the materiality of text (Lenoir) or code (Mackenzie). Rather than seeing interfaces as a technology to be optimized, interfaces have been conceptualized as a space of world-making (Holtrop) or as a site of aesthetic production worthy of critical attention (Ulrik and Pold). Also, in contrast to HCI approaches, the constitution of the agent, rather than servicing its needs, is made central to inquiry and design.

Digital humanities research also indicates the importance of the metaphor of surface to the shaping of interfaces, which affects not only the space of interaction but also the knower and what can be known. The flatness of most interfaces, without a vanishing point, tends to elide historical and cultural situatedness (Drucker, *Visualization*). This is also true for spatial situatedness since digital interfaces tend to define location not in terms of place but rather in relation to Cartesian coordinates that render spaces equivalent—but at the cost of erasing historical, cultural, and place-based understandings (Beaulieu, "A Space for Measuring Mind").

Finally, digital humanities propose a route to better interfaces through an examination of how different epistemic values might be inscribed in interfaces, enabling (but not determining) different interactions. Drucker (but see also Jue and McGillivray et al.) insists across her oeuvre on the need to inscribe ambiguity, complexity, and multiplicity of viewpoints to serve the interpretative agenda of the humanities. For her part, Brown stresses the need to consider how design fundamentally matters since it affects whether resources are considered useful and are actually used (Brown). Others have gone so far as to argue that argument and interface design are inseparable (Andrews and van Zundert).

This brief sketch provides some context to understand the contrast between interpretations of interfaces that foreground representation, immediacy, abstraction, and standardization to enable calculability and scalability, and interpretations that foreground the generative attributes of interfaces and support human attentiveness to embodiment, vitality, and materiality. The latter interpretations foreground interpretation and transformation, and correspondingly different approaches to interface design. Scholars have set out how the epistemic values of the humanities need to be supported through reflexive and critical infrastructure development (Liu, "Toward Critical Infrastructure Studies"; Drucker, *Visualization*) or how infrastructures can best serve diversity (Liu, "Towards a Diversity Stack") and access in digital humanities (Pawlicka-Deger). I contend that such reconfigurations are crucial, not only for the sake of providing better tools for the important work of the humanities, but also as a way to support a different relationship to the environment and enable liveable futures.

Digital Interfaces in the Anthropocene

Interfaces matter for the Anthropocene, for example in relation to growing datafication and digitization. To articulate new kinds of accountability and thereby stimulate more efficient supply chain logistics and carbon accounting, Paul Edwards suggests expanding how data exhaust is taken into account. Other lines of work argue that changing the scope of knowledge infrastructures (Beaulieu, *Revealing Relations*) and considering how more-than-humans are essential to the functioning of many knowledge infrastructures (Eren and Beaulieu) can offer new possibilities for knowledge for the Anthropocene. The multiplication of geolocation applications, such as radio frequency identification (RFID), tagged animals, and a wealth of other sensors, as well as the portability of screens, also mean that the digital and the natural are increasingly entwined in our tools and conceptualizations (Gabrys), a phenomenon sometimes labeled the "digital Anthropocene" (McLean). The design of better digital environments matters for modes of engagement with the living world and are relevant for wider lay audiences (Whitelaw and Smaill).

To simplify this complex discussion, rather than focus on specific technological implementations that may rapidly become obsolete, I identify core values to guide novel interfaces. First, there is a strong sense that moral order can be felt in the material state of the world and in the ontologies on which we elaborate in our knowledge systems (Tsing, *The Mushroom*; Bellacasa; Conty). This involves both a decentering of humans (Cielemęcka and Daigle) and renewed attention to how we position the material world ontologically and epistemologically. "New materialists" therefore argue for renewed attention to materiality as an alternative to the politics of representation. The resulting materialization of relations of knowing contrasts with narratives of commodification, scaling, and virtualization that have shaped digital technologies. It also draws into question the seamless and frictionless interface. If materiality matters, very literally, then interfaces that do justice to these sensibilities would make work, engagement, and friction prominent, radicalizing the current paradigm of tangible user interfaces where users are not merely presented with flat screens and surfaces or with sensors that disappear, but also are invited to engage with objects that have complex materiality and spatiality.

Another dimension of the kinds of relations needed for liveable futures questions the affectively neutral, detached interface. Detachment—of course also a form of affect—enables calculation (Turnhout and Purvis; Tichenor et al.), commensurability (Espeland and Stevens; Dijck, Poell, and Waal), aggregation (Alaimo and Kallinikos), and standardization (Beaulieu, "Voxels in the Brain"). So far, attempts to explore affect in relation to interfaces have been predominantly in the service of aggregation (the algorithmic treatment of "likes") or sentiment analysis (natural language processing of user-generated posts). While affective computing increases the range of interfaces, it views emotions as conduits of information by emphasizing multimodal interfaces, thereby aligning with the criterion of efficiency discussed in relation to HCI.

Clearly, if *affective* dimensions are to be integrated in our relationships to the world, be they care (Bellacasa) or curiosity (Lorimer, *Wildlife in the Anthropocene*), then a more radical approach to interfaces will be needed. Agostinho (2) asks how visualization might be rethought to enhance "dignity, ethical responsibility and care for what is not known." In addition, recent work on mediation in DH suggests that a focus on volumes and duration (rather than surface and immediacy) could also open our registers to affective interactions in the Anthropocene and, especially, expand our vocabulary for environmental justice (Jue).

While there are important differences in starting points and politics between scholars, the need for situated knowledges is prominently argued across many analyses. The consequences of lack of *situatedness* have been elaborated in relation to colonial practices of access to land (Liboiron); to bodies, populations, and their data (Murphy; Couldry and Mejias; Ruppert and Isin); and to history (Povinelli); as well as in relation to epistemic injustice with regard to indigenous knowledge (Todd; Turnbull). Across this work, the consequences of landless, bodiless, ahistorical, and abstracted knowledge are exposed as dispossession, genocide, and environmental degradation, and the often-undisputed desirability of access is problematized (Christen). If the erasure of provenance, context, and a variety of forms of ownership (besides economic) is implicated in the crisis of the Anthropocene, it makes sense to pursue a more situated and accountable way of knowing. This pleads for interfaces that would insist on assigning meaningful location, situation, and embodiment to data.

In critiques of modernist knowledge infrastructures, the dominance of individualized, discrete objects is also prominent (Christen; Bigo, Isin, and Ruppert). In contrast, many analysts plead for more attention to be paid to interactions and the importance of the *relational* for knowledge production for liveable futures. For example, if we shift the focus from nature as an external object and make relations to it central (Schroer), then to do justice to these relationships, interfaces should encourage composing practices within a preexisting and shifting web of relations (Arola). Interfaces that allow information seeking through "generous interfaces" (Whitelaw) or browsing (Ruecker, Radzikowska, and Sinclair), rather than solely retrieval systems based on "search," would contribute to better relations.

Yet another angle on the relational in terms of interfaces would be to enrich attempts at personalization and customization beyond a current focus on what is relevant for supporting transactions and consumption. Interfaces could allow for the enactment of a much wider range of relationships based on critical reflection. This could take the shape of interfaces designed to display relations for serendipitous discovery rather than information retrieval, as the Huni project aims to do (Verhoeven).[3] Or, if interfaces are also seen to have a duration (Jue), they could better support transformative encounters, spread over time, or closely entwined through digital copresence and liveness (Turnbull, Searle, and Adams). More attention to the relations of knowers to knowledge, therefore, could be a way to combat both testimonial and hermeneutic injustice (Fricker, 176–79). Since knowledge infrastructures, just like archives, rely on and reproduce particular socialities (Povinelli, 149), more attention to the social relations that undergird knowledge as collectively generated in infrastructures (Beaulieu and de Rijcke) also supports a less individualistic and atomistic sociality.

Finally, issues of *scale* are also powerfully identified across different analyses. When the Anthropocene is articulated as massive and planetary, the dynamics of globalization are reactivated (Tsing, *Friction*; Chakrabarty; Murphy) and contribute to the sense of "emergency" that may suspend the rights and freedoms of those already most disenfranchised, as we have seen in the recent responses to the Covid pandemic (Mattar et al.; Powers et al.). Again, ontology, epistemology, and politics of knowledge can be reconfigured by seeking to interface with issues of scale in mind. Interfaces that are oriented toward humility of intervention (Murphy) and support small-scale, intimate, and community-level developments would be invaluable.

These core values of interface design could contribute to better knowledge for the Anthropocene. They are relevant to developing a way of being in and knowing about the world that moves us beyond gesturing toward looming environmental and societal crises that our tools are insufficient to handle. They also enable us to avoid acting as accountants of death (who carefully monitor decline, as detached observers) and instead take more responsibility for the knowledge that we produce. Such interfaces could powerfully and iteratively connect knowledge to the possibility of identification, engagement, and meaningful action.

Interfacing Arctic Knowledge

This chapter has engaged with a wide range of scholarship and connected different kinds of concerns. It has set out many ambitions, calling for different ontologies, politics, and aesthetics that tangibly integrate attention to material, affective, situated, relational, and small-scale work for better interfaces. To explore how the current repertoire of interfaces can be related to these ambitions, I turn to a specific project focused on Svalbard, a Norwegian archipelago in the Arctic Ocean. This remote area of the world has a unique form of governance and welcomes a wide range of residents and visitors, both human and more-than-human, who immigrate or migrate to the archipelago. Many of these flows can be traced back over centuries, such that Svalbard is historically an important passage point for the exploitation of natural resources (e.g., hunting, mining) and the geopolitics of globalization (*terra nullius*, neutrality of trade). It is an area of intense interest for scientific research, being the object of a monitoring project called the Svalbard Integrated Arctic Earth Observing System (SIOS). Like so many other knowledge infrastructures, one of the driving principles of SIOS is to bring technologically mediated measurements together into a coherent and integrated observation and monitoring program.[4] The work of SIOS is informed by a natural science–based, Earth system approach. It monitors a wealth of factors, ranging from land and sea temperatures to snow depth, salinity, and ocean currents, to name but a few types of data. SIOS has a very thorough review and support structure that ensures that the data collected and consolidated demonstrates a high level of standardization, documentation, and interoperability. Analyses of how such values and practices shape knowledge have been elaborated on previously and will not be repeated here, but it is crucial to note the overarching paradigm of monitoring via abstraction that organizes this data on and about Svalbard—a paradigm touted and recognized by funders, researchers, and many agencies as the core strength of the initiative.

SIOS serves explicitly as a foil to a much smaller but very intensely pursued project, Understanding Resilience and Long-Term Environmental Change in the High Arctic: Narrative-Based Analyses from Svalbard (nicknamed SVALUR).[5] SVALUR seeks to examine the importance of narrative forms and of local and experiential knowledge in relation to the constitution and circulation of measurements, observations, and monitoring. It is fueled by the idea that important knowledge about Svalbard is elided by mainstream scientific knowledge, which the project terms *detached understanding*. Echoing the calls for different ways of knowing reviewed previously, SVALUR is concerned that in times of climate crisis, local communities of humans and nonhumans might not be optimally served by SIOS knowledge. The project initially set out to address this point by collecting accounts and interpretations of environmental change in context—namely, interviews, diaries, and testimonies mediated and preserved as audio recordings and their transcriptions. During the project, the SVALUR team also actively sought to remain in dialogue with SIOS through consultation and participation in SIOS events—unsurprisingly with varying degrees of success. Late in the project, as a member of SVALUR joined the Knowledge Infrastructures Department of Campus Fryslân, which I lead, a few members of the department started collaborating with the SVALUR team.

The work of SVALUR reveals much about dominant modes of interfaces and how they contrast with the aspirations of the SVALUR researchers. Two registers are repeatedly explored by these researchers as they seek to generate knowledge and engage others in it: experiencing and making visible. These registers are used in various subprojects to develop interfaces for SVALUR and serve here to make clear the epistemic tensions that arise around what can count as legitimate knowledge.

A first set of efforts in SVALUR sought to combine the epistemic strengths of SIOS (monitoring data) with the experience of generating this data and knowledge of the field, which is essential to properly understand it. Experiential and field knowledge is strongly elided from SIOS interfaces, although it is articulated by researchers in other settings and now is gathered by SVALUR team members. In seeking to order and make visible these other aspects, the project has engaged with Maptionnaire, an interface that uses maps to evoke data from participants and link it to specific coordinates.[6] Maptionnaire therefore makes room for experience and field knowledge through a mapping interface that situates, but also makes it possible to *commensurate* (make measurable by a common standard) with, the effect of rendering value (in all senses of the term) comparable. The space of the map is then the common denominator that coordinates input and potentially links the narratives of SVALUR and the observations of SIOS (see Figure 1.1). This is a way of turning experience into a format that is compatible with monitoring so these different types of data become potential layers to be integrated through reliance on a Cartesian space that becomes the universal translator for integration of wide-ranging data.[7] While this approach does make visible a wider range of types of knowledge, it typically does so by placing them next to each other as independent layers that can be clicked on and off. As such, it does not make visible the interdependence of these elements and elides the importance of experiential and embodied knowledge for the production and understanding of monitoring data. This interface does part of the job of bringing other modes of knowing into the realm of monitoring data, but because it does so through a logic of commensurability, making stories and measurements comparable, it comes at the cost of inserting the SVALUR narratives into a universalizing epistemology.

SVALUR researchers are committed to demonstrating how embodiment and field experience, as undocumented elements that affect measurements, are important for SIOS and for better knowing Svalbard. But this relationship is challenging

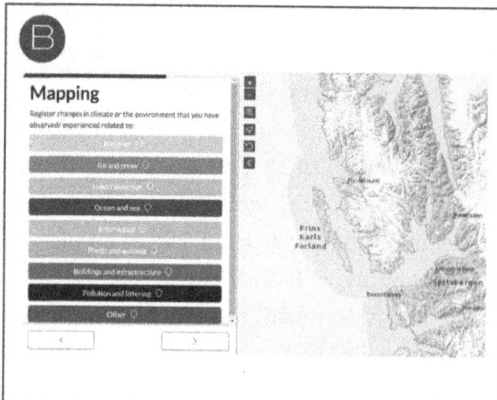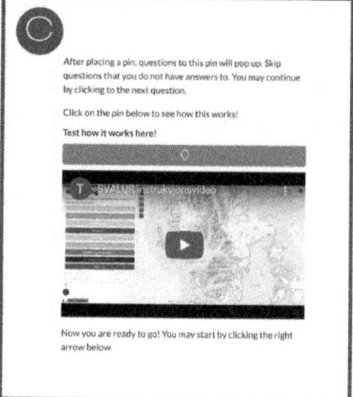

Figure 1.1. Elements of the SVALUR Maptionnaire interface included in a project publication. Reproduced with permission from Ann Eileen Lennert. Copyright Ann Eileen Lennert, all rights reserved.

to make visible. Members of the SVALUR team experienced different interfaces as reinforcing a dichotomy by pairing experience with narrative and empirical truth with numerical measurement data. To avoid subsuming what the team sees as valuable knowledge to a framing of experience as independent (if not completely extraneous) to monitoring, it has turned to an entirely different set of interfaces to foreground experience: artistic encounters on Svalbard and workshops at scientific conferences, along with opinion pieces in scientific journals. The hope is that these forms better convey the complexity, interplay, entanglement, and fluidity of field experience and its relationship to monitoring data. While these interfaces enable more intimate and coherent interactions among modes of knowing, the project team saw them as a different kind of output: opinion pieces were positioned at a somewhat different level than the empirical publications; satellite workshops at conferences were not viewed as being quite as weighty as presentations in the main program; and live events were less enduring than digital infrastructure. To do justice to the entwinement of narrative and measurement, other formats seemed necessary, but they were felt to be outside the mainstream.

In a third line of work, researchers in SVALUR examine their own knowledge production by expanding the range of interfaces that they work with. The team has experimented with the use of Obsidian to document the research process, led by one member's interest in this note-taking tool. Obsidian is designed to highlight connections among items and to map out complex sets of relations, and it is descended from knowledge systems that have index cards at their heart.[8] It supports a multiplicity of sources and relations and makes both searching and browsing possible via both textual and graphical interfaces. In that sense, the platform welcomes the range of formats that SVALUR wants to embrace and supports the creation of limitless relations among all objects. Obsidian can also display information in countless ways that are not predetermined, and the platform's software is open to enable further design improvements. But this openness and malleability also give rise to uncertainty about what Obsidian makes visible, as researchers struggle to make sense of how to use this extensive source of documentation. They see the risks: sharing so much of the workflow of the project can lead to user overload or lead to overexposure of the cracks in the project. Obsidian is known as a personal knowledge management tool, or more colloquially as a "second brain"—qualifications that stress the intimate, personal purpose of the platform. Sharing interdependence and complexity can make projects and researchers vulnerable, even as it showcases precious insights about the entwinement of experience and outcomes of research.

Each of these lines of work has particularities that require tailored approaches, resulting in SVALUR struggling with the overarching logics of interfaces that impose strong divides: public and private/personal, time/space, specific/generalizable, observation/experience, and measurement/account. In developing the project, the team has felt a tension between stories and data and sought not to repeat that problem, based on the conviction that monitoring is also full of experiential knowledge and inscribed in a particular set of narratives common to environmental data collection.

Two elements are worth noting here. First, the data of SIOS would not make sense without strongly internalized—and externally broadcast—narratives about the value of mechanical objectivity, data sharing, disciplined observation, empiricism, and long-term monitoring. The unearthing and critique of these dominant but unmarked narratives may be necessary to arrive at interfaces that do justice to a reconfiguration of knowledge that troubles the alignment of data-scientific-global and experience-narrative-local. Second, as increased diversity of knowledge is on every agenda, from posthuman and new materialist (Conty) to Earth system approaches (Anderies, Mathias, and Janssen), this may be achieved only through intense attention to novel design practices since existing supports and platforms tend to reinscribe the status quo. A pilot project called Contaminating Encounters, in whose development I participated, explores such alternatives (see Figure 1.2). As Tomás Sánchez Criado formulates it, any real attempt at "creating a lasting epistemic transformation in these contexts of asymmetric knowledge politics—so those neglected knowledges can matter—would explicitly require an immense experimental investment in forms of designing careful conditions" (67). As discussed in relation to SVALUR, the challenge should not be underestimated, but interfaces are a good starting point to move toward the kinds of knowledge needed for the Anthropocene.

The challenge of developing the knowledge needed for the Anthropocene does not lie in acquiring more data, but in engaging in different relations. This is why interfaces are a good place of intervention and why digital humanities can be such a powerful source of inspiration, know-how, and conceptual innovation for the formalizations needed for production (van Zundert et al.). With regard to interfaces specifically, digital humanities expertise can help create better interfaces that

Figure. 1.2. Screenshot of an experimental interface from the Contaminating Encounters project, which embraces digital forms for storytelling and relation-making (https://www.contaminating-encounters.nl/). Copyright 2024 Zdenka Sokolickova and Joshua Schaubel.

instantiate and support other relations to data beyond search, retrieval, and computation. They can enable a focus on meaning in relation to measurements of objects, on topologies rather than location, and on possibility and potential rather than calculation and probability. These are neither interesting but trivial design tweaks, nor the addition of enriching layers, but rather deeply transformative ways of looking at interfaces and knowledge infrastructures. If we take seriously these alternative relations while developing interfaces, this approach can help us relate to data in a mode that is problematizing ("speculative") rather than declarative ("truth-claiming") (Johansson and Stenlund, 74). Finally, interfaces where different models of agency that would enhance awareness of collectivities as constitutive of our selves and our material worlds would open different ways of knowing and acting on pollution, soil health, or land use.

This is not an easy project, and there are difficulties to be expected. For example, we should expect to fight those who claim that there is no viable alternative (TINAs). In this battle, having a good set of exemplars can help, and the present volume contains a wealth of successful and sustainable projects that can serve this purpose. There will also be organizational and material consequences in developing new kinds of interfaces, on both the front and back end—think of relations between client and servers and the alignment of hardware and infrastructures to serve different interfaces, as well as the implications for the materials and energy needed to run these infrastructures.[9] At present, these are aligned to search logics, and our current World Wide Web is optimized for retrieval based on search.

New ways of designing interfaces will require different skills (Whitelaw) and the ability to interact in even more diverse teams. This is no small task, and anyone having worked in inter/multi/transdisciplinary projects in universities can testify to the misalignment of administrative and reward structures to such ways of working. At the same time, there are many lines of work and activism that can fuel the project described here. Allies are essential and can be sought in movements for openness, public internet, digital literacy, and leadership, and in the strong currents of science policy, development work, and citizen science that seek and support alternatives to a regime of monitoring and indicators.

NOTES

1. I would like to thank the members of the Knowledge Infrastructures Department, the members of the SVALUR team, and in particular the reviewers and editors of this volume for valuable feedback on this text.

2. This term is rightly criticized along different axes: its anthropocentrism (Brannen) and its flattening of the huge differences in the relation, responsibility, and vulnerability of different parts of Anthropos (Chakrabarty; Haraway; Todd and Davis). Also see Lorimer, "The Anthropo-Scene," for an insightful overview of different lines of work that use Anthropocene as a label.

3. On HuNI, see also chapter 12 in this volume by Verhoeven, Jones, Burrows, and Borda.

4. https://sios-svalbard.org/.

5. https://www.slu.se/svalur/.

6. https://www.maptionnaire.com/.

7. See efforts to redesign map interfaces specifically in the Iigliinit project (Gearheard et al. 2011).

8. https://obsidian.md/.

9. On the energy requirements of information technology (IT), see chapter 7 in this volume by Cha and Miller.

BIBLIOGRAPHY

Agostinho, Daniela. "The Optical Unconscious of Big Data: Datafication of Vision and Care for Unknown Futures." *Big Data & Society* 6, no. 1 (2019). https://doi.org/10.1177/2053951719826859.

Alaimo, Cristina, and Jannis Kallinikos. "Computing the Everyday: Social Media as Data Platforms." *The Information Society* 33, no. 4 (2017): 175–91. https://doi.org/10.1080/01972243.2017.1318327.

Andrews, Tara L., and Joris van Zundert. "What Are You Trying to Say? The Interface as an Integral Element of Argument." In *Digital Scholarly Editions as Interfaces*, edited by Roman Bleier, Martina Bürgemeister, Helmut W. Klug, et al., 3–33. Books on Demand, 2018.

Anderies, John, Jean-Denis Mathias, and Marco A. Janssen. "Knowledge Infrastructure and Safe Operating Spaces in Social–Ecological Systems." *Proceedings of the*

National Academy of Sciences 116, no. 12 (2019): 5277–84. https://doi.org/10.1073/pnas.1802885115.

Aouragh, Miriyam, Seda Gürses, Helen Pritchard, et al. "The Extractive Infrastructures of Contact Tracing Apps." *Journal of Environmental Media* 1, no. 2 (2020): 9.1–9.9. https://doi.org/10.1386/jem_00030_1.

Arola, Kristin L. "Indigenous Interfaces." In *Social Writing/Social Media: Publics, Presentations, and Pedagogies,* edited by Douglas M. Walls and Stephanie Vie, 209–24. University Press of Colorado, 2017. https://doi.org/10.37514/PER-B.2017.0063.2.11.

Beaulieu, Anne. "Organising Knowledge for Sustainable Futures." In *Handbook of the Anthropology of Technology,* edited by Maja Hojer Bruun, Dorthe Brogaard Kristensen, Rachel Douglas-Jones, et al., 355–77. Palgrave, 2022.

Beaulieu, Anne. *Revealing Relations: Knowledge Infrastructures for Liveable Futures.* Bristol University Press, 2026.

Beaulieu, Anne. "A Space for Measuring Mind and Brain: Interdisciplinarity and Digital Tools in the Development of Brain Mapping and Functional Imaging, 1980–1990." *Brain and Cognition* 49, no. 1 (2002): 13–33. https://doi.org/10.1006/brcg.2001.1461.

Beaulieu, Anne. "Voxels in the Brain: Neuroscience, Informatics and Changing Notions of Objectivity." *Social Studies of Science* 31, no. 5 (2001): 635–80. https://doi.org/10.1177/030631201031005001.

Beaulieu, Anne, and Sarah de Rijcke. "Networked Knowledge and Epistemic Authority in the Development of Virtual Museums." In *Museums in a Digital Culture: How Art and Heritage Become Meaningful,* edited by Chiel van den Akker and Susan Legene, 75–92. Amsterdam University Press, 2016. http://muse.jhu.edu/book/66531.

Beaulieu, Anne, and Sabina Leonelli. *Data and Society: A Critical Introduction.* Sage Publications, 2021.

Bellacasa, María Puig de la. *Matters of Care: Speculative Ethics in More than Human Worlds.* 3rd ed. University of Minnesota Press, 2017.

Bigo, Didier, Engin Isin, and Evelyn Ruppert. "Data Politics." In *Data Politics: Worlds, Subjects, Rights,* edited by Didier Bigo, Engin Isin, and Evelyn Ruppert, 1–17. Routledge, 2019.

Bowker, Geoffrey, and Susan Leigh Star. *Sorting Things Out: Classification and Its Consequences.* MIT Press, 1999.

Brannen, Peter. "The Anthropocene Is a Joke." *The Atlantic.* August 13, 2019. https://www.theatlantic.com/science/archive/2019/08/arrogance-anthropocene/595795/.

Brown, Susan. "Remediating the Editor." *Interdisciplinary Science Reviews* 40, no. 1 (2015): 78–94. https://doi.org/10.1179/0308018814Z.000000000106.

Brown, Susan, Tany Clement, Laura Mandell, et al. "Creating Feminist Infrastructure in the Digital Humanities." In *Digital Humanities 2016: Conference Abstracts,* 47–50. Jagiellonian University and Pedagogical University, 2016.

Chakrabarty, Dipesh. "The Human Condition in the Anthropocene." *The Tanner Lectures on Human Values.* Lecture, Yale University, February 18–19, 2015. https://tannerlectures.utah.edu/_resources/documents/a-to-z/c/Chakrabarty%20manuscript.pdf.

Christen, Kimberly. "Relationships, Not Records: Digital Heritage and the Ethics of Sharing Indigenous Knowledge Online." In *The Routledge Companion to Media Studies and Digital Humanities*, edited by Jentery Sayers, 403–13. Routledge, 2018.

Cielemęcka, Olga, and Christine Daigle. "Posthuman Sustainability: An Ethos for Our Anthropocenic Future." *Theory, Culture & Society* 36, no. 7–8 (2019): 67–87. https://doi.org/10.1177/0263276419873710.

Contaminating Encounters. 2024. Zdenka Sokolickova and Joshua Schauble (creators), in collaboration with Anne Beaulieu, Hendrik Elzinga, Martinus van Hoorn, et al. University of Groningen, https://www.rug.nl/research/arctisch-centrum/?lang=en.

Conty, Arianne Françoise. "The Politics of Nature: New Materialist Responses to the Anthropocene." *Theory, Culture & Society* 35, no. 7–8 (2018): 73–96. https://doi.org/10.1177/0263276418802891.

Couldry, Nick, and Ulises A. Mejias. *The Costs of Connection: How Data Is Colonizing Human Life and Appropriating It for Capitalism*. Stanford University Press, 2019.

Cramer, Florian, and Matthew Fuller. "Interface." In *Software Studies: A Lexicon*, edited by Matthew Fuller, 149–52. MIT Press, 2008. http://direct.mit.edu/books/book/1924/chapter/52857/Interface.

Criado, Tomás Sánchez. "Anthropology as a Careful Design Practice?" *Zeitschrift für Ethnologie* 145 (2020): 47–70.

Dijck, José van, Thomas Poell, and Martijn de Waal. *The Platform Society: Public Values in a Connective World*. Oxford University Press, 2018.

Drucker, Johanna. *Graphesis: Visual Forms of Knowledge Production*. Harvard University Press, 2014.

Drucker, Johanna. *Visualization and Interpretation: Humanistic Approaches to Display*. MIT Press, 2020.

Edwards, Paul N. "Knowledge Infrastructures for the Anthropocene." *The Anthropocene Review* 4, no. 1 (2017): 34–43. https://doi.org/10.1177/2053019616679854.

Eren, Selen, and Anne Beaulieu. "Intermittent Care in Conservation Science: From Individual Birds in the Hand to Species Data in the Bank." *Theory, Culture & Society*, 2023. https://doi.org/10.1177/02632764231187584.

Espeland, Wendy Nelson, and Mitchell L. Stevens. "Commensuration as a Social Process." *Annual Review of Sociology* 24, no.1 (1998): 313–43. https://doi.org/10.1146/annurev.soc.24.1.313.

Fricker, Miranda. *Epistemic Injustice: Power and the Ethics of Knowing*. Oxford University Press, 2009.

Gabrys, Jennifer. "Sensing a Planet in Crisis." *Media+Environment* 1, no. 1 (2019): 10036. https://doi.org/10.1525/001c.10036.

Galloway, Alexander R. 2012. *The Interface Effect*. Polity.

Gearheard, Shari, Claudio Aporta, Gary Aipellee, et al. "The Igliniit Project: Inuit Hunters Document Life on the Trail to Map and Monitor Arctic Change." *The Canadian Geographer / Le Géographe Canadien* 55, no. 1(2011): 42–55. https://doi.org/10.1111/j.1541-0064.2010.00344.x.

Gobbo, Federico, and Federica Russo. "Epistemic Diversity and the Question of Lingua Franca in Science and Philosophy." *Foundations of Science* 25, no. 1: 185–207. https://doi.org/10.1007/s10699-019-09631-6.

Graeber, David, and David Wengrow. 2021. *The Dawn of Everything: A New History of Humanity*. McClelland & Stewart.

Haraway, Donna. "Anthropocene, Capitalocene, Plantationocene, Chthulucene: Making Kin." *Environmental Humanities* 6, no. 1: 159–65. https://doi.org/10.1215/22011919-3615934.

Harrison, Steve, Phoebe Sengers, and Deborah Tatar. "The Three Paradigms of HCI." *Proceedings of Alt.Chi,* 2007. https://sciencetechnologystudies.journal.fi/article/view/56743.

Holtrop, Tjitske. "6.15%: Taking Numbers at Interface Value." *Science & Technology Studies* 31, no. 4 (2018): 75–88. https://doi.org/10.23987/sts.56743.

Hookway, Branden. *Interface*. MIT Press, 2014.

Jensen, Casper Bruun, and Atsuro Morita. "Introduction: Infrastructures as Ontological Experiments." *Ethnos* 82, no. 4 (2017): 615–26. https://doi.org/10.1080/00141844.2015.1107607.

Jue, Melody. *Wild Blue Media: Thinking through Seawater*. Illustrated ed. Duke University Press, 2020.

Lennert, Ann Eileen, René van der Wal, Jasmine Zhang, et al. "Rich Local Knowledge Despite High Transience in an Arctic Community Experiencing Rapid Environmental Change." *Humanities and Social Sciences Communications* 10, no. 1 (November 4, 2023): 1–15. https://doi.org/10.1057/s41599-023-02310-9.

Lenoir, Timothy. *Inscribing Science: Scientific Texts and the Materiality of Communication*. Stanford University Press, 1997.

Liboiron, Max. *Pollution Is Colonialism*. Duke University Press, 2021.

Liu, Alan. "Toward Critical Infrastructure Studies." April 21, 2018, https://cistudies.org/wp-content/uploads/Toward-Critical-Infrastructure-Studies.pdf.

Liu, Alan. "Toward a Diversity Stack: Digital Humanities and Diversity as Technical Problem." *PMLA* 135, no. 1 (2020): 130–51. https://doi.org/10.1632/pmla.2020.135.1.130.

Lorimer, Jamie. "The Anthropo-Scene: A Guide for the Perplexed." *Social Studies of Science* 47, no. 1 (2017): 117–42. https://doi.org/10.1177/0306312716671039.

Lorimer, Jamie. *Wildlife in the Anthropocene: Conservation after Nature*. University of Minnesota Press, 2015.

Mackenzie, Adrian. *Machine Learners: Archaeology of a Data Practice*. MIT Press, 2017.

Maptionnaire. Platform to design and manage citizen engagement. Accessed September 26, 2023, https://www.maptionnaire.com/.

Mattar, Sennan D., Tahseen Jafry, Patrick Schröder, et al. "Climate Justice: Priorities for Equitable Recovery from the Pandemic." *Climate Policy* 21, no. 10 (2021): 1307–17. https://doi.org/10.1080/14693062.2021.1976095.

McGillivray, Barbara, Beatrice Alex, Sarah Ames, et al. "The Challenges and Prospects of the Intersection of Humanities and Data Science: A White Paper from the Alan Turing Institute." August 2020. https://doi.org/doi:10.6084/m9.figshare.12732164.

McLean, Jessica. "Feeling the Digital Anthropocene." In *Changing Digital Geographies: Technologies, Environments and People,* edited by Jessica McLean, 159–75. Springer International Publishing, 2020. https://doi.org/10.1007/978-3-030-28307-0_8.

Murphy, Michelle. "Against Population, Toward Afterlife." In *Making Kin, not Population,* edited by Adele E Clarke and Donna J Haraway, 102–24. Prickly Paradigm Press, 2018.

Obsidian. Note-taking platform and writing app. Accessed September 26, 2023, https://obsidian.md/.

Pawlicka-Deger, Urszula. "Infrastructuring Digital Humanities: On Relational Infrastructure and Global Reconfiguration of the Field." *Digital Scholarship in the Humanities,* 37, no. 2 (2021): 534–550. https://doi.org10.1093/llc/fqab086.

Povinelli, Elizabeth A. *Geontologies: A Requiem to Late Liberalism.* Duke University Press, 2016. https://doi.org/10.1215/9780822373810.

Powers, Martha, Phil Brown, Grace Poudrier, et al. "COVID-19 as Eco-pandemic Injustice: Opportunities for Collective and Antiracist Approaches to Environmental Health." *Journal of Health and Social Behavior* 62, no. 2 (2021): 222–29. https://doi.org/10.1177/00221465211005704.

Ruecker, Stan, Milena Radzikowska, and Steven Sinclair. *Visual Interface Design for Digital Cultural Heritage: A Guide to Rich-Prospect Browsing.* Routledge, 2016.

Ruppert, Evelyn, and Engin Isin. "Data's Empire: Postcolonial Data Politics." In *Data Politics: Worlds, Subjects, Rights,* edited by Didier Bigo, Engin Isin, and Evelyn Ruppert, 207–27. Routledge, 2019.

Schroer, Sara Asu. "The Arts of Coexistence: A View from Anthropology." *Frontiers in Conservation Science* 2 (2021). https://www.frontiersin.org/article/10.3389/fcosc.2021.711019.

SIOS. Svalbard Integrated Arctic Earth Observing System Project. Accessed September 26, 2023, https://sios-svalbard.org/.

Johansson, Veronica, and Jörgen Stenlund. "Making Time/Breaking Time: Critical Literacy and Politics of Time in Data Visualisation." *Journal of Documentation* 78, no.1 (2021): 60–82. https://doi.org/10.1108/JD-12-2020-0210.

SVALUR: Understanding Resilience and Long-Term Ecosystem Change in the High Arctic. Accessed September 26, 2023, https://www.slu.se/svalur.

Tichenor, Marlee, Sally E Merry, Sotiria Grek, et al. "Global Public Policy in a Quantified World: Sustainable Development Goals as Epistemic Infrastructures." *Policy and Society* 41, no. 4 (2022): 431–44. https://doi.org/10.1093/polsoc/puac015.

Todd, Zoe. "An Indigenous Feminist's Take on the Ontological Turn: 'Ontology' Is Just Another Word for Colonialism." *Journal of Historical Sociology* 29, no. 1 (2016): 4–22. https://doi.org10.1111/johs.12124.

Todd, Zoe, and Heather Davis. "On the Importance of a Date, or, Decolonizing the Anthropocene." *ACME: An International Journal for Critical Geographies* 16, no. 4 (2017): 761–80.

Tsing, Anna. *Friction: An Ethnography of Global Connection.* Princeton University Press, 2005.

Tsing, Anna. *The Mushroom at the End of the World: On the Possibility of Life in Capitalist Ruins.* Princeton University Press, 2015.

Turnbull, David. "Futures for Indigenous Knowledges." *Futures* 41, no. 1 (2009): 1–5. https://doi.org/10.1016/j.futures.2008.07.002.

Turnbull, Jonathon, Adam Searle, and William M. Adams. "Quarantine Encounters with Digital Animals: More-than-Human Geographies of Lockdown Life." *Journal of Environmental Media* 1, no. 1 (2020): 6.1–6.10. https://doi.org/10.1386/jem_00027_1.

Turnhout, Esther, and Andy Purvis. "Biodiversity and Species Extinction: Categorisation, Calculation, and Communication." *Griffith Law Review* 29, no. 4 (2020): 669–85. doi: 10.1080/10383441.2020.1925204.

Ulrik, Christian, and Soeren Bro Pold. "Introduction." In *Interface Criticism: Aesthetics Beyond Buttons,* edited by Christian Ulrik and Soeren Bro Pold. Aarhus University Press, 2011.

Verhoeven, Deb. "HuNI: Helping Humanities Researchers Get Lucky." YouTube, October 22, 2014, https://www.youtube.com/watch?v=qiFj7JjlaSQ.

Whitelaw, Mitchell. "Generous Interfaces for Digital Cultural Collections." *DHQ: Digital Humanities Quarterly* 9, no. 1 (2015) https://www.digitalhumanities.org/dhq/vol/9/1/000205/000205.html.

Whitelaw, Mitchell, and Belinda Smaill. "Biodiversity Data as Public Environmental Media: Citizen Science Projects, National Databases and Data Visualizations." *Journal of Environmental Media* 2, no. 1 (2021): 79–99. https://doi.org/10.1386/jem_00041_1.

Zundert, Joris van, Smiljana Antonijevic, Anne Beaulieu, et al. "Cultures of Formalization: Towards an Encounter between Humanities and Computing." In *Understanding Digital Humanities: The Computational Turn and New Technology,* edited by David M. Berry, 279–294. Palgrave Macmillan, 2011.

PART I][*Chapter 2*

Replatforming

SUSAN BROWN

Digital studies increasingly recognize the importance of platforms from the web as a whole, through internet-based services and specific sites, to hardware/software couplings such as gaming systems. Among *platform*'s earliest senses in the *OED* is "A raised level surface on which people or things can stand" to enable speech, performance, and political activity. The word's meanings then extended as technologies developed—from ships, railways, and oil rigs to its computational sense of "a standard system architecture; a (type of) machine and/or operating system, regarded as the base on which software applications are run" (*OED*, s.v. "Platform").

Platforms are sites of power and potential for political change. Associated with the rise of political protest and democracy, platforms as technologies for embodied speech were most famously institutionalized in 1872 at Speakers' Corner in Hyde Park, London (Meisel 224, 236, 273). Corporate web platforms, which are now naturalized as the default locations for most online speech, are currently undergoing intense scrutiny due to their immense economic and social impact. While many believe that "platform alternatives" (Lovink 207) are elusive, others are more optimistic. Jennifer Wemigwans (Anishnaabekwe [Ojibwe/Potawatomi], from Wikwemikong First Nation) argues that Indigenous platforms can propel new forms of relation and knowledge creation, "contributing to a radical Indigenous resurgence" (32–33). Platforms are sites of contestation, their significatory slipperiness amplified by the discourse of global technology companies that strategically cast themselves as neutral facilitators of others' speech to obscure "real and substantive interventions into the contours of public discourse" (Gillespie, "Politics," 348).[1]

Platforms in one sense are foundational components underlying a layered computing system or stack (Montfort and Bogost). However, large-scale platforms are now understood as messy totalities (Lovink 188), sociotechnical entities characterized by Anne Helmond as "the dominant infrastructural and economic model of the social web" ("Platformatization") and linked by Nick Srnicek and Shoshana Zuboff to new capitalist forms that derive value from user behavior and data (*Platform*

Capitalism; The Age of Surveillance Capitalism). Platforms have massive social implications in areas as diverse as mental health, labor relations, and political processes. Their material components are enmeshed in the geopolitics of colonial extractive capitalism and energy consumption.[2] The circulation of human expression through digital platforms is thus imbricated with the most profound challenges of our era.

This chapter probes scholarly platforms through the concept of *replatforming*, a critical inversion of what social media calls *deplatforming*. Deplatforming bans individuals from platforms, whereas *replatforming* invokes collective iterative activity. Replatforming prompts inquiry into what it means to provide and sustain platforms to advance digital scholarship and online diversity. My argument has three sections discussing the benefits and risks of scholarly platforms (*re:platforming*); the critical implications of republishing data (*re-platforming*); and the challenges of ongoing repair and rebuilding (*replatforming*). While they differ in scale and purpose from the sites of platform capitalism, digital humanities (DH) platforms nevertheless warrant examination. Intersectional feminist, Indigenous, critical race, and archival studies suggest how infrastructure can advance social justice and mitigate harm, underscoring the need for scholarly involvement in the ongoing relational work of creating and sustaining knowledge platforms.

Re:platforming

"Platform thinking" is characteristic of DH, argues Steven E. Jones, stressing the human collaborations behind "layered systems, built and rebuilt by maker-scholars as starting places, platforms on which to build future things" (16, 174). I focus here on such scholar-led infrastructure that emerges from researchers' practices (Anderson, 10), including three interlinked DH projects, one that stumbled into creating infrastructure and the others funded as such, that I have codeveloped (Orlando Project, "Credits"; Canadian Writing Research Collaboratory, "Credits and Acknowledgements"; Linked Infrastructure for Networked Cultural Scholarship, "Our Team"). The first, the Orlando Project, an experiment since 1995, recently redesigned its original publication platform, but before that had retired its homegrown production backend in favor of the second, the Canadian Writing Research Collaboratory (CWRC, pronounced "quirk"), which launched in 2016 and relaunched in 2025 after a full rebuild. CWRC serves as a virtual research environment for many individual projects built on it. The third, the Linked Infrastructure for Networked Cultural Scholarship (LINCS), provides tools for converting and publishing datasets as linked open data. They are illustrations, accompanied by other projects that I will mention, of the complexities of what Alan Liu has called the diversity stack: "a fused techno-ideological apparatus—a platform in all senses" (133).

The born-digital scholarship of the Orlando Project aimed to platform marginalized voices through detailed treatment of British women's writing. Orlando now profiles 1,444 writers using 3 million semantic tags embedded in 9 million words.

Of course, such counting is not in itself meaningful, and this summary misrepresents the ethos of a project devoted to qualitative data. In evaluating feminist digital archives, Jacqueline Wernimont cautions against reliance on "patriarchal tropes of size, mastery, and comprehensive collection" that reinforce funders' notions of impact (par. 4, par. 5). Yet these numbers are important as measures of inclusion and a recognition that formation of the literary canon in digital space stems as much from inequitable access to infrastructure as from abstract measures of value. On the early do-it-yourself (DIY) web, Amy Earhart notes, digital literary scholarship "reinserted women, queers, and people of color into the canon" (*Traces*, 66). However, "the democratization of knowledge made possible by the developing technological infrastructure" nosedived as web publishing became less accessible and much early recovery work was lost, along with the ability to sustain platforms ("Can information?"). Miriam Posner therefore assigns an ethical mandate to those who operate DH platforms: "it is incumbent on all of us (but particularly those of us who have platforms) to push for the inclusion of underrepresented communities in digital humanities work, because it will make all of our work stronger and sounder" (39). Scholarly platforms can fight injustice by amplifying marginalized voices and nonhegemonic perspectives.[3]

The Orlando Project dedicated its semantic markup to platforming difference while seeking to diversify online content. From what is now called an intersectional feminist approach, Orlando devised a means of encoding differences, refusing, for instance, to let whiteness operate as a silent and unmarked norm against which a smaller number of Black, Indigenous, and people of color (BIPOC) writers were defined, and encoding ethnicity, religion, class, and gender in conjunction with race. It was activist in avowing the constructedness of its markup and the ideological investments of its knowledge representation. Embedded semantics helped to counter some of the problematic implications of platforming individualist authorial profiles—for instance, by enabling materials to be retrieved by the semantic tag for "destruction of work" instead of only by text search, revealing the recurring suppression of women's writing by male relatives and authorities. The tagging works across individual profiles to illuminate systemic patterns and processes of change. Orlando's ability to platform difference depended in the early 2000s upon building a multilayered platform comprised of a markup schema or data model, a backend collaborative editing system, and an experimental front-end that leveraged tagging.[4] This platform allowed Orlando to register as a significant intervention that shifted how one thinks (Nixon), modeled intersectional knowledge representation (Risam), and changed the study of women's writing (Bowers).

The CWRC emerged from rebuilding Orlando's aging infrastructure, aiming to increase the diversity of scholarly web content and empower those with modest technical expertise to adopt DH practices. CWRC hosts 400,000+ objects (e.g., texts, images, videos, and bibliographic records). The thirty-five projects that it houses include Karen Skinazi's collection of writings by Winnifred Eaton (pen name

Onoto Watanna), the first Asian North American novelist; *Canadian Jewish Women Writers,* directed by Catherine Caulfield; and *The People and the Text,* a history of Indigenous texts directed by Métis/Cree scholar Deanna Reder. For CWRC as a settler-led project to collaborate with *The People and the Text* in platforming Indigenous scholarly and literary voices has been a meaningful way to work toward truth and reconciliation.

Indigenous infrastructures lead the way in decolonizing information systems and centering sustainability. For some content, both data and platform sovereignty are essential to Indigenous sovereignty. The OCAP® principles of "ownership, control, access, and possession" assert First Nations control over data collection and use, and with the CARE principles for Indigenous data governance (*C*ollective benefit, *A*uthority to Control, *R*esponsibility, *E*thics), complement the FAIR (*F*indable, *A*ccessible, *I*nteroperable, *R*eusable) principles for open data (GO FAIR) in recognizing the tension between openness and data sovereignty (Carroll). Assumptions about universal openness as an unquestioned good are challenged when ownership is understood as collective and access as more complicated than all or nothing. Traditional Knowledge (TK) labels, as developed by Jane Anderson, Kim Christen, and colleagues (Anderson and Christen, "'Chuck a Copyright on It'") meet the infrastructural need to inscribe protocols developed in consultation with Indigenous communities. Platforms such as Mukurtu operationalize the metadata represented by TK labels to guide who should have access to what content under what circumstances (Christen, "Does Information?"). Such systems reinforce other scholars' insights that people, communities, and relationality are the heart of infrastructure (Endings; Edwards et al.; Star).

The state's "power to name" was key to how railways, as communications infrastructure, advanced colonialism (Duarte and Belarde-Lewis, 681, quoting Olson), so Indigenous approaches to situating knowledge through metadata can model responsible platforming. For Wemigwans, a "digital bundle" of data and its cultural protocols provide an infrastructure for "decolonizing the digital" by countering decontextualization (*A Digital,* 42, 43; "Digital Bundles").[5] Imaginings grounded in Indigenous knowledges and decolonizing methodologies that accommodate "a plurality of knowledge systems" (Duarte and Belarde-Lewis, 678) can help overwrite colonial classifications, but they themselves require platforms. The Respectful Terminology Platform Project, led by Stacy Allison-Cassin (Métis Nation of Ontario) and Camille Callison (Tahltan Nation), will enable the development of a dynamic, multilingual set of terminologies to replace inappropriate metadata and model strategies for countering harmful data legacies.

Parallel concerns emerge when considering how platforms can promote diversity and epistemic justice in collaboration with affected communities (Fricker; Poirier and Costelloe-Kuehn). T. L. Cowan and Jasmine Rault's Cabaret Commons, a community-based and -sourced project in CWRC initially devoted to ephemera from the queer cabaret scene, highlights the complexity of platforming data involving living persons. Their speculative exploration, continuing through the Digital Research Ethics Collaboratory, produced a closed archive in which data initially uploaded in a pilot to CWRC was kept dark due to privacy and permissions concerns. The collaborative Feminist Data Manifest-No, with Cowan and Rault as coauthors, similarly creates a no-data anti-archive that treats data as fundamentally personal, tactical, and temporary: "*We refuse* the use of data about people in perpetuity. *We commit* to embracing agency and working with intentionality, preparing bodies or corpuses of data to be laid to rest when they are not being used in service to the people about whom they were created" (Cifor et al.). The Manifest-No collaborative builds on a long tradition of feminist understanding of the imbrication of processes of knowing with power and social relations (Scott). Focused on the data extraction, dataveillance, and vulnerability of minoritized groups associated with corporate platforms, it stresses the harm inflicted by data on persons treated as research objects rather than coresearchers, and insists upon the right to be forgotten (Cifor et al.; Scott).

Platforming thus evinces online infrastructure's dynamic presentism. Facing current scholarly and social debates, resurgent and emergent epistemologies, and evolving understandings of responsibility and accountability requires that scholarly platforms be grounded in the relationality of care and repair for people and communities, as well as for machines and software. Like large corporate platforms (Helmond, "Facebook's Evolution," 125), but for other reasons, scholarly platforms engage in ongoing updates and iteration as resources permit, resituating content (*re-platforming*) and renewing systems (*replatforming*)

Re-platforming

While deplatforming typically limits platform access to mitigate harm (Thomas and Wahedi), the concept of *re-platforming* raises questions about the potential harm involved in republishing online already public information or migrating online content to a different platform.[6] Re-platforming emphasizes the importance of asking whether and how to republish existing data, its hyphen symbolizing the pause that should accompany such decisions, even if the data was previously published online. *Deplatforming*, as noted previously, came to the fore in connection with social media platforms, and, like *platform*, the word is etymologically rooted in live speech (Perlman).[7] Objections to deplatforming in a social media context typically critique limitations on speech, whereas arguments for it stress the public good and denounce the erosion of the information commons by private interests (Gillespie, "Politics"; Jhaver et al.; Rogers). Debate informed by postcolonial, queer, and feminist critiques addressing the ethics of what I am terming re-platforming has been extensive within the archival community (Agostinho; Dalgleish; Povinelli). It deserves more discussion within DH, particularly since sustainability, preservation, and access considerations are making re-platforming increasingly common activities in the field.

Re-platforming can seem straightforward, especially with respect to historical data. For instance, CWRC re-platformed Thomas B. Vincent's CD-ROM *Index to Pre-1900 Canadian Cultural and Literary Magazines* (originally published in 1994) as the *Early Canadian Cultural Journals Index*.[8] This index constitutes a valuable resource for researching early Canadian culture because it indexes many journals found online in the *Canadiana* collection, which is among the largest collections of digitized Canadian documentary heritage in the world. *Canadiana* provides the full text of many of these magazines but lacks detailed indexing, whereas Vincent indexed editorial content, provided keywords, and designed the dataset to support future linking ("Opportunities"). Online and downloadable, with a Creative Commons license, *Early Canadian Cultural Journals Index* is more accessible and supports analysis in ways that go beyond the FAIR principles because it is not only archived for preservation with other CWRC data but also accessible alongside related content through a dynamic interface.

There are problematic angles to data rescue, however. Marika Cifor draws on Sara Ahmed, Lauren Berlant, and Ann Cvetkovich to argue that archives create affective relations with social justice implications, "creating, documenting, maintaining, reconciling and (re)producing such relations—between records and people, ideologies, institutions, systems and worlds." As noted, Vincent's original *Index* recorded much of the content that would later be put online in *Canadiana*, whose home page warns that "The Canadiana Collections contain content created over five centuries that tell an incomplete, oftentimes distorted and, sometimes harmful, story of Canada. The content, metadata, and resource descriptions in the Canadiana Collections contain language that reflects the biases, norms, and perspectives of the time in which they were created. This includes harmful and offensive wording, cultural references, and stereotypes."

This warning is relevant to Vincent's *Index*. Counting item titles and keywords, the word "Indian" or "Indians" appears in 436 of the *Index*'s records as the predominant term for Indigenous peoples in the area now known as Canada. Moreover, titles such as "The Indian Problem and the Washakada Indian Home, Elkhorn, Man" refer to residential schools that caused individual and collective traumas. The CWRC home page for the re-platformed *Index* displays the same warning about problematic historical content. In re-platforming the data again for the updated platform, CWRC is also adopting measures followed by memory institutions, including *Canadiana*'s host (the Canadian Research Knowledge Network), in revising the most egregious terms in the *Index*, and it hopes to engage Indigenous experts, scholars of early Canadian writing, and library partners on how best to handle other potentially problematic terms that require deeper knowledge to address respectfully.

Re-platforming existing data, then, even when that data does not represent living persons, can re-present epistemic violence and perpetuate trauma associated with racism, colonialism, and other forms of oppression. Dialogue with evolving standards and values and affected communities is therefore crucial. Jessica Marie

Johnson lays out the affective tensions between the historical record and present needs to critique the complicity of digital design and implementation in commodifying Black bodies. Yet despite the horror invoked in her essay, she notes how the Trans-Atlantic Slave Trade Database, which is a central exemplum of her argument, was able to reshape the study of slavery by filling in, albeit with the smallest traces of material existence, the gap left by women and children excluded from the historical record. Johnson calls for a "Black digital practice [that] requires researchers to witness and remark on the marked and unmarked bodies, the ones that defy computation, and finds ways to hold the null values up to the light" (70–71).

Not only does the colonial archive suffer from harmful gaps and silences—untold numbers of unnamed human data points and "missing" datasets—but it also suffers from the reinforcement of the imperial gaze that can occur without careful contextualization of the cultural work done by "digital artifacts of a sensitive and dehumanizing nature" (Odumosu S298). Careful re-platforming can mitigate harm by negotiating the complexity of networked archives in the era of big data (Agostinho; Carter; Hartman; Onuoha). Alternatively, disrupting the algorithmic logic of "search, find, identify, claim, or steal," which keeps "the colonial episteme intact," Temi Odumosu argues, may require denying an artifact further reproduction, which is to say, refusing re-platforming (S298).

Re-platforming, then, risks perpetuating the extractive colonial logic that initially populated the collections of memory institutions through the further dissemination of representations of violence, violation, and atrocity, whether via images of enslaved persons, diagrams of slave ships, or representations of sacred Indigenous ceremonies. Decontextualized, often misattributed, or simply reproduced, such materials participate in what Susanne Kappeler describes as the pornography of representation, which reproduces and legitimizes the unequal power relations embedded in the artifact's creation, against which social justice strives (Anderson and Christen, "Decolonizing"; Duff; Callison et al.; Guiliano and Heitman; Kappeler).[9] As with the University of Nebraska's decision not to publish online a digitized archive of the satirical *Awgwan* student paper from the early 1900s, which was rife with discriminatory and derogatory content, sometimes outright refusal to re-platform is the answer (Brink et al.) and can serve as a generative stance (Tuck and Yang).

When digital content references living persons, as indicated earlier, complexities multiply. The digitization of *On Our Backs,* a pornographic lesbian magazine pivotal to the twentieth-century feminist sex wars, is a powerful example of the need for care in platforming formerly published content. Reveal Digital, a company that crowdfunds through libraries to digitize collections and make them open access, re-platformed the journal in 2012 as part of its "Independent Voices" collection. Reveal Digital withdrew *On Our Backs* in 2016 for multiple reasons, including concern about outing individuals represented in the collections. With respect to the latter, librarian Tara Robertson initiated the ethical debate in a blog post on her diversity, equity, and inclusion consultancy website, highlighting the tension between access and consent: "Consenting to a porn shoot that would be in a queer [print] magazine is a different thing to consenting to have your porn shoot be available online." Digital preservation and remediation often blur into re-platforming. Indeed, platform limitations (namely, the inability to restrict access by age to comply with pornography laws in some states) were a determining factor in the decision to pull *On Our Backs*, even though the periodical's content had been reframed as "alternative press" by its placement within the Independent Voices collection (Reveal Digital). That reframing remediated not only the content, but also "scale, audience, and interaction" through the JSTOR platform (Groeneveld).

These kinds of affordances are included in Jay Bolter and Richard Grusin's notion of remediation (*Remediation*). However, considering re-platforming raises the question of who is accountable for the distribution of remediated materials. Major social media platforms deny responsibility for hosted content (Gillespie, "Regulation," 666; "Politics"), a position contradicted by the practice of deplatforming. The mandate advocated here for scholarly platforms is an ethical one, based on the insight that the politics of digitization are inseparable from platform and cataloging choices that determine access.[10] Unpacking the tensions associated with such responsibility is crucial to understanding both the implications of how research platforms work and how we might build them otherwise.

Parallelling Indigenous articulations of data sovereignty are arguments that vulnerable groups need to control their own means of digital reproduction in ways that go beyond conventional understandings of copyright or consent (Boyles). As Elizabeth Groeneveld argues, corporate and academic archives whose "digitization frameworks have not been developed with the specific needs of minoritized communities in mind" operate with different values from "a grassroots community archive that emerges directly out of queer/feminist principles" (78). Exemplifying the magnitude of this challenge is Cowan and Rault's search for "media and infrastructures of risk and care" that produce a Cabaret Commons that refuses extractive practices and exploitation by transforming "institutional and platform logics" (122).

For Johnson, practices of hacking, remixing, and institutional resistance contribute to "an effort to dismantle the residue of commodification that is slavery's legacy" and an ongoing source of trauma. Such practices compel designers to "collaborate with the living descendants of the enslaved, who still claim as ancestor and kin those who can only be rendered in databases as '1' or a single pièce d'Inde" (Johnson 66, 71). The Digital Oral Histories for Reconciliation (DOHR) project, a platform for first-person voices of survivors of the Nova Scotia Home for Colored Children, hints at the possibilities. Collaborators Jennifer Roberts-Smith, Kristina Llewellyn, and members of the Victims of Institutional Child Exploitation Society (VOICES) incorporated "relational presence," which foregrounds difference rather than immersion, into the design of DOHR's virtual reality experience based on values that include "affective dissonance" (Smith et al.; Roberts-Smith). This platforming emerged from a community-based restorative justice process that supported intense design collaboration and the embedding of digital artifacts within a curricular process supported by trained facilitators—a whole-project rather than solely an infrastructure initiative. Moreover, VR permitted a degree of control over access and context impossible for web platforms. Scaling up and opening up on the web entail significant challenges, many associated with the linking that is the web's foundational affordance.

In fact, ordinary linking on the web needs considerable supplementation through linked data to overcome such challenges. Linking in the latter sense—that is, the disambiguation of data through reconciliation against shared and persistent vocabularies—is identified by Donald J. Waters as a crucial component of "emerging digital infrastructure for research in the humanities" (87). Re-platforming existing datasets in new forms, such as linked open data, makes data FAIR and yet demands critical consideration. Representing people and other entities unambiguously on the web and interlinking them has immense potential for advancing knowledge. For instance, linking named to unnamed subjects provides one approach to addressing the null values invoked by Johnson, opening up positions for the unnamed within historical networks of relationships. Indeed, the project *Enslaved: Peoples of the Historical Slave Trade* is re-platforming historical records related to enslaved persons as linked open data for inclusive and reparative ends.

Yet disambiguation can be deeply problematic in the context of queer identities. For instance, reusing unique identifiers to link people from *On Our Backs* to current identities or to link two discrete personas from different contexts could be ruinous or traumatic for those involved. As Cowan and Rault observe of past queer cabaret performers, "even the research required to find these folks and ask for permission can have the unintentionally harmful consequences of detonating a former name in a search engine, algorithmically linking that name, gender, sex, and/or other marker to a current identity that the person has worked hard to delink" (125).

LINCS recognizes that open is not an indisputable good, and that linking may harm. This suite of tools for remediating and re-platforming datasets aims to support the use of linked data in reciprocal and respectful ways as a bridge to closed or sensitive data. Enabling such use includes developing processes for flagging problematic data and honoring requests to anonymize or deprecate data, notwithstanding the challenges of excising portions of interlinked datasets. What problematic or sensitive data often requires, however, is context, which is greatly determined by platform configurations and affordances. This means that although domain experts bear responsibility for materials that they select or create, scholarly platforms have an obligation to address concerns that arise from re-platforming those materials[11] and to support contextualization. LINCS has sought to ensure that its core data structures—the ontologies that govern relationships between entities, determining what it is possible to represent structurally—support contextualization through the adoption of two core ontology structures, the CIDOC-CRM ontology and the Web Annotation Data Model. These two ontologies situate linked data assertions in ways that reduce the risk of decontextualization and help address the ethical concerns associated with re-platforming (Canning et al.; Brown, Canning et al.). LINCS maintains living policy documents to communicate platform values and to work with researchers toward a shared understanding of our mutual responsibilities. For instance, Figure 2.1 shows how the platform's ontology policies adapt ideas from Catherine D'Ignazio and Lauren Klein's *Data Feminism* (218–19) to lay out strategies for countering systemic inequality.

As LINCS grows, it will monitor whether attempts to convey values and critical data awareness through policy actually shape practices. Similarly, the ability of linked data to structure data in ways that support diverse epistemologies can be evaluated only when there is a critical mass of data representing different ways of knowing.

Re-platforming involves the "capacity of data to move on from the original signature and produce new possibilities" (Agostinho et al., 438). Those possibilities are exciting but entail risk. At times, not platforming may be more appropriate than platforming or re-platforming. CWRC has declined to host projects that its Research Board deemed incongruent with the platform's values, including a case where categories associated with the representation of Indigenous persons were not aligned with current vocabularies. Likewise, LINCS chose not to re-platform a social media dataset involving vulnerable subjects. Platforming becomes complicated when a dataset is no longer active, since domain experts are no longer engaged in caretaking related to sensitive content. It becomes a question of whether a platform has the expertise and resources to undertake remediative labor alongside all the other forms of care and repair required to sustain itself, the final sense of replatforming that I explore next.

Bias	Method
Ableism	Develop tools with accessibility considerations included from the point of inception.
Cissexism	Center trans perspectives in discussions of the gender binary; use trans-inclusive language.
Classism	Develop solutions that can be used by partners or others without requiring additional expensive technologies; provide training and education resources.
Colonialism	Adopt or develop ontologies that promote equal views of epistemological differences and/or promote Indigenous ways of knowing.
Discrimination	Promote understanding of different worldviews, along with histories, cultures, and contexts; use justice-oriented language.
Heteronormativity	Resist assumptions about family structures, sexual and romantic preferences and identities, and gender roles.
Racism	Center the perspectives of people of color in discussions of race.

Figure 2.1. Excerpt from the LINCS Ontology Adoption Policy (Canning and Brown et al.).

Replatforming

There is rising awareness within infrastructure studies of the critical importance of maintenance rather than innovation (Russell and Vinsel), and within DH of the importance for sustainability of factors such as people, communities, and the environment (Neylon; Nowviskie "Capacity," "Digital Humanities"; Drucker; Sample; Goddard and Seeman). Replatforming in this maintainer's and caretaker's sense involves the repetitive processes of operating, repairing, and upgrading scholarly platforms. For Steven J. Jackson, *care* is an ethical proposition that breaks down subject-object relations between humans and things. These blurred boundaries, coupled with the "gendered infrastructural imaginaries" (Agostinho and Thylstrup, 754) associated with container technologies (Sofia), lead to infrastructure and operations being devalued as support or service in contrast to higher values placed on innovation and production (Anderson; Brown, "Delivery Service"). Not only is the "repair or replacement of broken infrastructure . . . necessary for any form of sociality to extend itself," Lauren Berlant argues, but the extension of infrastructure can also be sufficiently generative to open up utopian possibilities of a renewed commons (393).

The desire to experiment and renew, of course, pulls against the need to standardize and stabilize for sustainability. Orlando had ambitions well beyond what its initial interface offered, planning affordances for exploring intersectional identities, analyzing markup patterns, and enabling user contributions. Some aspirations were realized, but with limitations; most resources were instead absorbed in fixing, tweaking, and upgrading the initial platform in ways that, while making a difference, were not transformative. After publication in 2006, Orlando's growth reflected the types of resources available; support in the form of research assistance enabled ongoing content production. Experimental interfaces created for the project stimulated insights (Holland and Elford; Brown, Ruecker et al.), but resources could not stretch to building production versions.[12] Sustainability needs thus pull against infrastructural ambitions; yet the paucity of funding for operations, as opposed to innovation, pushes platforms toward overextension rather than improvement of core functionality. In a DH context, this story of possibilities curtailed by sustainability needs is unusual only in that Orlando's longevity speaks to its phenomenal good fortune in being able to operate for as long as it has and rebuild both its front and back ends.[13]

Care and repair can be generative and rewarding. However, the term *replatforming* stresses the Sisyphean aspect of platform maintenance, even when conducted according to best practices within stabilizing institutional frameworks (Smithies et al.). The quotidian task of monitoring and maintaining a platform with scant resources so it is always accessible is shadowed by awareness that systems need major repairs, often suddenly and urgently, and eventual rebuilds. Even conceptualizing systems as distinct from people is wrongheaded (Ciula and Smithies), as becomes glaringly obvious when DH infrastructure fails because a sole sustainer disappears. People critical to the task of sustaining systems include professional technical staff, students, librarians, and/or faculty, whose labor may be divided, soft-funded, ill defined for this purpose (in relation to their job as a whole), or voluntary.

The precarity of scholar-led research platforms can exact high levels of administrative work in the perpetual production of grant applications to keep platforms going and, if the applications are successful, in project tracking and reporting. Such precarity consumes labor and creativity that could otherwise be directed toward generating infrastructural insights for platform renewal. Design thinking is easier at early stages, as is evident even in the case of the innovative Scalar platform for multimodal long-form arguments, which built front-end customizability into its architecture from the start (McPherson, *Feminist*). The more established an infrastructure, the greater the friction associated with design changes it increases once a tool shifts from being a purely lab-based solution or tactical object for thinking (Galey and Ruecker) to operational use. It is hardly surprising, given his expertise, that Bruno Latour built and then "gracefully degraded" (Nowviskie and Porter) rather than tried to sustain, the bespoke web platform for contributing content to his An Inquiry into Modes of Existence project. The platform, an act of critical making, was revised toward the end of its funding period, prioritizing normative platform criteria (speed, stability, usability, reusability, and durability) and mundane maintenance ("the host of problems such as errors, backtracking, modifications, and bugs that are all inherent to this kind of adventure") while dispensing with more innovative, labor-consuming features for engaging "co-inquirers" and active contributors (Ricci 50; Latour; Brown, "Taking AIME"). The major outcomes of Latour's project were disseminated in a monograph report, articles, and a website with reduced functionality.

Platforming with a Difference

I favor verb forms of *platform* because, as Susan Leigh Star and Karen Ruhleder emphasize, "infrastructure is something that emerges for people in practice, connected to activities and structures" (379). Platforming is processual and performative, played out each time functionality is invoked (Hayles, 274). Replatforming, as explored here, performs knowledge production, elevating voices through iterative acts that embody value. The political stakes of platforms can be elicited in "the ways in which social forms are written into the technological scaffolding of information, and how they reflect and materialize power dynamics, thereby structuring the possibilities for social action" (Agostinho and Thylstrup, 750). In recognizing that platforms are imbricated with power, academics and professionals can work to push scholarly platforms beyond established practices and reshape the intimate space of information exchange (Christen and Anderson, "Slow Archives").

Researchers are ethically responsible for how they interact with archives (Moravec), and yet how they do so is massively shaped by the platforms with which they interact as consumers, remediators, and producers of knowledge online. Envisioning how an archive of colonial violence might signal "where and how sensitivity is required, not as an optional stance but as a prerequisite for the digital encounter" (Odumosu, S298) suggests how platforms might perform what Anjali Arondekar calls "a counterrecord of that history" of the colonial moment (12). For Odumosu, metadata is a quiet "undercommons reconfiguring the digital thoroughfares (associations, keywords, hyperlinks) that bring a public into encounters with challenging histories" and "capable of narrating in full an object's life and afterlife, . . . making that known to users with each right-click and download" (S299). As her invocation of affordances offered through hyperlinks and clicks indicates, this vision conjoins metadata with platforming as an active, ethical intervention in embodied knowledge spaces. Johnson similarly envisions a subversive Black digital practice that "engages data promiscuously, across multiple platforms, taking up the nearest tools at hand to defy, dismiss, jeer, and sneer at the presumed legitimacy claimed by institutional structures and categories of analysis generated by the Ivory Tower" (71). The extent to which Indigenous, feminist, Black, queer, and trans values challenge normative institutional logics and epistemologies underscores the impediments to (re)platforming with a difference. As Cowan and Rault warn, it is a long game: "this will be a collective struggle that may take more than a generation to reframe" (136). Yet shifting slightly, iteratively, is still movement.

Jones characterizes DH platforms as "Frankenstein's monsters that we stitch together ourselves and for which we take responsibility" (174). Thinking through platforming helps identify the messy overlapping areas of collective and individual responsibilities involved in the creation, situation, and dissemination of networked content. The ongoing, iterative processes of platforming materials, re-platforming them in new forms, and renewing platforms in pursuit of better ways of knowing is rife with precarity but also with empowerment and promise. Understanding the relationships and responsibilities bound up in these multifaceted processes is key to pursuing the potential that is glimpsed, but only partially and unevenly realized, in scholarly platforming to date: a path to better situated and more ethical knowledge production and circulation.

Replatforming invokes the inevitable human processes whereby analog or digital materials are resituated in the process of migrating from one platform to another, and whereby scholarly platforms require care, repair, and eventual retirement or renewal. The challenges of replatforming with a difference are vast and varied because even platforms that offer modest alternatives to mainstream platforms defy logics that are fundamental to the economic and ideological conditions of our historical moment. For that reason, replatforming must be understood as a crucial component of ongoing efforts with DH to embody cultural critique and advance justice in the knowledge that we network and the systems that we build.

NOTES

1. Notably, after Google's 2006 acquisition of YouTube consolidated its power, YouTube swapped out the terms *website, service, forum,* and *community* in favor of describing itself as a "distribution platform for original content creators and advertisers large and small," thus aiming to "strike a regulatory sweet spot between legislative protections that benefit them and obligations that do not, and to lay out a cultural imaginary within which their service makes sense" (Gillespie, "Politics," 348). See Helmond et al. on how Facebook transformed from social media site to platform. See also Bratton (34).

2. Thus "Earth" is the name of the fundamental layer of Benjamin Bratton's *The Stack,* and of the opening chapter in Kate Crawford's *The Atlas of AI,* which stresses that the cloud is "made of rocks and lithium brine and crude oil" and is voraciously consuming rare "elements that required billions of years to form inside the earth" (31).

3. Earhart stresses the need for infrastructure to support small as well as large, high-profile projects; Moya Bailey highlights the challenges for community members of engaging with such infrastructure (cited in Boyles, 123).

4. The Orlando Project, a generously funded project directed by Patricia Clements (1995–2007), Susan Brown (2008–2015), and Brown, Katherine Binhammer, and Isobel Grundy (2016–present), originally presumed that available solutions would serve for project publication; therefore, nothing was budgeted for programming. The project was fortunate to be able to build both back-end content management and front-end web publication systems with in-kind resources, grant funding from the Canada Foundation for Innovation (secured by project member Susan Hockey), and a loan from the University of Alberta.

5. FourDirectionsTeachings.com, developed by Wemigwans in reciprocity and relationship with Indigenous Elders, "speaks back to dominant colonial systems of knowledge" through multimediality that expresses Indigenous Knowledge "in ways that contribute to the reflection of Indigenous ontology and values on the World Wide Web" (Four Directions Teachings.com; Wemigwans, *A Digital Bundle,* 2, 42).

6. Other senses of *replatforming* include shifting an application from one platform to another, typically cloud-based, platform, and Julia Ebner's use of the term to discuss the parallel online worlds of right-extremist alternative social media.

7. Deplatforming is most strongly associated with attempts to reduce hate speech and harassment associated with groups and individuals on the political right (e.g., Khoo), but affects others as well. For instance, mainstream journalists critical of Elon Musk were removed from Twitter after he acquired that platform and renamed it X; and, in another instance, feminist academics in the United Kingdom were allegedly silenced for their stance on gender self-identification legislation, inspiring the Cambridge Radical Feminist Network's Spring 2021 "Replatforming Deplatformed Women" series. Lovink doubts that deplatforming will "fundamentally change the ways entertainment and distraction are organized" (*Stuck,* 144).

8. This massive bibliographical dataset, published in 1993 as a CD-ROM (Vincent, *Index; Early*), contains 137,943 records of prose, poetry, drama, fiction, and other items

from 203 nineteenth-century periodicals published in Canada, many now digitized and available online through the *Canadiana* collections. It was re-platformed on CWRC with Vincent's permission.

9. Kappeler's choice of a photograph of white men displaying the body of a Black man whom they murdered as a paradigm of the representational structures she seeks to reframe anticipates Tonia Sutherland's analysis of the traumatic impacts for the black community in the United States of the repeated online trafficking of images of Black bodies in the absence of reparative and transitional justice: "While communities of color have long engaged in ritual practices of (re)membering and bearing witness to violent acts as modes of resistance and mourning; in digital spaces these practices have been appropriated to reinforce systems of white supremacist power and racial inequality, re-inscribing structural and systemic racism" (33).

10. Zaagsma's summary of the politics of digitization nicely articulates these imbrications.

11. Tara McPherson's handling of complaints about privacy violations in a Scalar project that had harvested social media content illustrates the tension between ensuring that re-platformed content aligns with a platform's ethical commitments and assigning researchers responsibility for the content they produce ("From *Vectors*," 48, 56).

12. Orlando's platform remained substantially unchanged from 2006 to 2022, a span that attests to the design foresight, talents, and dedication of technical-team lead Jeffery Antoniuk in keeping aging production and publication platforms functioning while juggling other responsibilities.

13. With its rebuilt interface launched in 2022, the team again hopes to extend functionality to do better justice to the data but struggles to meet fundamental production requirements. For a more detailed account in the Canadian funding context of infrastructure challenges (including for Orlando and for CWRC), see Susan Brown, Kim Martin, and Asen Ivanov.

BIBLIOGRAPHY

Agostinho, Daniela. "Archival Encounters: Rethinking Access and Care in Digital Colonial Archives." *Archival Science* 19 (2019): 141–65. https://doi.org/10.1007/s10502-019-09312-0.

Agostinho, Daniela, Catherine D'Ignazio, Annie Ring, et al. "Uncertain Archives: Approaching the Unknowns, Errors and Vulnerabilities of Big Data through Cultural Theories of the Archive." *Surveillance and Society* 17, no. 3/4 (2019): 422–41.

Agostinho, Daniela, and Nanna Bonde Thylstrup. "'If Truth Was a Woman': Leaky Infrastructures and the Gender Politics of Truth-Telling." *Ephemera* 19, no. 4 (2019): 745–75.

Anderson, Jane, and Kimberly Christen. "'Chuck a Copyright on It': Dilemmas of Digital Return and the Possibilities for Traditional Knowledge Licenses and Labels." *Museum Anthropology Review* 7, no. 1–2 (2013): 105.

Anderson, Jane, and Kimberly Christen. "Decolonizing Attribution: Traditions of Exclusion." *Journal of Radical Librarianship* 5 (2019): 113–52.

Anderson, Sheila. "What Are Research Infrastructures?" *International Journal of Humanities and Arts Computing* 7, no. 1–2 (2013): 4–23.

Arondekar, Anjali. "Without a Trace: Sexuality and the Colonial Archive." *Journal of the History of Sexuality* 14, no. 1/2 (2005): 10–27.

Berlant, Lauren. "The Commons: Infrastructures for Troubling Times." *Environment and Planning D: Society and Space* 34, no. 3 (2016): 393–419.

Bogost, Ian, and Nick Montfort, ed. "Platform Studies: A Book Series from The MIT Press." n.d. http://platformstudies.com/.

Bolter, Jay David, and Richard Grusin. *Remediation: Understanding New Media*. MIT Press, 2000.

Bowers, Toni. "Exploring the Richardson Circle Using the Orlando Database." *The Scriblerian and the Kit-Cats* 45, no. 1 (2012): 56–58.

Boyles, Christina. "Intersectionality and Infrastructure: Towards a Critical Digital Humanities." In *People, Practice, Power: Digital Humanities Outside the Center*, edited by Anne McGrail, Angel David Nieves, and Siobhan Senier, 118–26. University of Minnesota Press, 2021. https://www.jstor.org/stable/10.5749/j.ctv2782dmw.11.

Bratton, Benjamin H. *The Stack: On Software and Sovereignty*. MIT Press, 2016.

Brink, Peterson, Mary Ellen Ducey, and Elizabeth Lorang. "The Case of the *Awgwan*: Considering Ethics of Digitization and Access for Archives." *The Reading Room: A Journal of Special Collections* 2, no. 1 (2016): 1081.

Brown, Susan. "Delivery Service: Gender and the Political Unconscious of Digital Humanities." *Bodies of Information: Intersectional Feminism and Digital Humanities*, edited by Elizabeth Losh and Jacqueline Wernimont, 261–85. University of Minnesota Press, 2018.

Brown, Susan. "Taking AIME at Face Value: An ANT-like Crawl Through the Digital Project." *Resilience: A Journal of the Environmental Humanities* 4, no. 1 (2016): 130–36.

Brown, Susan, Erin Canning, Kim Martin, et al. "Ethical Considerations in the Development of Responsible Linked Open Data Infrastructure." In *Ethics in Linked Data*, edited by B. M. Watson, Alexandra Provo, and Kathleen Burlingame, 297–324. Library Juice Press, 2023.

Brown, Susan, Patricia Clements, and Isobel Grundy, ed. *Orlando: Women's Writing in the British Isles from the Beginnings to the Present*. Cambridge University Press, 2006–2024.

Brown, Susan, Kim Martin, and Asen Ivanov. "Linking Out: The Long Now of DH Infrastructures." In *Canadian Digital Humanities: Future Horizons*, edited by Paul Barrett and Sarah Roger, 315–47. University of Ottawa Press, 2023.

Brown, Susan, Stan Ruecker, Jeffery Antoniuk, et al. "Reading Orlando with the Mandala Browser: A Case Study in Algorithmic Criticism via Experimental Visualization." *Digital Studies/Le champ numérique* 2.1 (2010). https://www.digitalstudies.org/article/id/7226/.

Cabaret Commons. Home page. Directed by T. L. Cowan and Jasmine Rault. n.d. Canadian Writing Research Collaboratory. https://web.archive.org/web/20250214222024/https://cwrc.ca/project/cabaret-commons.

Callison, Camille, Ann Ludbrook, Victoria Owen, and Kim Nayer. "Engaging Respectfully with Indigenous Knowledges: Copyright, Customary Law, and Cultural Memory Institutions in Canada." *KULA: Knowledge Creation, Dissemination, and Preservation Studies* 5, no. 1 (2021). https://doi.org/10.18357/kula.146.

Cambridge Radical Feminist Network. Profile page. Facebook. https://www.facebook.com/camradfems/. Accessed November 19, 2022.

Canadian Jewish Women Writers. Home page. Directed by Catherine Caufield. n.d. Canadian Writing Research Collaboratory. https://cwrc.ca/project/canadian-jewish-women-writers.

Canadian Writing Research Collaboratory (CWRC). Home page. Directed by Susan Brown. n.d. https://cwrc.ca.

Canadian Writing Research Collaboratory (CWRC). "Credits and Acknowledgements." n.d. https://cwrc.ca/about/credits-and-acknowledgments.

Canadiana. Home page. 2023. Canadian Research Knowledge Network. https://www.canadiana.ca/.

Canadian Research Knowledge Network. "Decolonizing Canadiana Metadata: An Overdue Step in Removing Harmful Subject Headings." January 25, 2022. https://www.crkn-rcdr.ca/en/decolonizing-canadiana-metadata-overdue-step-removing-harmful-subject-headings.

Canning, Erin, and Susan Brown, with Kim Martin, Alliyya Mo and Sarah Roger. "LINCS Ontologies Adoption & Development Policy." Zenodo, February 11, 2022. https://doi.org/10.5281/zenodo.6047748.

Canning, Erin, Susan Brown, Sarah Roger, et al. "The Power to Structure: Making Meaning from Metadata Through Ontologies." *KULA: Knowledge Creation, Dissemination, and Preservation Studies,* 6, no. 3 (2022). https://doi.org/10.18357/kula.169.

Carroll, Stephanie Russo, Ibrahim Garba, Oscar L. Figueroa-Rodríguez, et al. "The CARE Principles for Indigenous Data Governance." *Data Science* 19, no 1. (2020): 43. http://doi.org/10.5334/dsj-2020-043.

Carter, Rodney G. S. "Of Things Said and Unsaid: Power, Archival Silences, and Power in Silence." *Archivaria,* 2006, 215–33.

Christen, Kimberly. "Does Information Really Want to Be Free? Indigenous Knowledge Systems and the Question of Openness." *International Journal of Communication* 6 (2012). https://ijoc.org/index.php/ijoc/article/view/1618.

Christen, Kimberly, and Jane Anderson. "Toward Slow Archives." *Archival Science* 19, no. 2 (June 1, 2019): 87–116. https://doi.org/10.1007/s10502-019-09307-x.

CIDOC-CRM Conceptual Reference Model. Home page. n.d. International Committee for Documentation, International Council of Museums. https://cidoc-crm.org/.

Cifor, Marika. "Affecting Relations: Introducing Affect Theory to Archival Discourse." *Archival Science* 16, no. 1 (2016): 7–31.

Cifor, Marika, Patricia Garcia, T. L. Cowan, et al. "Feminist Data Manifest-No." 2019. https://www.manifestno.com/.

Ciula, Arianna, and James Smithies. "Sustainability and Modelling at King's Digital Lab: Between Tradition and Innovation." In *On Making in the Digital Humanities: The Scholarship of Digital Humanities Development in Honour of John Bradley,* edited by Julianne Nyhan, Geoffrey Rockwell, Stéfan Sinclair, and Alexandra Ortolja-Baird, 78–104. UCL Press, 2023. https://library.oapen.org/handle/20.500.12657/62237.

Crawford, Kate. *The Atlas of AI: Power, Politics, and the Planetary Costs of Artificial Intelligence.* Yale University Press, 2021.

Cowan, T. L., and Jasmine Rault. "Onlining Queer Acts: Digital Research Ethics and Caring for Risky Archives." *Women & Performance: A Journal of Feminist Theory* 28, no. 2 (2018): 121–42.

Dalgleish, Paul. "The Thorniest Area: Making Collections Accessible Online While Respecting Individual and Community Sensitivities." *Archives and Manuscripts* 39, no. 1 (2011): 67–84.

Digital Oral Histories for Reconciliation (DOHR). Home page. n.d. https://dohrprojectca.wordpress.com/.

Digital Research Ethics Collaboratory. Directed by T. L. Cowan and Jas Rault. Home page. n.d. http://www.drecollab.org/.

D'Ignazio, Catherine, and Lauren Klein. *Data Feminism.* MIT Press, 2020.

Drucker, Johanna. "Sustainability and Complexity: Knowledge and Authority in the Digital Humanities." *Digital Scholarship in the Humanities* 36, Supplement_2 (2021): ii86–94. https://doi.org/10.1093/llc/fqab025.

Duarte, Marisa Elena. *Network Sovereignty: Building the Internet Across Indian Country.* Indigenous Confluences. University of Washington, 2017.

Duarte, Marisa Elena, and Miranda Belarde-Lewis. "Imagining: Creating Spaces for Indigenous Ontologies." *Cataloging & Classification Quarterly* 53, no. 5–6 (July 4, 2015): 677–702. https://doi.org/10.1080/01639374.2015.1018396.

Duff, Wendy M., Andrew Flinn, Karen Emily Suurtamm, et al. "Social Justice Impact of Archives: A Preliminary Investigation." *Archival Science* 13, no. 4 (2013): 317–48.

Earhart, Amy E. "Can Information Be Unfettered? Race and the New Digital Humanities Canon." In *Debates in the Digital Humanities,* edited by Matthew K. Gold, 309–18. University of Minnesota Press, 2012. https://dhdebates.gc.cuny.edu/read/untitled-88c11800-9446-469b-a3be-3fdb36bfbd1e/section/cf0af04d-73e3-4738-98d9-74c1ae3534e5#ch18.

Earhart, Amy. *Traces of the Old, Uses of the New: The Emergence of Digital Literary Studies.* University of Michigan Press, 2015.

Early Canadian Cultural Journals Index. Created by Thomas B. Vincent. Canadian Writing Research Collaboratory, 2022. https://cwrc.ca/project/early-canadian-cultural-journals-index.

Ebner, Julia. "Replatforming Unreality." *Journal of Design and Science* 6 (2019). https://jods.mitpress.mit.edu/pub/k109mlhg/release/1.

Edwards, Paul N., Steven J. Jackson, Geoffrey C. Bowker, et al. "Understanding Infrastructure: Dynamics, Tensions, and Design." 2007. https://deepblue.lib.umich.edu/handle/2027.42/49353.

Endings Project. Home page. n.d. https://endings.uvic.ca/.

Enslaved: Peoples of the Historical Slave Trade. Home page. n.d. https://enslaved.org/.

Fitzpatrick, Kathleen. *Planned Obsolescence: Publishing, Technology, and the Future of the Academy.* New York University Press, 2011.

First Nations Information Governance Centre (FNIGC). Home page. 2023. https://fnigc.ca/.

Floridi, Luciano. "Trump, Parler, and Regulating the Infosphere as Our Commons." *Philosophy & Technology* 34, no. 1 (2021): 1–5.

FourDirectionsTeachings.com. Producer Jennifer Wemigwans. Home page. n.d. https://fourdirectionsteachings.com/.

Fricker, Miranda. *Epistemic Injustice: Power and the Ethics of Knowing.* Oxford University Press, 2007. https://doi.org/10.1093/acprof:oso/9780198237907.001.0001.

Galey, Alan, and Stan Ruecker. "How a Prototype Argues." *Literary and Linguistic Computing* 25, no. 4 (2010): 405–24. https://doi.org/10.1093/llc/fqq021.

Gillespie, Tarleton. "The Politics of 'Platforms.'" *New Media and Society* 12, no. 3 (2010): 347–64. https://doi.org/10.1177/1461444809342738.

Gillespie, Tarleton. "Regulation of and by Platforms." In *The SAGE Handbook of Social Media,* edited by Jean Burgess, Alice Marwick, and Thomas Poell, 254–78. SAGE, 2018.

Goddard, Lisa, and Dean Seeman. "Negotiating Sustainability: Building Digital Humanities Projects That Last." In *Doing More Digital Humanities,* edited by Constance Crompton, Richard J. Lane, and Ray Siemens, 38–57. Routledge, 2019.

GO FAIR. "FAIR Principles." n.d. https://www.go-fair.org/fair-principles/. Accessed May 20, 2022.

Groeneveld, Elizabeth. "Remediating Pornography: The *On Our Backs* Digitization Debate." *Continuum* 32, no. 1 (January 2, 2018): 73–83. https://doi.org/10.1080/10304312.2018.1404677.

Guiliano, Jennifer, and Carolyn Heitman. "Difficult Heritage and the Complexities of Indigenous Data." *Journal of Cultural Analytics* 4, no. 1 (2019). https://doi.org/10.22148/16.044.

Hartman, Saidiya. "Venus in Two Acts." *Small Axe: A Caribbean Journal of Criticism* 12, no. 2 (2008): 1–14.

Hayles, N. Katherine. "Translating Media: Why We Should Rethink Textuality." *Yale Journal of Criticism* 16, no. 2 (2003): 263–90.

Helmond, Anne. "The Platformization of the Web: Making Web Data Platform Ready." *Social Media + Society,* vol. 1, no. 2 (2015). https://doi.org/10.1177/2056305115603080.

Helmond, Anne, David B. Nieborg, and Fernando N. van der Vlist. "Facebook's Evolution: Development of a Platform-as-Infrastructure." *Internet Histories* 3, no. 2 (2019): 123–46.

Holland, Kathryn, and Jana Smith Elford. "Textbase as Machine: Graphing Feminism and Modernism with *OrlandoVision.*" In *Reading Modernism with Machines,* edited by Veronica Alfano and Andrew Stauffer, 109–34. Palgrave Macmillan, 2016.

Jackson, Steven J. "Rethinking Repair." *Media Technologies: Essays on Communication, Materiality, and Society,* edited by Tarleton Gillespie, Pablo J. Boczkowski, and Kirsten A. Foot, 221–39. MIT Press, 2014.

Jhaver, Shagun, Christian Boylston, Diyi Yang, et al. "Evaluating the Effectiveness of Deplatforming as a Moderation Strategy on Twitter." *Proceedings of the ACM on Human-Computer Interaction* 5, no. CSCW2 (October 18, 2021): Article no. 381:1–30. https://doi.org/10.1145/3479525.

Johnson, Jessica Marie. "Markup Bodies, Black [Life] Studies and Slavery [Death] Studies at the Digital Crossroads." *Social Text* 36, no. 4 (137) (2018): 57–79. https://read.duke upress.edu/social-text/article/36/4%20(137)/57/137032/Markup-BodiesBlack-Life-Studies-and-Slavery-Death.

Jones, Steven E. *The Emergence of the Digital Humanities.* Taylor & Francis, 2013.

Kappeler, Susanne. *The Pornography of Representation.* John Wiley & Sons, 2013.

Khoo, Cynthia. "Deplatforming Misogyny." Women's Legal Education and Action Fund (LEAF), 2021. Accessed December 3, 2022. https://www.leaf.ca/publication/deplatforming-misogyny/.

Latour, Bruno. "A New Website." An Inquiry into Modes of Existence (blog), October 19, 2015. http://modesofexistence.org/a-new-website/.

Linked Editing Academic Framework (LEAF-VRE). Home page. 2023. https://www.leaf-vre.org/.

Linked Infrastructure for Networked Cultural Scholarship (LINCS). Home page. 2023. https://lincsproject.ca.

Linked Infrastructure for Networked Cultural Scholarship (LINCS). "Team." 2023. https://lincsproject.ca/docs/about-lincs/people/team.

Liu, Alan. "Toward a Diversity Stack: Digital Humanities and Diversity as Technical Problem." *PMLA* 135, no. 1 (2020): 130–51. https://doi.org/10.1632/pmla.2020.135.1.130.

Lovink, Geert. *Stuck on the Platform.* Valiz, 2022.

McPherson, Tara. *Feminist in a Software Lab: Difference + Design.* metaLABprojects. Harvard University Press, 2018.

McPherson, Tara. "From Vectors to Scalar: A Brief Primer for Applied Media Studies." In *Applied Media Studies,* edited by Kirsten Ostherr, 48–59. Routledge, 2018.

Meisel, Joseph S. *Public Speech and the Culture of Public Life in the Age of Gladstone.* Columbia University Press, 2001. https://doi.org/10.7312/meis12144.

MIT Press. "Platform Studies." 2023. https://mitpress.mit.edu/series/platform-studies/.

Montfort, Nick, and Ian Bogost. "Series Foreword." In *Racing the Beam: The Atari Video Computer System,* by Nick Montfort and Ian Bogost, vii–viii. MIT Press, 2009.

Moravec, Michelle. "Feminist Research Practices and Digital Archives." *Australian Feminist Studies* 32, no. 91–92 (April 3, 2017): 186–201. https://doi.org/10.1080/08164649.2017.1357006.

Neylon, Cameron. "Sustaining Scholarly Infrastructures Through Collective Action: The Lessons That Olson Can Teach Us." *KULA: Knowledge Creation, Dissemination, and Preservation Studies* 1, no. 1 (December 27, 2017): 3. https://doi.org/10.5334/kula.7.

Nixon, Cheryl. "A Sampling of Results." In Ros Ballaster, Laura McLean, Matthew Risling, Jennifer Currin, Betty A. Schellenberg, and Cheryl Nixon. "The Orlando Project." *Eighteenth Century Fiction* 22, no. 2 (2009): 371–79.

Nowviskie, Bethany. "Capacity Through Care." In *Debates in the Digital Humanities 2019*, edited by Matthew K. Gold and Lauren Klein, 424–26. University of Minnesota Press, 2019. https://dhdebates.gc.cuny.edu/read/untitled-f2acf72c-a469-49d8-be35-67f9ac1e3a60/section/3a53cbc1-5eee-421a-a4f6-82bb5dfb1c17.

Nowviskie, Bethany. "Digital Humanities in the Anthropocene." *Digital Scholarship in the Humanities* 30, Supplement_1 (2015): i4–i15.

Nowviskie, Bethany, and Dot Porter. "The Graceful Degradation Survey: Managing Digital Humanities Projects Through Times of Transition and Decline." In *Digital Humanities 2010 Conference Abstracts*, edited by Elena Pierazzo, 192–93. King's College London, 2010. https://dh2010.cch.kcl.ac.uk/academic-programme/abstracts/papers/html/ab-722.html.

Odumosu, Temi. "The Crying Child: On Colonial Archives, Digitization, and Ethics of Care in the Cultural Commons." *Current Anthropology* 61, no. S22 (2020): S289–S302.

Olson, Hope A. "The Power to Name: Representation in Library Catalogs." *Signs* 26, no. 3 (2001): 639–68.

Onuoha, Mimi. "The Library of Missing Datasets." 2016. https://mimionuoha.com/the-library-of-missing-datasets.

Orlando Project. "Credits." https://orlando.cambridge.org/about/credits.

Orlando Project. Home page. Edited by Susan Brown, Patricia Clements, and Isobel Grundy. Cambridge University Press, 2006–2022. https://www.artsrn.ualberta.ca/orlando.

Oxford English Dictionary. Edited by John Simpson et al. Oxford University Press, 1989.

The People and the Text. Home page. Directed by Deanna Reder. n.d. Canadian Writing Research Collaboratory (CWRC). https://thepeopleandthetext.ca/.

Perlman, Merrill. "The Rise of 'Deplatform.'" *Columbia Journalism Review,* February 4, 2021. Accessed November 19, 2022. https://www.cjr.org/language_corner/deplatform.php.

Poirier, Lindsay, and Brandon Costelloe-Kuehn. "Data Sharing at Scale: A Heuristic for Affirming Data Cultures." *Data Science Journal* 18, no. 1 (September 30, 2019): 48. https://doi.org/10.5334/dsj-2019-048.

Posner, Miriam. "What's Next: The Radical, Unrealized Potential of Digital Humanities." In *Debates in the Digital Humanities 2016,* edited by Matthew K. Gold and Lauren Klein, 32–41. University of Minnesota Press, 2016. https://dhdebates.gc.cuny.edu/read/untitled/section/a22aca14-0eb0-4cc6-a622-6fee9428a357.

Povinelli, Elizabeth A. "The Woman on the Other Side of the Wall: Archiving the Otherwise in Postcolonial Digital Archives." *Differences* 22, no. 1 (May 1, 2011): 146–71. https://doi.org/10.1215/10407391-1218274.

Respectful Terminology Platform Project. Home page. Led by Camille Callison and Stacy Allison-Cassin. n.d. National Indigenous Knowledge & Language Alliance/Alliance nationale des connaissances et des langues autochtones (NIKLA-ANCLA). https://www.nikla-ancla.com/respectful-terminology.

Reveal Digital. *Independent Voices.* Home page. n.d. JSTOR. https://www.jstor.org/site/reveal-digital/independent-voices.

Ricci, Donato, Robin De Mourat, Christophe Leclercq, and Bruno Latour. "Clues. Anomalies. Understanding. Detecting Underlying Assumptions and Expected Practices in the Digital Humanities Through the AIME Project." *Visible Language* 49, no. 3 (2015).

Risam, Roopika. "Beyond the Margins: Intersectionality and the Digital Humanities." *Digital Humanities Quarterly* 9, no. 2 (2015). https://www.digitalhumanities.org/dhq/vol/9/2/000208/000208.html.

Roberts-Smith, Jennifer, Justin Carpenter, Kristina R. Llewellyn, et al. "'Relational Presence': Designing VR-Based Virtual Learning Environments for Oral History-Based Restorative Pedagogy." *Journal of Interactive Technology and Pedagogy* 17 (May 20, 2020). https://jitp.commons.gc.cuny.edu/relational-presence-designing-vr-based-virtual-learning-environments-for-oral-history-based-restorative-pedagogy/.

Robertson, Tara. "Digitization: Just Because You Can, Doesn't Mean You Should." *Tara Robertson Consulting*, March 21, 2016. https://tararobertson.ca/2016/oob/.

Russell, Andrew L., and Lee Vinsel. "After Innovation, Turn to Maintenance." *Technology and Culture* 59, no. 1 (2018): 1–25. https://doi.org/10.1353/tech.2018.0004.

Sample, Mark. "The Black Box and Speculative Care." *Debates in the Digital Humanities* (2019): 445–48. https://dhdebates.gc.cuny.edu/read/untitled-f2acf72c-a469-49d8-be35-67f9ac1e3a60/section/3aa0b4f4-bd72-410d-9935-366a895ea7a7#ch43.

Scott, Joan W. "The Evidence of Experience." *Critical Inquiry* 17, no. 4 (1991): 773–97.

Skinaszi, Karen, ed. Winnifred Eaton/Onoto Watanna—The Alberta Years. Home page. n.d. Canadian Writing Research Collaboratory (CWRC). https://cwrc.ca/project/winnifred-eaton-onoto-watanna-alberta-years.

Smith, Tony, Gerald Morrison, Tracy Dorrington-Skinner, et al. "Digital Oral Histories for Reconciliation: The Nova Scotia Home for Colored Children History Education Initiative (DOHR)." *Journal of the Canadian Association for Curriculum Studies* 18, no 1. (2020): 60–61. https://jcacs.journals.yorku.ca/index.php/jcacs/article/view/40582.

Smithies, James, Carina Westling, Anna-Maria Sichani, et al. "Managing 100 Digital Humanities Projects: Digital Scholarship & Archiving in King's Digital Lab." *Digital Humanities Quarterly* 13, no. 1 (2019). http://www.digitalhumanities.org/dhq/vol/13/1/000411/000411.html.

Sofia, Zoë. "Container Technologies." *Hypatia* 15, no. 2 (ed 2000): 181–201. https://doi.org/10.1111/j.1527-2001.2000.tb00322.x.

Srnicek, Nick. *Platform Capitalism*. Theory Redux. Polity Press, 2017.

Star, Susan Leigh. "The Ethnography of Infrastructure." *American Behavioral Scientist* 43, no. 3 (1999): 377–91.

Star, Susan Leigh, and Karen Ruhleder. "Steps Toward an Ecology of Infrastructure: Design and Access for Large Information Spaces." In *Boundary Objects and Beyond: Working with Leigh Star*, edited by Geoffrey C. Bowker, Stefan Timmermans, Adele E. Clarke, and Ellen Balka, 377–415. MIT Press, 2015.

Sutherland, Tonia. "Making a Killing: On Race, Ritual, and (Re) Membering in Digital Culture." *Preservation, Digital Technology & Culture* 46, no. 1 (2017): 32–40.

Thomas, Daniel Robert, and Laila A. Wahedi. "Disrupting Hate: The Effect of Deplatforming Hate Organizations on Their Online Audience." *Proceedings of the National Academy of Sciences* 120, no. 24. (13 June 2023): e2214080120.

Tuck, Eve, and K. Wayne Yang. "Unbecoming Claims: Pedagogies of Refusal in Qualitative Research." *Qualitative Inquiry* 20, no. 6 (2014): 811–18. https://doi.org/10.1177/1077800414530265.

Vincent, Thomas B. *Index to Pre-1900 English Language Canadian Cultural and Literary Magazines.* Computer disk and quick reference guide. Optim Corporation, 1994.

Vincent, Thomas B. "Opportunities and Pitfalls: Observations on the Application of Database Software to Bibliographical Activities." *Papers of The Bibliographical Society of Canada* 28, no. 1 (1989), 17–18.

Waters, Donald J. "The Emerging Digital Infrastructure for Research in the Humanities." *International Journal on Digital Libraries* 24 (2022): 87–102.

Wemigwans, Jennifer. *A Digital Bundle: Protecting and Promoting Indigenous Knowledge Online.* University of Regina Press, 2018.

Wemigwans, Jennifer. "Digital Bundles: Creating Cultural Space for Indigenous Knowledge through the Use of New Technologies." Critical Digital Humanities Initiative International Conference 2022, University of Toronto, September 30, 2022.

Wernimont, Jacqueline. "Whence Feminism? Assessing Feminist Interventions in Digital Literary Archives." *DHQ: Digital Humanities Quarterly* 7, no. 1 (2013). https://www.digitalhumanities.org/dhq/vol/7/1/000156/000156.html.

Williams, Raymond. *Keywords: A Vocabulary of Culture and Society.* Oxford University Press, 1983.

World Wide Web Consortium (W3C). "Web Annotation Data Model." Edited by Robert Sanderson, Paolo Ciccarese, and Benjamin Young. World Wide Web Consortium. February 23, 2017.

Zaagsma, Gerben. "Digital History and the Politics of Digitization." *Digital Scholarship in the Humanities* 38 (2023): 830–51. https://doi.org/10.1093/llc/fqac050.

Zuboff, Shoshana. *The Age of Surveillance Capitalism: The Fight for a Human Future at the New Frontier of Power.* Public Affairs, 2020.

PART I][Chapter 3

Networking the Nation
Settler Colonialism as an Analytic in Critical Infrastructure Studies

SARAH MONTOYA

> The "real" of settler colonial society is built on the violent erasures of alternative modes of mapping and geographic understandings. The Americas as a social, economic, political, and inherently spatial construction has a history and a relationship to people who have lived here long before Europeans arrived. It also has a history of colonization, imperialism, and nation-building.
>
> —Mishuana Goeman, *Mark My Words: Native Women Mapping Our Nations*

At the core of discussions of information and communications technology (ICT) infrastructures are land and property ownership, the understandings of which are shaped by settler colonialism in the United States and administered by a constellation of global settler state formations. Settler colonialism names an orientation to the world and life; it is a way of being and relating to land, peoples, objects, and knowledge that prioritize the liberal humanist citizen-subject and the creation and acquisition of property and possession, and it is deeply characterized by white supremacy and enforced through various forms of targeted violence (Estes et al., 59, 107). When utilized as an analytical approach, settler colonial studies insists on making legible the relationship between land, infrastructure, private property, and the project of settler statecraft. This framework extends research in electronic and digital colonialism and makes explicit the spatial and racial violence required to maintain settler society.

Bringing forth the connections between the settler state, surveillance projects realized through the military-industrial complex, and the impact of commercial interests in the shaping of the use of internet technology, this chapter argues that a web of networked settler logics emerges through a critical reading of ICT infrastructure that links the development of the telegraph, telegraphone, and the internet to settler colonial property regimes in the United States. While this chapter applies settler colonial studies as a lens through which to analyze ICTs in the United States, it is important to approach other geopolitical sites with attention to their specific imperial and colonial histories and understand that occupation is quotidian, ongoing, and altered by local conditions. The goal of this chapter is to confront and dismantle narratives of settler technological supremacy and consider anticolonial and decolonial approaches in critical infrastructure studies (CIS).

Analyzing and Challenging Settler Colonialism

Settler colonialism names a worlding technique as much as it names historically specific flows of power and possession. Although this work focuses on U.S. settler society, it indicates only one of many areas marked by violence: settler colonial societies share common characteristics.[1] The characteristics noted here do not represent an exhaustive list; rather, they should be understood as rationales and functions that take different shapes in different geopolitical contexts. They all guarantee the same outcome, however: the establishment and dominance of a settler society. This pattern of practices includes the acquisition of territory, resource and labor extraction, the legal codification of private property, and the erection of a network of legal and administrative settler state legal structures used to manage subordinate populations. Such practices are predicated on the production of raced and gendered hierarchies, which strengthens the dominance of settler supremacy.

The imposition of a solipsistic, self-authorizing juridical and administrative settler state legal structure is crucial to the colonial project. Alyosha Goldstein describes modern U.S. colonialism as the "administration of populations [which] operates in tandem with the juridical, political, military-strategic, economic, and cultural production and control of property, territory, and resources" (8). Similarly, Goeman argues that within the settler configuration of land as an exploitable resource, Indigenous peoples are flattened into "flora and fauna" (18), turned into objects to be managed. Managerial techniques are evidenced in reservations, residential schools, and in the enslavement of Native women and children as domestic labor (Estes, 29, 55). As the settler state's formation solidifies, the mechanisms for discipline and "the logic of elimination" oscillate between the overt targeting of racialized communities, resulting in death, and a range of techniques including the use of carceral spaces and methods of slow death (Wolfe, "Settler Colonialism and the Elimination of the Native," 387; Razack, 916). Here, the settler state paternalistically deploys itself to manage targeted communities under the guise of "civilizing" projects. Settler invasion and the construction of settler supremacy are iterative, performative processes that must be enacted and reenacted to establish and entrench their legitimacy. Globally, nation-state structures function to affirm the formation of the nation-state and its authority. Settler colonialism is understood here as a foundational structure to globalization.

Settler property regimes are intimately linked with settler geographic knowledges, or ways of coming to know and understand place and space. We see this in spatial metaphors that reveal underlying assumptions and reify forms of spatial and geographical knowledge. The pervasiveness of the frontier narrative in the United States is a testament to its effectiveness and malleability; it has endured because the notion of a "frontier" is an established but also evolving national myth. Consider the framing of cyberspace-as-frontier space. In the 1990s, when such phraseology emerged, the language did the dual work of representing a land that promises to be conquered by an already-triumphant settler society and reflecting a now-familiar spatial order.[2] The term invokes the memory of an unruly land and peoples that pose a threat to settler order but will ultimately be ordered, catalogued, and transformed. Constituent forms of spatial and geographic knowledge are displaced and erased. White Earth Ojibwe historian Jean O'Brien terms this practice "firsting," which "asserts that non-Indians were the first people to erect the proper institutions of a social order worthy of notice" (xiv). This myth-making process permits settlers to perpetually supplant and/or evacuate Native and Indigenous peoples from narratives of social, cultural, and technological development.

Possession and ownership are inherently racial logics in settler property regimes. The possession of private property names a relationship with land. Cheryl Harris's formative work, "Whiteness as Property," traces the development of chattel slavery and the consequences of racialization in which whiteness is legally associated with property and whiteness itself becomes a form of property by exercising the rights of exclusion (1726, 1736). In the making of racial categories, whiteness was codified as settler law and guaranteed its primacy and social and legal protections. Quandamooka scholar Aileen Moreton-Robinson uses the concept of "possessive logics" to understand the link between the assertion of property rights by people with white, patriarchal settler identities and the project of establishing and maintaining the settler state. The possessive logic serves as an orientation and rationalization to name settler investments in "reproducing and reaffirming the nation-state's ownership . . . and circulate sets of meanings about ownership of the nation as part of common-sense knowledge" (12). Settler colonialism constitutes the fabric of social, economic, and political worlds so much that settler common sense displaces and replaces any constituent modes of relation and posits the dominance of settler society as inevitable.

The formulation of an ideal settler citizen-subject is accomplished through settler colonial epistemes and ontologies. Chickasaw scholar Jodi Byrd argues that the scope of settler knowledge production is enabled through an Indigenous identity rendered as an *interpretable body* that no longer exists in the real. It is through this malleability that a settler colonial society can simultaneously "forget" the violence done to Indigenous peoples even as it, at other times, uses the figure of "the Indian" as a hypervisible character in a celebratory historical narrative (19). Tiffany Lethabo King combines Black and Indigenous feminist work to "illuminate the ways that white humanity and its self-actualization require Black and Native death as its condition of possibility" and names this "conquistador humanism." Here, Black and

Indigenous peoples are "transformed into lesser forms of humanity—and, in some cases, become nonhuman altogether" (King, 16, 20). The violence of this transformation cannot be overstated because it authorizes the invocation of *terra nullius* and the enslavement of lesser nonhumans. Shiri Pasternak argues that land is viewed as unoccupied and available for seizure and settlement precisely because these racial Others do not register as human, nor do they enact administration over territory in a manner reflecting settler notions of private property (155–56). Settler citizenship is predicated on dispossession, and this violence is continually referenced and sanitized. Consider the desire for home ownership as a mundane part of a settler statecraft.

Settler colonial legal structures are consistently organized around a raced and gendered construction of the category of the human. This malleable construction defines who will have social and legal recognition within the settler state juridical system and can own property. Wolfe understands racialization as a central, organizing logic in which racial hierarchies are established to center white supremacist settler societies and produce both Indigenous dispossession and Black enslaved labor ("Settler Colonialism and the Elimination of the Native," 387–88). It is important to note that that Wolfe's narrow structuring of racial categories obfuscates both the enslavement of Indigenous peoples and the dispossession of African peoples from Africa (Kelley, 268). Heteropatriarchy is codified within the logics of property and settler state law based on male inheritance. For example, the 1887 Dawes Act, or General Allotment Act, which required naming a male head of household, radically altered tribal relationships and care networks and divested Native women from political representation (Goeman, 2013, 91–92). Interpreting settler colonialism as a gendered project also highlights the extensive intellectual genealogies of Black and Indigenous feminism. This work attests to the gendered racial violence of chattel slavery, sexual violence, and role of blood quantum, all of which rely on horrific violence visited upon racialized women's bodies.[3]

More recently, a significant body of work addresses how settler violence continues through a language of logistics. Deborah Cowen notes that "the supply chain of contemporary capitalism resonates so clearly with the supply line of the colonial frontier" and the "old enemies of empire—'indians' and 'pirates'—are among the groups that pose the biggest threats" (9). Racialized communities, intentionally divested of resources, are continually represented as antagonistic to the project of modernity and liberal progress. Pasternak and Dafnos argue that "Indigenous rights are increasingly problematized as threats to the existing infrastructures of supply chains, as well as to their future expansion" (747). We see this reflected in the sustained protective demonstrations by Native, Indigenous, First Nations, and Aboriginal communities globally over land, water, mineral, gas, and oil rights.[4] The strategic denial of basic human needs to communities targeted by the settler state represents ongoing settler colonial violence. As Wolfe notes, settlers "come to stay; invasion is a structure, not an event" ("Settler Colonialism and the Elimination of the Native," 388). Rather than understand colonial contact and violence as a singular event in the past, settler colonialism as a methodological framework connects our imperial, colonial, and settler colonial past to our settler colonial present and extends our critique of the nation-state.

Electronic and Digital Colonialism and Settler Colonialism

Settler colonial studies' emphasis on territoriality and the violent mechanisms used to sustain settler invasion and ensure settler supremacy nuances the formative and growing body of work in digital colonialism. It is worth noting some foundational assumptions here. Rather than conflate capitalist privatization with colonialism, I note that these are linked social and economic projects, but *not* equivalents. My use of the term "violence" is deliberate in its naming of a scale of violence bound up in colonial conquest and intentional erasures of brutal histories. Furthermore, settler colonialism names an ideological investment that many people uphold, including people of color, although the concept is associated with the primacy of a white, able-bodied, heterosexual, and male/masculine subject. It is also important to acknowledge that conceptions and experiences of imperialism, colonialism, and settler colonialism shift according to the geopolitical context. Finally, although representative in important ways, the United States should not be understood as a proxy for all settler societies or a stand-in for specific and complex imperial and colonial histories. In India and Africa, the terms *postcolonial* and *decolonial* often have different meanings than in the U.S. context. In the United States, the use of the term *postcolonial* to describe sociopolitical issues is contested, as settler colonial studies claims that we live under occupation and postcolonial theory was originally oriented toward the critique of colonialism (Byrd, xxxiii–xxxvi). None of this terminology (or its application) is free from debate; language is fraught as we try to describe our specific histories and name the deep and lasting impact that colonial occupation created and maintains.

Marxist approaches to information studies often focus on disparities of socioeconomic power via critiques of ownership and infrastructure in the First World/Global North and Third World/Global South and can be strengthened by settler colonial studies' critique. Herbert Schiller's foundational work highlights the use of communication infrastructures to disseminate Western ideologies in media; for example, he argues that this form of cultural imperialism was part of a deliberate effort undertaken by U.S. military-communication conglomerates to maintain dominance commercially and politically (9). His characterization expanded to describe media domination and Third World dependence on First World communications infrastructures. This relationship entrenches economic dependence on First World governments, which dictate the consequences of use as the infrastructure's owners both surveil and profit from the data, which ensures perpetual market dominance. Contemporary discourse follows a similar approach; in "Data Colonialism through

Accumulation by Dispossession," Thatcher et al. argue that data colonialism "occurs through asymmetrical relations between data producers (end-users) and data collectors and owners (corporate entities) that mirror processes of primitive accumulation or accumulation by dispossession that occur as capitalism colonizes previously non-commodified, private times and places" (7). I am wary of the use of colonialism as a metaphor for capitalist privatization or definitions that are bloodless, are inattentive to raced and gendered violence, or assume equivalence between Indigenous dispossession and broad practices of data exploitation. Problems also exist with definitions that assume that occupation and dispossession are largely historical issues. Many authors are quick to note the existence of sustained colonial legacies, but they are less likely to consider how contemporary rulings undermine tribal sovereignty and jurisdiction—functioning in similar ways to the land grabbing techniques of "historical colonialism" (Couldry and Mejias, 70–4).

Recent work on digital colonialism grapples with a legacy of colonial and settler colonial ways of interpreting, cataloguing, and owning data that dehumanize racialized communities. Roopika Risam identifies colonial archival violence by noting the way that data is circulated through colonial ideologies and ontologies, thus reaffirming colonial structures (48–49). Similarly, Morehshin Allahyari defines digital colonialism as "a framework for critically examining the tendency for information technologies to be deployed in ways that reproduce colonial power relations" and stipulates that the term refers specifically to the use of digital technologies and is differentiated from "material or object-based colonialism" (Sharma et al., 201). Allahyari critiques the use of three-dimensional (3D) printers and scanners by "Western institutions and digital archaeology spaces in Eurocentric countries and North American countries . . . [who] go to the Global South and [3D] scan historical and cultural sites. Their claim is that they will save cultural heritage that we all share" (Sharma et al., 197). This enacts the same colonial violence that many institutions now repudiate by offering "less invasive" methods of sacred site and grave desecration. Such practices are part of a long continuum of the settler colonial violence that King terms "conquistador humanism" (16). In the violent cataloging and categorizing of subhuman and nonhuman flesh, knowledge is bound up in conquest, and violence is sanitized, as institutions claim that the work is for the public good.

Settler Colonialism as Method: Reapproaching Telecommunications Infrastructures

Understanding the central logics of settler colonialism allows us to grasp the historical foundations and political stakes of telecommunications structures in the United States. My work borrows from scholars tracing the relationship between technologies and empire, and builds on work interrogating the relationships among the nation-state, the military-industrial complex, and surveillance technologies.[5] State-created telecommunications infrastructures and technologies in the United

States have fundamentally enabled and been realized through frontier violence and continue to be used to surveil communities of color.[6] As an exhaustive history is impossible, I briefly touch on specific areas of telegraph, telephone, and early internet development and consider how the infrastructure and deployment of these technologies sustain the racial hierarchies that are integral to settler colonial societies.

Telegraph infrastructure represented westward colonial expansion visually, materially, and ideologically. James Schwoch's *Wired into Nature* positions telegraphy as a form of electronic communication that was central to the development of national security frameworks. The work describes the role of the U.S. Military Telegraph Corps and the U.S. Army Signal Corps in the establishment and maintenance of telegraph technology and the completion of the transcontinental telegraph system. Schwoch notes the connections between the collection of bioinformation and the use of the telegraph as a "military asset to wage long-term low-scale continuous asymmetric warfare against Native Americans" (10). The U.S. military and settler state was absolute in its recognition of the importance of communications infrastructure to accomplish this targeting. So critical were telegraph structures to the state that when Native peoples sabotaged telegraph poles in response to the Sand Creek Massacre, the army led a scorched-earth campaign and set a wildfire in January 1865 (122–23).

Settler geographic knowledge was codified in no small part due to environmental and climatological surveys. Sara Grossman's *Immeasurable Weather* examines how climatological knowledge understood as "multiregional settler data was key to the story and enactment of settler environmental dominance" and entrenched a settler relation to land and nature bound to ownership and mastery (39). Critical geographer Nicholas Blomley argues that army surveys, like the ones that guided the construction of the telegraph system, function as a mechanism to establish settler legitimacy. Blomley notes that "space is simultaneously a means of disciplining the performances that are possible within it. These social performances are citational, reiterating past performances and thus reproducing dominant norms and practices at the same time as they diverge from them" (122). The coupling of "nature" and "Native" as things to be surveilled and tracked constitutes the epistemological violence that Goeman mentioned; Native communities were cast as part of the landscape, simultaneously objectified and transformed into natural history. Schwoch notes that the telegraph helped to establish geodetic knowledge and the codification of increasingly standardized latitudinal and longitudinal coordinates, which "enhanced the ability of the federal government to contain Native Americans within the perimeters of geodetically plotted boundary stakes" (32). Today, reservation boundaries remain policed and Native women continue to be murdered and missing. Settler colonial surveys attempted to upend Indigenous geographic knowledge and relationship with place, and represent a pattern of violence long enacted by the state.

The importance of such communications infrastructures is well documented in the legal language of acts that cede right-of-way to the federal government. The

1862 Pacific Railway Act and the Dawes Act were central to the act of dispossession and the systematic diminishment of Native lifeways and sovereignty. The creation of public lands through federal power appears in legislation from the Land Ordinance of 1785 and well into the late 1800s, with the Homestead Act of 1862. Section 2 of the Pacific Railway Act granted right of way through tribal lands and codified the state's right to "extinguish as rapidly as may be the Indian titles to all lands falling under the operation of this act" (National Archives). Similarly, the Dawes Act stated that the federal government maintained the "right of way through any lands granted to an Indian, or a tribe of Indians, for railroads or other highways, or telegraph lines, for the public use" and pairs it with the demand that allotted Indigenous peoples "[adopt] the habits of civilized life" (National Archives). We see here the entwined demands to diminish Indigenous lifeways and ensure settler domination by shoring up the power of the settler state and its public. Equally important to note is how foundational such legal precedent remains in lawsuits, including the Keystone XL pipeline project, which was eventually terminated (Native American Rights Fund). Viewed through the lens of Moreton-Robinson's possessive logics, the erection of the telegraph pole functioned as a representation of a unified nation. Like railroads and roadways, telegraph poles marked the land as part of the project of colonization and offered the visage of national unity and security.[7]

Telephone poles represented a different moment in the settler colonial project, though no less violent, and can be viewed as a way station toward digital communication infrastructures. The relationship between telegraph and telephone infrastructures echoes the eventual use of telephone lines for dial-up internet: the technological and conceptual structures were grafted onto each other. Where telegraph lines were sought as a means of protecting settler safety, telephone poles were regarded as aesthetically displeasing; however, once the increased speed of communication was realized, the telephone infrastructure flourished. While dispossession remained coded in the infrastructure, telephone poles also functioned as gallows in the lynching of Black people. Eula Biss notes that her search for the phrase "telephone pole" in *The New York Times* archives from 1880 to 1920 resulted in 370 results showing these instances of racial terror (89). As the telephone infrastructure grew more robust, so did the U.S. government's ability to surveil. This is glaringly apparent in the use of wiretaps and phone calls by the Federal Bureau of Investigation (FBI) in the COINTELPRO files, which targeted Black communities organizing against social and state violence.[8] As the telegraph supplied the state with information on surveilled communities, so would the telephone. Although it is important to remember the historically situated context of telegraph and telephone infrastructures (there are discontinuities as well as continuities in their respective histories), it is important to acknowledge the inequitable deployment of telecommunications technologies and the way that they have helped to ensure that settler hegemony remains intact.

A substantial body of work traces internet history from its bourgeoning moments within the military-industrial complex through to its commercialization.

Abbates's *Inventing the Internet* describes the military roots of internet and computational technologies. As dial-up internet evolved to its current state, the nation-state and commercial entities guided the development of online protocols and extended the reach of private property law to infrastructure. Dan Schiller's *Digital Capitalism* traces the expansion of the internet network toward a for-profit model, noting that business networks effectively drove policy agendas with the Federal Communications Commission that allowed them to build proprietary systems and utilize extant telecommunications structures. He reports that "some 60 percent of the internet's host computers in early 1997 were located in the United States" (35). It is perhaps no surprise, then, how heavily governmental and commercial interests shaped international policy in the United States—all the way to outer space. After all, how do we convert outer space into the property of a nation-state without the parallel extension of communications infrastructures? Settler state expansion never ceased; now it needs to adapt legal language from airspace and outer space. James Hay sketches out such legalities, noting how the eventual legal constructions of outer space can be traced to precedents over airspace and the related industry of radio (26–29). Christy Collis's work traces the fraught legalities of the Geostationary Earth Orbit, otherwise known as the most expensive piece of outer space due to the position it guarantees satellites (61). As *terra nullius* undergirded the doctrine of discovery, *res nullius* undergirds the creation of territory in outer space.

When settler colonial studies are applied to critical infrastructure studies, the networked settler logics manifested and managed through telecommunications infrastructures become apparent. Dan Schiller's *Crossed Wires: The Conflicted History of US Telecommunications* examines how the establishment of the Post Office facilitated information flows and exchange "among dispersed and typically small units of capital . . . Thus the Post Office functioned as a two-stroke engine, firing on behalf of both territorial conquest and market intensification" (21). In many ways, these technologies are doing exactly what they were designed to do: advancing settler state interests and security. Schwoch offers an insightful reflection when he writes that "when the Transcontinental Telegraph relayed the messages that triggered the Great Prairie Fire, the Transcontinental Telegraph presaged the ultimate (but never issued) prime directive of ARPANET: relaying messages to deploy American weapons of mass destruction" (124).

Our work can refuse to pull a curtain before the historical colonial and ongoing settler colonial violence. Settler colonial studies as a method reminds us to look back and draw out these complex and violent histories and their many iterations.

Confronting and Refusing the Settler State

Methods attentive to settler colonialism ask us to reorient our analysis and address the limitations of state-based interventions. Where neoliberal approaches may address, for instance, how women *also* contributed to settler state projects, an approach attentive to settler colonialism enables an analysis that does not treat gender as an additive component of labor analysis. Instead, a robust analysis considers how the use of raced, sexed, and gendered labor is itself part of a settler colonial nation-building project. Rather than treat legal constructions of private property and ownership as givens, these become additional areas to be interrogated. More pointedly, we can see the limits of proposed solutions like increased information accessibility or open-access data and transparency when the settler state is tasked to be an arbiter. Solutions realized through settler state intervention reentrench the centrality of the state and allow it to act as a paternalistic benefactor. Hupa abolitionist feminist Stephanie Lumsden reminds us that settler law "must be understood as a tool of colonization rather than an instrument of justice. This means that the law has been implemented to maintain the violent settler colonial hegemony that undergirds the state" (86). Neoliberal approaches are inherently limited by their reliance on settler state juridical interventions, and the limited critique fails to address the systemic oppression and violence that so characterize U.S. settler society.

In practicing self-reflexive work, we must ask ourselves how our projects may perpetuate and sustain harm in a settler society. This allows us to address what Tuck and Yang term "settler moves to innocence," which "problematically attempt to reconcile settler guilt and complicity, and rescue settler futurity" (3). We see this resuscitation of the settler in discourses that cast the internet as a "public good," which assumes a shared set of cultural values and structures of equality and access that simply do not exist in the settler state; rhetoric that joins "democracy," "freedom," and "equal access" demonstrates an investment in upholding settler hegemony. Embedded in the notion of "the commons"—a category that presumes settler citizenship as its normative identity—there remains a refusal to grapple with settler violence. Extant frameworks do not always "look back" to consider how racial, spatial settler, colonial histories so intimately shape our settler colonial present. Instead, theorists attempt to salvage a colonial project through re-packaging; if a project can equally serve all populations, then there is no need to address the violent dispossession that made its very advent possible nor is there precedent to address ongoing settler colonial violence.

What if, instead of a worlding practice that is predicated on death to preserve settler subjectivity, we ask ourselves how we can create a world that promises protection and life to those currently rendered most vulnerable? Native, Indigenous, and First Nations communities are working on data sovereignty and data protocols in response to settler colonialism and digital colonialism. Cheyenne scholar Desi Rodriguez-Lonebear claims that "the data sovereignty revolution in Indian country is going to be built tribe by tribe and community by community. Reclaiming the right to understand the diverse realities of our peoples on our terms and to chart sustainable courses for future generations is a matter of contemporary survival for indigenous peoples" (268). Rather than seek sweeping or broad settler state legislative reforms, these approaches ask for nuance, specificity, and directives delineated by the communities themselves. There is no shortage of extant models for working toward a just future, beyond the state. Such mobilizations are part of a long continuum of Indigenous resistance. Beyond this, we can actively work toward decolonization, including the return of the land to Indigenous and ancestral relations, which is at once possible and necessary. This insistence is not born of naive idealism: in 2022, significant tracts of land were returned to the InterTribal Sinkyone Wilderness Council, the Wiyot, and the Rappahannock Tribe (Kunze, "After 350 Years, Rappahannock Tribe Gets Land Back"). Covid-19 and the supply chain crisis continue to show us the profoundly precarious status of all settler states. New methods can herald new ethics.

NOTES

1. For more on settler colonialism in varied global contexts, see Veracini's *Settler Colonialism* and *The Settler Colonial Present*; Wolfe's *Settler Colonialism and the Transformation of Anthropology*; Shalhoub-Kevorkian's *Security Theology, Surveillance and the Politics of Fear*; Barker's edited collection, *Sovereignty Matters*; Shigematsu and Camacho's edited collection, *Militarized Currents*; and Kanji's "Settler Moves to Innocence."

2. The use of the frontier metaphor is associated with the Electronic Frontier Foundation (EFF), founded in 1990. The invocation of the frontier in an American context was intentional, as the EFF sought to garner support from the public. See "Across the Electronic Frontier" by EFF cofounders Mitchell Kapor and John Perry Barlow.

3. For more on gendered racial violence, see Haley, *No Mercy Here*; Snorton, *Black on Both Sides*; TallBear, *Native DNA*; and Deer, *The Beginning and End of Rape*.

4. For more on Indigenous and Native demonstrations and protection movements, see Estes and Dhillon's edited collection, *Standing with Standing Rock* and the Wet'suwet'en land and water protectors at Raventrust.com.

5. See Karuka's *Empire's Tracks*; and Voyles's *The Settler Sea* and *Wastelanding*.

6. See Benjamin, *Captivating Technology*; and Dubrofsky and Magnet, *Feminist Surveillance Studies*.

7. For work focused on the role of the telegraph in nation building, settler colonial religious imaginaries, and its impact within the Oneida Nation, see Montgomerie, *When the Medium Was the Mission*.

8. See "Black Extremist" in the COINTELPRO archive.

BIBLIOGRAPHY

Abbate, Janet. *Inventing the Internet*. MIT Press, 1999.

Barker, Joanne, ed. *Sovereignty Matters: Locations of Contestation and Possibility in Indigenous Struggles for Self-Determination*. University of Nebraska Press, 2006.

Benjamin, Ruha, ed. *Captivating Technology: Race, Carceral Technoscience, and Liberatory Imagination in Everyday Life*. Duke University Press, 2019.

Biss, Eula. "Time and Distance Overcome." *Iowa Review* 38, no. 1 (2008): 83–9.

Blomley, Nicholas. "Law, Property, and the Geography of Violence: The Frontier, the Survey, and the Grid." *Annals of the Association of American Geographers* 93, no. 1 (2003): 121–41.

Byrd, Jodi A. *The Transit of Empire: Indigenous Critiques of Colonialism.* University of Minnesota Press, 2011.

COINTELPRO. FBI Records: The Vault. Accessed August 9, 3023, vault.fbi.gov/cointel-pro.

Collis, Christy. "The Geostationary Orbit: A Critical Legal Geography of Space's Most Valuable Real Estate." In *Down to Earth: Satellite Technologies, Industries, and Cultures,* edited by Lisa Parks and James Schwoch, 61–81. Rutgers University Press, 2012.

Couldry, Nick, and Ulises A. Mejias. *The Costs of Connection: How Data Is Colonizing Human Life and Appropriating It for Capitalism.* Stanford University Press, 2019.

Cowen, Deborah. *The Deadly Life of Logistics: Mapping Violence in Global Trade.* University of Minnesota Press, 2014.

Deer, Sarah. *The Beginning and End of Rape: Confronting Sexual Violence in Native America.* University of Minnesota Press, 2015.

Dubrofsky, Rachel E., and Shoshana Magnet, eds. *Feminist Surveillance Studies.* Duke University Press, 2015.

Estes, Nick, and Jaskiran Dhillon, eds. *Standing with Standing Rock: Voices from the #NoDAPL Movement.* Indigenous Americas. University of Minnesota Press, 2019.

Estes, Nick, et al. *Red Nation Rising: From Bordertown Violence to Native Liberation.* PM Press, 2021.

Goeman, Mishuana. *Mark My Words: Native Women Mapping Our Nations.* University of Minnesota Press, 2013.

Goldstein, Alyosha. "Towards a Genealogy of a U.S. Colonial Present." *Formations of United States Colonialism.* Duke University Press, 2014.

Grossman, Sara. *Immeasurable Weather: Meteorological Data and Settler Colonialism from 1820 to Hurricane Sandy.* Duke University Press, 2023.

Haley, Sarah. *No Mercy Here: Gender, Punishment, and the Making of Jim Crow Modernity.* University of North Carolina Press, 2016.

Harris, Cheryl I. "Whiteness as Property." *Harvard Law Review* 106, no. 8 (1993): 1707–91.

Hay, James. "The Invention of Air Space, Outer Space, and Cyberspace." In *Down to Earth: Satellite Technologies, Industries, and Cultures,* edited by Lisa Parks and James Schwoch, 19–41. Cultures. Rutgers University Press, 2012.

Kanji, Azeezah. "Settler Moves to Innocence: A Transnational Legal Glossary." *Yellowhead Brief,* no. 133, April 13, 2023. https://yellowheadinstitute.org/wp-content/uploads/2023/04/kanji-settler-moves-april2023.pdf.

Kapor, Mitchell, and John Perry Barlow. "Across the Electronic Frontier." The Electronic Frontier Foundation, July 10, 1990. https://www.eff.org/pages/across-electronic-frontier.

Karuka, Manu. *Empire's Tracks: Indigenous Nations, Chinese Workers, and the Transcontinental Railroad.* University of California Press, 2019.

Kelley, Robin D. G. "The Rest of Us: Rethinking Settler and Native." *American Quarterly* 69, no. 2 (2017): 267–76.

King, Tiffany Lethabo. *The Black Shoals: Offshore Formations of Black and Native Studies.* Duke University Press, 2019.

Kunze, Jenna. "After 350 Years, Rappahannock Tribe Gets Land Back." *Native News Online.* April 2, 2022. Accessed August 9, 2023, nativenewsonline.net/sovereignty/after-350-years-the-rappahannock-tribe-gets-land-back.

Kunze, Jenna. "50 Acres of Ancestral Homeland Repatriated to the Wiyot Tribe." *Native News Online.* August 24, 2022. Accessed August 9, 2023, nativenewsonline.net/sovereignty/50-acres-of-ancestral-homeland-repatriated-to-the-wiyot-tribe.

Kunze, Jenna. "Land Returned to Consortium of 10 Tribes." *Native News Online.* February 2, 2022c. Accessed August 9, 2023, nativenewsonline.net/sovereignty/land-returned-to-consortium-of-10-tribes.

Lumsden, Stephanie. "Missionization, Incarceration, and Ohlone Resilience." In *Counterpoints: Bay Area Data and Stories for Resisting Displacement*, 84–86.PM Press, 2021.

Moreton-Robinson, Aileen. *The White Possessive: Property, Power, and Indigenous Sovereignty.* University of Minnesota Press, 2015.

National Archives: Milestone Documents. "Dawes Act (1887)." Archives.gov/milestone-documents/dawes-act.

National Archives: Milestone Documents. "Pacific Railway Act (1862)." Archives.gov/milestone-documents/pacific-railway-act.

Native American Rights Fund. "Keystone XL Pipeline (Rosebud Sioux Tribe v. Trump)." Narf.org/cases/keystone.

O'Brien, Jean M. *Firsting and Lasting: Writing Indians out of Existence in New England.* University of Minnesota Press, 2010.

Pasternak, Shiri. "Jurisdiction and Settler Colonialism: Where Do Laws Meet?" *Canadian Journal of Law and Society* 29, no. 2 (2014): 145–61.

Pasternak, Shiri, and Tia Dafnos. "How Does a Settler State Secure the Circuitry of Capital?" *Environment and Planning D: Society and Space* 36, no. 4 (2018): 739–57.

Razack, Sherene H. "Memorializing Colonial Power: The Death of Frank Paul." *Law & Social Inquiry* 37, no. 4 (2012): 908–32.

Risam, Roopika. *New Digital Worlds: Postcolonial Digital Humanities in Theory, Praxis, and Pedagogy.* Northwestern University Press, 2019.

Rodriguez-Lonebear, Desi. "Building a Data Revolution in Indian Country." In *Indigenous Data Sovereignty: Toward an Agenda,* edited by Tahu Kukutai and John Taylor, 253–72. ANU Press, 2016.

Schiller, Dan. *Digital Capitalism: Networking the Global Market System.* MIT Press, 1999.

Schiller, Dan. *Crossed Wires: The Conflicted History of US Telecommunications from the Post Office to the Internet.* Oxford University Press, 2023.

Schiller, Herbert. *Communication and Cultural Domination.* International Arts and Sciences Press, 1976.

Schwoch, James. *Wired into Nature: The Telegraph and the North American Frontier.* University of Illinois Press, 2018.

Shalhoub-Kevorkian, Nadera. *Security Theology, Surveillance and the Politics of Fear.* Cambridge University Press, 2015.

Sharma, Sarah, et al. "3D Printing and Digital Colonialism: A Conversation with Morehshin Allahyari." In *Re-understanding Media: Feminist Extensions of Marshall McLuhan*, edited by Sarah Sharma and Rianka Singh, 192–207. Duke University Press, 2022.

Shigematsu, Setsu, and Keith L. Camacho eds. *Militarized Currents: Toward a Decolonized Future in Asia and the Pacific.* University of Minnesota Press, 2010.

Snorton, C. Riley. *Black on Both Sides: A Racial History of Trans Identity.* University of Minnesota Press, 2017.

Supp-Montgomerie, Jenna. *When the Medium Was the Mission: The Atlantic Telegraph and the Religious Origins of Network Culture.* NYU Press, 2021.

TallBear, Kimberly. *Native American DNA: Tribal Belonging and the False Promise of Genetic Science.* University of Minnesota Press, 2013.

Thatcher, Jim, et al. "Data Colonialism Through Accumulation by Dispossession: New Metaphors for Daily Data." *Environment and Planning D Society and Space* 36, no. 6 (2016): 990–1006.

Tuck, Eve, and K. Wayne Yang. "Decolonization Is Not a Metaphor." *Decolonization: Indigeneity, Education & Society* 1, no. 1 (2012): 1–40.

Veracini, Lorenzo. *Settler Colonialism: A Theoretical Overview.* Palgrave Macmillan, 2010.

Veracini, Lorenzo. *The Settler Colonial Present.* Palgrave Macmillan, 2015.

Voyles, Traci Brynne. *The Settler Sea: California's Salton Sea and the Consequences of Colonialism.* University of Nebraska Press, 2021.

Voyles, Traci Brynne. *Wastelanding: Legacies of Uranium Mining in Navajo Country.* University of Minnesota Press, 2015.

"Wet'suwet'en Campaign." RAVEN: Respecting Aboriginal Values and Environmental Needs. Accessed August 9, 2023, Raventrust.com.

Wolfe, Patrick. "Settler Colonialism and the Elimination of the Native." *Journal of Genocide Research* 8, no. 4 (2006): 387–409.

Wolfe, Patrick. *Settler Colonialism and the Transformation of Anthropology: The Politics and Poetics of an Ethnographic Events.* Continuum, 1999.

PART I][Chapter 4

Manifesting Connection
Digital Humanities for the Critical Study of Logistics

MATTHEW HOCKENBERRY

I am hurt all over, but I cannot tell the full extent yet, because the doctor is not done taking inventory. He will make out my manifest this evening. However, thus far he thinks only sixteen of my wounds are fatal. I don't mind the others.

—Mark Twain, "Niagara," 1875

Manifest entered the English vernacular from either French or Italian influences. The meaning in old French was of something "evident" or "palpable," and indeed it is that latter understanding that most accurately conveys the sensorial associations of the word. For something to "be manifest" implies its inhabitance in a space—not only physical presence, but the capability for physical apprehension. The word's more distant Latin origins seem—at first—to make this clear. *Manifestus* derives, in part, from *manus*, hand—an etymology familiar in words like *manual* and *manufacture*. But there is less consensus on the rest. If *manifestus* is taken to mean "caught by hand" (and "to make manifest"—consequently, to be "caught in the act"), then perhaps it also derives from *infestare*—to attack, disturb, or "seize." And there is something powerful that this other etymology offers. The manifest was, after all, a document of violence.

A "manifest," as an early reference from Edward Hatton's *Merchant's Magazine* defined it in 1697, "is a Transcript of a Master of a Ship's Cargo, showing what is due to him for Freight from each person to whom the Goods in his Ship belong" (Hatton, xiv, 234/1). This definition economically embeds the manifest within seventeenth-century notions of property and commerce. For a thing to appear in a manifest, it explains, it must be owned, and that thing must have value. And so it was that this form carried out the unforgiving ends of its dictates. In the pages of the manifest, resources could be stripped out of the world of nature and brought into a world of commerce. The manifest thus deprived enslaved persons—at least for its readers—of their humanity and subjectivity; and while it may not have been the only tool carving up the geographies of Indigenous lands, it was the one responsible for recording the riches of Indigenous material cultures, preparing them for ships bound across the ocean, and placing them in the storehouses of colonial encampments. What's more, contrary to its more manual etymology, the manifest had become transcribed—no longer fixed in hand, but now abstracted in ink on linen paper. Digitization aside, that is where it has stayed. It is a means for marking relations—economic, legal, and otherwise. While for most, *manifest* is nothing other than an archaic, officious synonym for a list of materials, its historic association with transportation (as in a ship's or plane's manifest) remains.

My personal history in manufacturing these sorts of listings goes back to when I created a project called Sourcemap at the Center for Civic Media at the Massachusetts Institute of Technology (MIT) as an early attempt to document product supply chains.[1] The idea was not to make a list. It was to make a map with a calculator—a map, to see where the parts of a product were coming from; and a calculator, to quantify the impact of those piecemeal geographies. We had little context for our work. In the United States, the Digital Humanities Initiative had been announced only recently by the National Endowment for the Humanities, and the idea of the supply chain—at least as an object of discourse—remained centered on the impact of globalization. Our primary interest was as the "makers" of various other hardware projects, and there was a palpable resonance in knowing how far the materials that we sourced for these projects had traveled and what they had cost—not in dollars, but in social and environmental footprints. It was not that we set out to create listings. Indeed, we were trying to move *away* from listings—from stacks of catalogues that assembled components as placeless lines on pages, absent the textures that constitute the reality of human production. This is just to say that, when listings did appear in our project, it was not that we sought them out, but rather that they seemed a necessary precondition of what (we supposed) were other, more interesting things.

This chapter takes up the idea of the manifest to survey how the digital humanities (DH) and allied approaches like human-computer interaction (HCI), data visualization, and others engage with the critical study of logistics, especially with regard to the spatial and nonspatial visualization of logistical networks, media representations and aesthetic imaginations of logistical sites, and close readings of logistical training regimes and logistical software systems. In examining the projects in this survey, I consider how approaches such as the digital mapping of infrastructures for global logistics are shared. Given that industry software and management pedagogy have historically treated the spatialization of logistical processes as largely supplemental to the management of those processes, and that—when geographic perspectives have been adopted—they have been deployed in the service of self-interested corporate discourse on supply chain *transparency* and *resilience*, I argue that special care must be taken in implementing digital approaches. Drawing from my own work on Manifest, a new tool for visualizing, analyzing, and documenting supply chains and trade networks (further discussed later in this chapter), I argue for models centered around the act of "manifesting connection"—not only of logistical networks, but of the external points of contact that situate these networks in broader patterns of sociality, while also outlining the ethical dimensions at stake and the future challenges that will emerge.

Supply chains are not natural objects. They are abstractions, models, for a form of human sociality—and rather recent ones at that (Cowen, "The Deadly Life"; Farris). In many respects, the idea of the supply chain seems to have coincided with the development of supply chain management, as if the whole of this new object emerged, more or less, alongside the discipline that purported to study it. If logistics concerns flows of materials, the term *supply chain management* describes a specific system for managing those flows. When in a 1982 meeting with the consumer electronics manufacturer Philips, business consultant Keith Oliver proposed the idea of modeling separate systems of production, marketing, distribution, sales, and finance "as though" they were a single entity, the result was a singular and total unit of managerial analysis: the supply chain (Laseter and Oliver; see also Cowen, "The Deadly Life"). While the elements that the supply chain operated on were not in themselves new, the changes brought by "supply chain capitalism" lay in the infrastructural and managerial model that it set out; it was a network that could be analyzed through techniques of network optimization (Tsing). Extending the work of *systems,* from Taylorism and Fordism to Toyotism, contemporary production would no longer consist of siloed sites of assembly; rather, they became, as Marc Steinberg suggests, a "platform"—a "networked enterprise" that couples suppliers and distributors to maximize efficiency. Supply chain capitalism defined an order of operation with associations formed by arrangements of subcontracting and outsourcing, a new mobility of labor, and an overriding logic of flexibility and interchange (Tsing, 149).

That the lines and nodes of this network overlaid real geographies, with real people and real materials, was—for its architects—immaterial. From ports and warehouses to cargo containers and delivery trucks, the sites of logistical operation enroll a diverse assembly of software, hardware, and physical architectures in networks that are both spatially distant and temporarily transient. But flattened, it was all reduced to rows in tables, where abstract calculations could determine opportunities for optimization, risk, redundancy, and resilience. This is the violence of the manifest writ large: a form of abstraction operating on the elimination of differences such as class, race, and gender, while being ever more precisely attuned to the potential for exploiting them. At the birth of modern logistics, it was the abstractive power of the "space in the hold"—and the manifest that recorded it—Christina Sharpe writes, that removed not only personhood but sheer nominality (94). Shiri Pasternak and Tia Dafnos note how settler colonial regimes could configure "new imperatives of capital flow" across borders, such that something like "the domestic 'problem of Indians' in Canada" could be reimagined as "the international problem of supply chain management" (Pasternak and Dafnos, 741). How could one attempt, as Brenna Bhandar asks, to "reconceive place, territory, land, or property when it appears settled, firmly ensconced in real estate and financial markets organized according to capitalist rationalities that bear the mark of historically embedded processes of abstraction?" (182).

My interests are not those of supply chain management, but those of the critical study of logistics, an emerging field that examines the impact of supply chains on society (Hockenberry et al., "Assembly Codes"; Brown). Increasingly, scholars working from the perspectives of media studies, science and technology studies, and DH have set out to "trace the vast networks of human labor, data, and natural resources that fuel our digital lives," attempting to reassemble the "metals, refineries, factories, shipping containers, and warehouses that not only manufacture and deliver our electronics but form the infrastructure that organizes our society." According to these "supply studies," it is only in "distill[ing] and mak[ing] legible these global networks" that we can account for the "overflows" of the supply chain and assess what forms of social, environmental, and cultural "harm" hide in its abstractions (Brown).

Substitution and Simulation

There are difficulties in attempting to survey the DH for the critical study of logistics. The most pressing of these is that there are simply not many digital projects that both explicitly address logistics and imagine themselves operating in DH. Jussi Parikka has asked: "Does digital humanities have things to say about, for example, postcolonial issues in digital culture and the strange planetary ties of digital infrastructures, whether those of satellite realities of visual culture, of supply chains of materials, of resource extraction, or of electronic waste as the residual level of dead media?" (Parikka, 451). Seeking projects that both address logistics and identify with DH, we are left with only a handful of examples. In some sense, this is not so surprising. While the first wave of work in DH was concerned with the application of digital methods to areas that lacked them, contemporary logistics was already a digital domain (Berry). Everything from purchasing to production scheduling was governed by enterprise resource planning (ERP) systems produced by companies like SAP, IBM, and Oracle. And conducting DH research *inside* this domain was difficult. Although examining proprietary ERP packages could provide valuable insights about their operation, such packages have largely inaccessible and opaque interfaces. Even if one could gain access or afford the licensing costs, these systems are meaningless outside their working implementations.

Nor is the data held in logistical systems trivial for researchers to access. Subsumed under an ever-expanding state security apparatus, a proliferation of legal designations, and the formalization of third-party logistics, data regarding traditional sites of material traffic has become inaccessible to the public. Ports, shipyards, and railways—once part of the city—now stand outside it. Warehouses and factories are no longer commercial establishments, but gigantic megastructures set in industrial geographies. Any intervention to render data about this infrastructure accessible must compete with systems that not only possess enhanced access but are materially responsible for the sequestration of this infrastructure for use in the network of global supply. In "the absence of access," we are left with "a kind of substitute interface . . . to the software actually used in logistical industries" (Rossiter, "Materialities").

But as Nicole Starosielski has noted, recognizing the "resource operation of digital media" means acknowledging the capacity of DH and allied approaches "both to reconfigure informational resources and to facilitate new extractions of meaning" (402). Some DH projects have thus attempted to offer such reconfigurations, gathering and visualizing data on contemporary port operations, warehouse traffic, and similar areas of internal logistical operation. The following section details distinct categories of projects that offer such substitutions for the critical study of logistics, providing interactive visualizations of logistical processes, new sources of logistical data, and spatial representations of supply chains and other networks.

Developed as part of the Transit-Labour research network, a collaborative project investigating changing patterns of labor and mobility in Asia, the Port Botany Visualisation software application is a representation of "wharf activity and truck turnover times at Port Botany, Australia" (Neilson, Ang, and Rossiter; also see Figure 4.1).[2] Informed by a domain knowledge understanding of the "underlying dynamics of system operations" at ports, its goal is to study changes in port productivity contextualized in terms of their connection to "economic events, labour relations and technological change." Indeed, upon launching the application, we are told that while "ports are logistical machines for creating productivity," this productivity "rests on events and disturbances that elude integration" (Port Botany Visualisation; Neilson et al.). Users are introduced to a graphical table separated into land-side (truck) and wharf-side (ship) transportation. The data—from government reports and interviews with port authorities and cargo companies—is grouped by year and quarter. It is also partial and incomplete. In some respects, such abridgment is typical. Not only is logistics data—which might include both broad trade statistics and highly specific operation records—affected by changes in standards, it is often subject to varying levels of accessibility depending on the administrative authority overseeing it.

The display for each quarter provides both numeric data and graphic representations, with most of these representations showing observed data against a maximum value. The wharf-side data, for example, includes gauges for information like crane rate and the variability of the stevedores (dockworkers, longshoremen) responsible for unloading cargo. The display also includes graphics of stacked containers that visualize data such as throughput or the vessel's working rate. The landside data is similarly organized, and for years after 2008, it even shows a detailed breakdown of daily trucking slots. While the project includes documentation, the measures that it treats are quite technical. One would need to know, for example, that

Manifesting Connection [63]

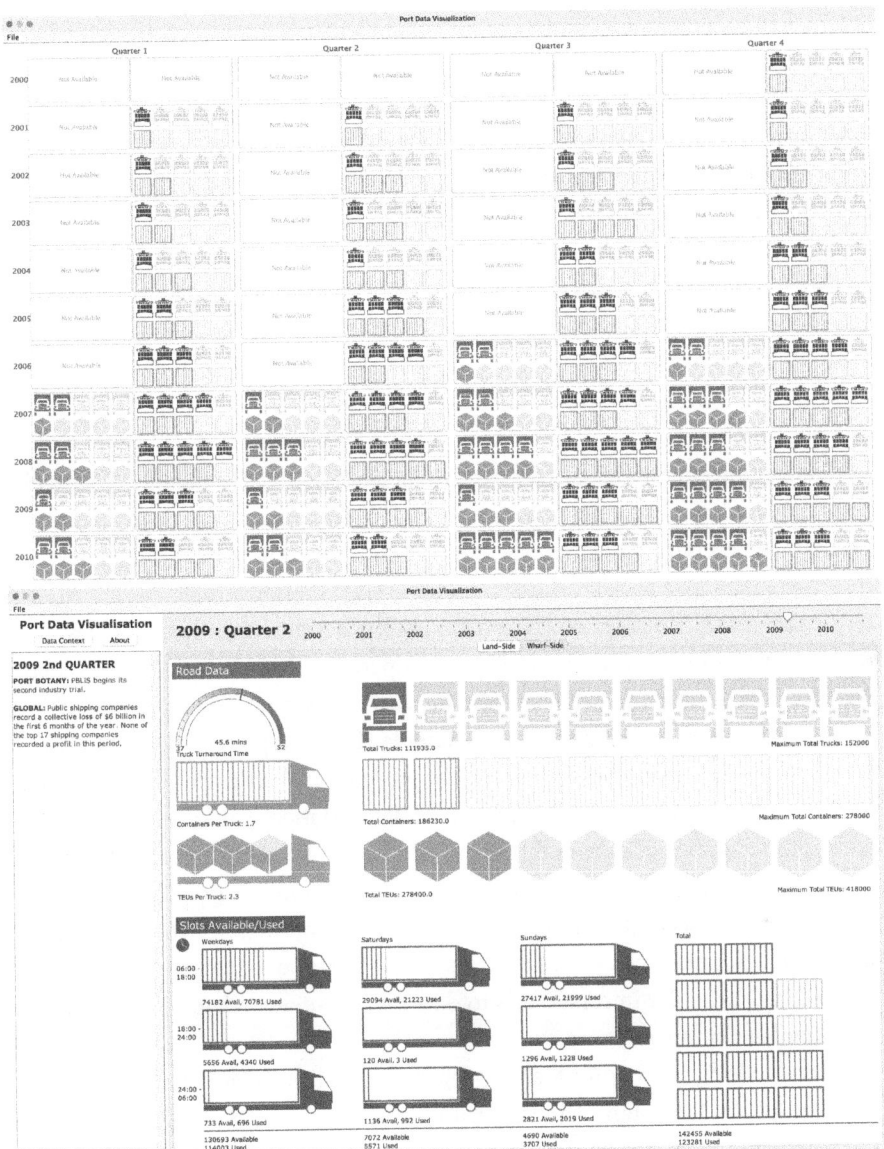

Figure 4.1. From top to bottom: screenshots from the Port Botany Visualisation showing the main browsing interface organized by year (2000–2010) and quarter, with icons indicating rough productivity measures (top section), and the page for land-side data from the second quarter of 2009, with relevant news from that quarter loaded in the side panel and graphical indicators showing detailed comparisons of truck and container throughput (bottom section).

"ship rate" compares crane rate and crane intensity—the number of allocated hours divided by the labor time of workers from ship boarding to departure, minus delays. Still, there is value just in seeing a holistic comparison of productivity across quarters. Not only does the interface make visible the ebb and flow of cargo over time, but the side panel with quarterly news and information also provides connections that at least begin to integrate qualitative accounts of the surrounding economic, political, and logistical circumstances. Indeed, Ned Rossiter notes that while "the aesthetic logic of the visualization is not markedly different from the many visualizations developed in digital humanities," its deployment here highlights the "multiplicity of logistical forces exerted upon labour." As a substitute media form "that [makes] visible the pressures on labour within the shipping and transport industries," it effects its own "algorithmic architecture," which is entirely "distinct from those available through proprietary licenses" (Rossiter, "Materialities," 222, 224, 233). The Port Botany Visualisation is not an application that coordinates supply chains, but "an analytical medium through which to register the interrelations between logistical infrastructure, algorithmic rules, labor practices, and supply chain assemblages" (Rossiter, *Software*, 69).

As an emblem for a class of digital—though not necessarily digital *humanities*—tools for which the primary purpose is indeed to *make visible* logistics data, the Port Botany Visualisation is distinguished by its humanistic interest in labor, infrastructure, and qualitative social context. But like all such tools, it is also characterized by its focused (but limited) quantitative data, a lack of actionable end points (these tools disassemble, rather than assemble, logistical operations), and a constrained range of visualization techniques. While geographic visualization may be a function of these tools, it is rarely the primary one. For example, Trase—a project by the Stockholm Environmental Institute and Global Canopy—provides a supply chain visualization allowing users to examine measures of deforestation, land use, and emissions for a select number of commodities and countries.[3] While its historic data varies, and while Trase does include a map of producing regions, geographic visualization in the project is secondary to providing graphs of production over time, importing countries, ports, and so on. Perhaps Trase's most valuable feature is a section connecting particular interest views to data for potentially related measures, commodities, and countries.

Trase, along with comparable projects like Visualizing Supply Chains for Eco-conscious Redesign (ViSER), primarily engage with logistics data from the perspectives of economics, industrial design, and environmental science (Bernstein et al.). With conflicting standards for representing supply chains, projects often adopt relatively flat and inflexible data models—though a few, like ViSER, employ relational data formats that can be used to generate graphlike visualizations of connections. ViSER goes so far as to present itself as a general-purpose dashboard with a variety of visualization panels and support for arbitrary datasets.[4] But while this dashboard interface may be a common metaphor used in industry software, it is critical to recognize that this is *not* industry software. Even smaller industry packages and providers like Panopticon, InetSoft, and Log-hub, though lacking the operational capabilities of platforms like SAP, are not bound by the constraints of the substitutes offered by academic projects like ViSER or Trase.[5] Panopticon, for example, has the potential to connect to live databases and subscribe to logistical event streaming feeds from systems like Apache's Kafka—a functional impracticality for the scale of a DH project.

Rather than attempt to offer a substitute for industry software, some projects subvert them. Ian Cook et al.'s followthethings, for example, presents what Miriam Matthiessen and Anne Lee Steele describe as an "Amazon underbelly of sorts, designed to match the feel of online shopping but linking the visitor instead to scholarship, films, stories, reporting on a given product" (Matthiessen and Steele, 12; see also Cook).[6] Other efforts have attempted directly to analyze their functions and effects. In 1996, for example, anarchist programmers affiliated with LabourNet cracked SAP's software to identify lines of codes "that would affect workplace activity in detrimental ways," and Ned Rossiter argues that, while less dramatic, similar "methods developed within the digital humanities . . . have an important role to play in the critique of logistical power" (*Software*, 55). Still, Miriam Posner is quick to note that, in many respects, "the term 'software' is misleading" when applied to industry systems. These are not singular applications, but complex "suite[s] of tools joined together through a shared database" with a modular nature that means that many "functions are 'black-boxed.'" Posner writes, "In a company that makes use of an integrated SAP suite for supply chain planning, planners pass data along a chain," moving, "with increasing granularity, down the line, until it reaches workers on the factory floor" (Posner, "Breakpoints"). This close reading nevertheless allows her to demonstrate how SAP's specialization allows operations on different time scales ("time horizons") simultaneously, and to show that, despite appearances, features like the "Supply-Chain Cockpit"—a graphic representation of the geography of factory locations, shipping routes, and warehouses—are "not dynamic," but "a moment of the planning process, frozen in time" (Posner, "The Software").

There are also those projects that attempt to provide access to the logistical data produced by corporations, nonprofit and activist organizations, and administrative government records by developing end points for new kinds of users to access that data.[7] Although much of this data is, as Rossiter puts it, more "remodeled" than raw, there can be novel derived measures (*Software*, 62). Part of the work of the Helena Kennedy Centre, for example, involves cross-referencing import records and manufacturing data to determine likely pathways of forced labor (Helena Kennedy Centre); the University of Delaware's project on the "Global Illicit Trade in Energy-Critical Materials" uses machine learning and remote sensing to map illegal extraction (Klinger et al.); The Prepared initiative's Dong Xi project mined companies and locations from PDFs of corporate supplier reports (Wright). But despite the best efforts of projects like the Open Supply Hub, Wikirate, and others, the open-data ecosystem facilitating such study remains limited.[8] While U.S. customs records, for example, are ostensibly public reports, they are buried behind administrative hurdles. Companies like Panjiva, ImportGenius, and MarineTraffic, which offer end points to logistics datasets, have made access increasingly restrictive or expensive over time, and projects like Importyeti and PortExaminer, which currently provide free queries for this kind of data but without well-defined application programming interfaces or direct access, have uncertain futures.[9] But in all the projects creating substitutions or simulations of logistics information, there remains a disconnect—an abstraction—between the screen and the real people and places that comprise the landscape of logistics.

Spaces of Possibility

"Logistics space," Deborah Cowen writes, is the network cartography defined by the global circuits of circulations, the deliberate result of "thinking and calculating space anew." But "there is a dearth of scholarship on the representations of [this] space" (Cowen, "The Deadly Life," 33, 48). Cartographic representations require a different kind of manifest, one that anchors abstract flows to material ones. As Johanna Drucker et al. note, maps require reconciling what might exist only in relative scale and space by organizing (or reorganizing) actors, events, and relationships: "forcing" this geographical representation, then, can represent a "profound, even violent, intervention" (Drucker et al., 57). But indeed, as I have already suggested, to map the supply chain is to respond to the violence of listing already embedded in its origins—to ground the network of the supply chain in those very geographies that it has already abstracted and de-territorialized. This may be why maps are rarely central to industry applications. Perhaps they are inherently "counter-logistical" pursuits (Bernes).

Supply-chain mapping projects employ a mix of commercial platforms and open-source software, including geographic information systems such as ArcGIS and QGIS, web mapping platforms such as Carto, Mapbox, and Leaflet, and spatial databases such as OpenStreetMap. Such projects tend toward two distinct areas of intervention: narrating accounts of commodities or representing landscapes of logistical infrastructures. The former is common to projects created by journalists and activists, whose stories gather insights by revealing the geographies traversed by commodities. Examples of academic projects in this area include mixed media initiatives such as Commodity Histories or the Charnley-Persky House Archaeological Project, which connects past consumer products to their wider logistical networks—listing, mapping, and historicizing the use and supply of household goods to "give voice to those who are not typically in the documentary record" (Charnley-Persky House Archaeological Project, "Archaeology").[10] But the direct connection to the manufacturer on the label of commodities represents only a partial perspective, and researchers cannot follow everything. Even in activist projects, high-profile companies like Apple and Nike must stand in for problems endemic to entire sectors, with articulations of those problems varying widely across numerous firms with sometimes divergent operational strategies. Often absent from these accounts are the ports, warehouses, and other logistical intermediaries, along with the suppliers and partners that may have contributed to their material inhabitance at locations such as Charnley House in Chicago. As a result, also absent are the pathways for the flows of the materials themselves.

The latter sort of intervention can be seen in projects such as the Empire Logistics research initiative, which addresses the representation of absent infrastructure directly, making the scale of network space commensurate to its real geographies to "articulate the infrastructure and [its] 'externalized cost'" (Empire Logistics, "About").[11] Focused on warehouses and logistical infrastructure in the Inland Empire in southern California, the project's "Supply Chain Infrastructure" map (Figure 4.2) includes information on railways, ports, and warehouses, identifying specific distribution centers for companies such as Amazon, Costco, and Walmart, and thus showing the broader context of technology, logistics, and finance.

Epitomized here is the power of listing: the logistical network of the Inland Empire gathered together and placed in a public geography of roads, airports, neighborhoods, and cities that makes manifest the spaces "where proletarian solidarity has the greatest possibility to spread up and down the chain" ("About"). But as a "counter-logistical" project (Bernes), Empire Logistics also uses its maps to open the choke points of the networks, thus making possible new—and given the sensitive nature of these choke points, its own potentially violent—forms of resistance. There are other examples of projects addressed to questions of logistical power, each with their own ethical or political stakes: the Abandoned Seafarer Map, which documents workers whom corporations have left behind at sea (Ader, Bolton, and Matthiessen); the Reducing Opportunities for Unlawful Transport of Endangered Species (ROUTES) Dashboard, which records wildlife trafficked through airports; and the Feral Trade project, which records transactions of a collective freight network, albeit one organized outside conventional logistical channels.[12] Other investigations reveal unfamiliar logistical infrastructures to draw attention to their geopolitical complications, material fragility, and complex entanglements. This includes Nicole Starosielski, Erik Loyer, and Shane Brennan's Surfacing project, which documents undersea fiber-optic communication cables, as well as work like Kate Crawford and Vladan Joler's Anatomy of an AI System and Anna Tsing, Jennifer Deger, Alder Keleman Saxena, and Feifei Zhou's Feral Atlas.[13]

Projects looking at earlier historical periods face different challenges. Not only does the reconstruction of historic trade networks depend on the availability of archival materials, but these networks exist within political landscapes and physical geographies that may be largely unrecognizable. Projects like the Viabundus Premodern Street Map, for example, attempt to make historical trade networks legible to contemporary viewers.[14] Digitizing the routes recorded in Friedrich Bruns and Hugo Weczerka's 1962 *Hansische Handelsstraßen*, the street map cleverly juxtaposes a modern interface for trip planning with directions from centuries past to make this trade network accessible in ways that a purely textual account cannot. Other projects, like Old Weather, recover maritime pathways from ships' logs to enable new "perspectives on the movement patterns of early American capitalism" (Schmidt);

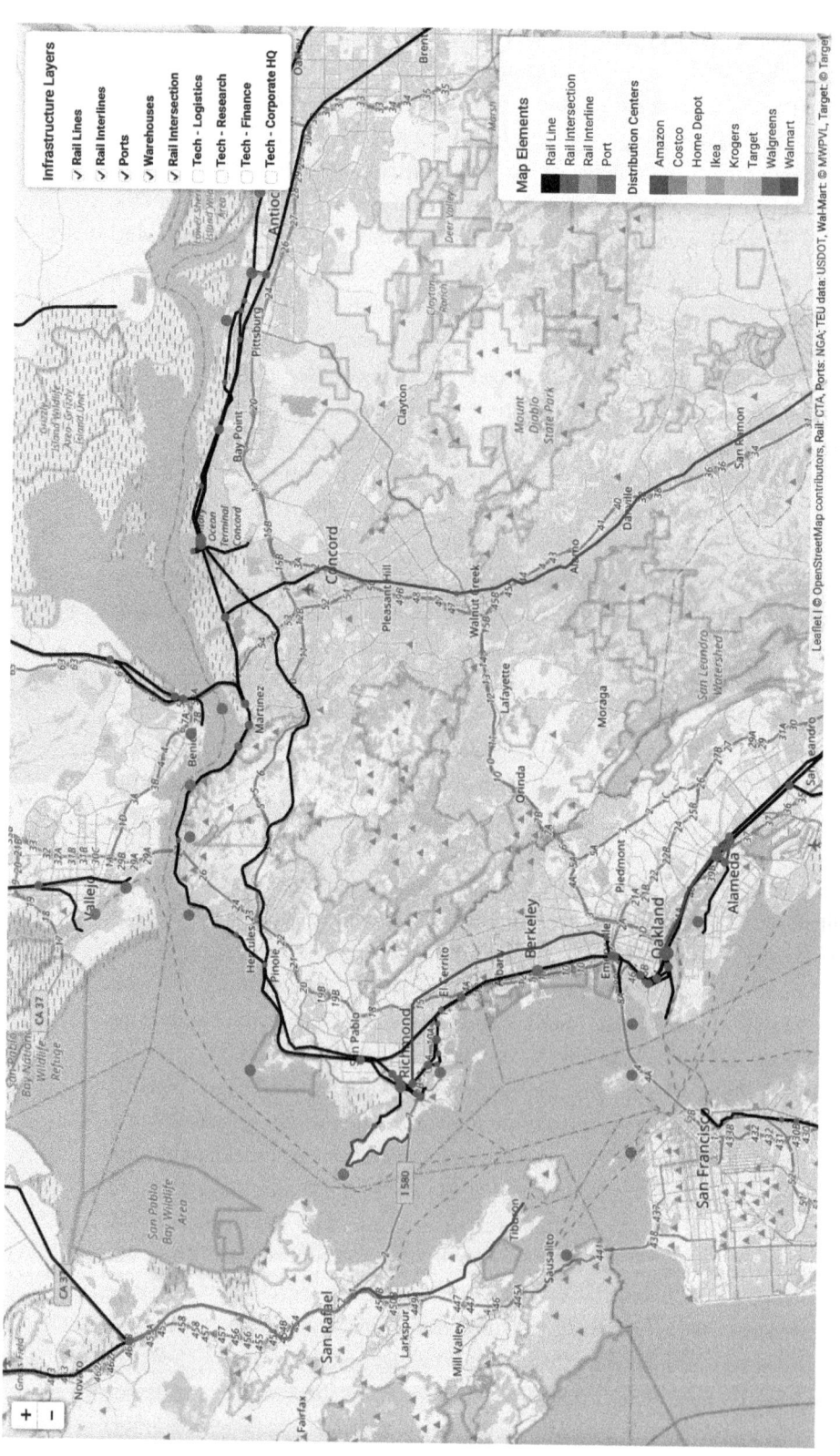

Figure 4.2. Screenshot of the Empire Logistics map of southern California's Inland Empire, with infrastructure layers for rail lines, ports, and warehouses, with indication of the companies associated with specific distribution centers (such as Amazon, Costco, IKEA, Target, and Walmart). Copyright 2020, Empire Logistics.

and Trading Consequences reconstructs networks of trade, economy, and environment in the British Empire during the nineteenth century by mining over 200,000 documents for references to commodities or locations.[15] The visualizations that resulted from this latter trade "search engine" allowed one to "correlate information extracted for one commodity with that of others," and thus to analyze the flows of materials and configurations of networks across decades (Trading Consequences). But one must be cautious about such historical representations. As Ben Schmidt reflects, visualizations of historical data do not remove us from problem of grappling with the historical context, the underlying archive, and the constraints under which information about the past was collected. While such visualizations can be valuable tools, one must be cautious about reading current concerns into them. There were no supply chains for the Hanseatic League or for whalers in the nineteenth century, at least as we know them.

There are, of course, digital projects in other domains that intersect with the critical study of logistics. "Although they diverge in their object of study," Starosielski writes, projects in ecological digital media such as Growing Blue, Aqueduct, Bodily Natures, and Protest are similarly "invested in digital technologies" and offer their own logistical insights (402). For more artistic inventions, we might consider projects like Georgina Voss and Wesley Goatley's *Familiars,* an immersive installation embedding the viewer in a mapped representation of intercepted logistical signals broadcast via radio (Goatley); the interactive game *Cargonauts*; or Jesse LeCavalier's architectural exploration *Landscapes of Fulfillment*.[16] And there are countless other such projects to which we might look for inspiration.

Manifesto

This final section discusses the Manifest project (Figure 4.3), outlining how its technical and ethical framework has been informed by the earlier projects that I have described here, before concluding with future challenges for DH in the critical study of logistics.[17] Manifest is based on some of the concepts behind Sourcemap, the web platform that I created to integrate social and environmental impact assessment (carbon footprinting, water, and energy usage) with geospatial visualizations of supply chains.[18] While originally intended as a tool of journalistic or activist investigation and scholarly research, Sourcemap was ultimately commercialized in 2011 as an industry platform by other members of the project team. Prior to this, the project published over ten thousand user-submitted supply-chain documents.

Released in 2021, Manifest is "an investigative toolkit intended for researchers, journalists, students, and scholars interested in visualizing, analyzing, and documenting supply chains, production lines, and trade networks" ("About Manifest"). Its goals are to develop a common data format for work in the critical study of logistics that can incorporate information relevant to a range of research projects and to provide an extensible visualization framework for studying supply chains. It has support

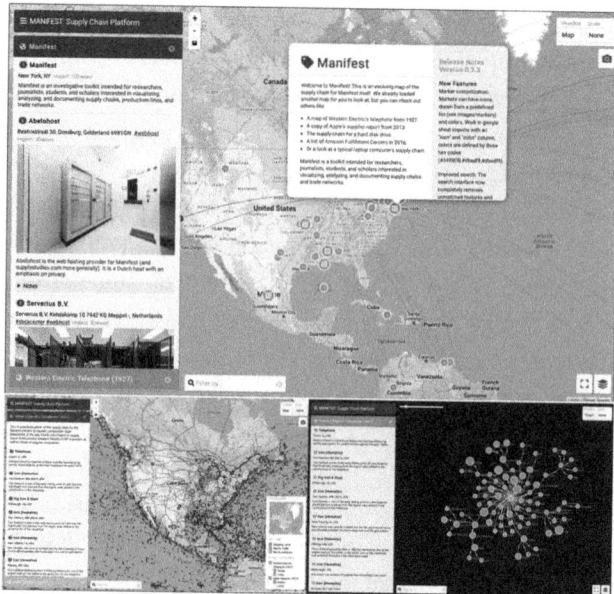

Figure 4.3. Counterclockwise from top: A spatial (map) visualization of two documents rendered simultaneously—namely, the hosting infrastructure of the Manifest project and the production supply chain of the telephone manufacturer Western Electric in 1927; a spatial (map) visualization of two documents rendered simultaneously—namely, Western Electric's production supply chain and Apple's 2013 worldwide supplier list, with these documents set over data layers for global shipping lanes, railway connections, and live reports of global marine traffic; a nonspatial (graph) visualization of the production supply chain of Western Electric in 1927. Copyright 2024, Manifest Project/Matthew Hockenberry.

for both spatial and nonspatial representations, as well as an interface that allows for the aggregation of multiple concurrent datasets, such as multiple discrete supply chain documents, infrastructural data layers like port locations or railways, and live data sources such as container ships or trucks. Figure 4.3, for example, shows a series of screenshots from Manifest that illustrate both map and graph visualizations of several supply chains overlaid on multiple infrastructural data layers and live data sources. The project is supported by a Digital Humanities Advancement Grant from the National Endowment for the Humanities (NEH), which provides funding for both overall technical development and the publication of a series of Manifest "case studies" produced by external contributors (Hockenberry and Perold); for an example of a Manifest case study, see Schwartz et al.[19]

From a technical perspective, Manifest documents in simple JavaScript Object Notation (JSON) files represent trade networks, supply chains, and similar accounts; and the tool allows users to construct, visualize, and analyze multiple such documents concurrently. Each document contains a list of nodes, including information like a name, an address, coordinates, connections to other nodes, and user-defined text, categorizations, imagery, and impact measures, along with arbitrary key-value pairs (which could include citation information, data generated by other applications, or other links and references not directly useful to Manifest). For example, the Western Electric supply chain shown in Figure 4.3 includes an entry for "Iron," with an address listed as "Hibbing, MN, USA," coordinates of 47.4271546°N latitude and -92.937688700002°W longitude, a connection to Western Electric's factory in Hawthorne, Illinois, a text entry noting that "the Hull-Rust-Mahoning Mine in Hibbing, Minnesota is one of the largest open pit iron mines in the world," categorization as a "raw material," and a measure that provides the approximate tonnage of iron extracted. In any visualization, these nodes can be filtered by category or keyword, or scaled to show the relative difference of the quantitative measures. The map visualization is customizable: colors, sizes, and icons can be set, and a variety of base tiles—the series of composited images that comprise the visual appearance of the map—like satellite imagery can be selected and layered with user-selectable supplemental datasets such as the shipping lanes, rail networks, and maritime traffic data previously described. Manifest provides extensible plug-in support for other visualizations as well, currently including a spreadsheet view, a chart view, the graph view shown in Figure 4.3, and flow and chord diagrams. It also has capabilities for sharing documents and generating static images of visualizations. While a public installation is available on the Manifest home page, anyone can set up an instance from the project's GitHub repository.[20]

Manifest diverges from its predecessor, Sourcemap, in several critical ways. The map is still present. But although Manifest supports visualizing quantitative data, the calculator is less so. This is partly because while approximate information proved to be meaningful in geographic visualization, approximate calculations did not. This is one of the reasons why Sourcemap was reproduced as a tool for business rather than maintained as the messy means of journalistic and activist "muckraking" it was originally intended to be. As the projects detailed in this chapter demonstrate, logistics data is often partial and incomplete, with corporate actors the most likely ones to have detailed quantitative data (or the power to compel its collection).

Manifest is also less invested than Sourcemap in centralizing supply chain data or providing comprehensive accounts of supply chains. From a practical perspective, this is due to the constrained nature of most supply chain investigations. Indeed, one of the greatest obstacles in the classroom is a concern about limited data. Manifest was not designed to be a clearinghouse for complete accounts of logistical operation, but rather an interface for assembling multiple supply chains (or partial supply chains) to understand the relationships among them. In Manifest, a map of three nodes can be as meaningful as one with three hundred, since that map can be brought into alignment with an arbitrary number of other Manifest documents and data layers. This decentralization also speaks to the finite lifespan of digital projects in general (Meneses). Some of the projects that I have surveyed here, like Port Botany and ViSER, continue to be usable but are no longer under development; others, such as Trading Consequences and Empire Logistics, are all but inoperable; while still others, like ImportYeti, seem destined for increased commercialization. Even more fundamental than the ephemerality of digital projects, of course, is the fact that the nature of global supply itself is ever changing. Projects not intended to be historical work will soon become so. Port Botany, for example, makes visible the labor of workers who are no longer employed. Empire Logistics shows infrastructure that has come and gone. And while some projects leave their data accessible (Port Botany's, for example, is available through the Australian Research Data Commons) or are committed to making regular updates (as in the case of the Abandoned Seafarer Map), many do not allow for sustainability. Fewer still share their code.

In Manifest, the data format is divorced from the interface. The data format not only privileges simple textual descriptions and clearly calculated measures, it supports arbitrary data that Manifest neither interprets nor visualizes. This leaves open the possibility that data produced for use in Manifest can be meaningful for other tools past or present and can be imported into general data science systems or even read in a basic text editor. Reliant on few external services, Manifest documents can be edited in a basic interface that requires users to geocode locations manually. Even the live map tiles can be exchanged for a static world map image.

But issues of functionality aside, the ethical framework that Manifest sets forth is important because of its endeavor to acknowledge the power of listing, working against a landscape of industry software best suited to those who have the ability not just to take, but to make, inventory. In this regard, Manifest does not just act counter to the prevailing logistical order but also as an inquiry into the fundamental structure that such order necessitates. Rejecting structures that seem to be collected and complete in favor of the distributed, the partial, and the temporary, Manifest advocates the creation of a scaffolding for collecting fragments of logistical networks. It is not a database, nor does it aspire to control or own its data—in part because this is exactly what has allowed for the abstraction of logistics space by enterprise software systems. The supply chain, after all, is a structure that "ravenously consumes other data" (Posner, "The Software"). Indeed, the very politics of supply chain transparency that Sourcemap once pioneered has become a central component of corporate social responsibility.

In the epigraph that starts this chapter, the manifest is described as an accounting of injuries. Despite our meager sympathies for Twain's protagonist, I think that this perspective provides a better model for unraveling the global supply chain than transparency. Rather than allow transparency to remain as a form of corporate responsibility, with "mapping the supply chain" an exercise in corporate power, "making out its manifest" might now be the ideal that accounts for our value, as well as our injuries. Such an accounting records the places where labor has been exploited, the Earth has been plundered, waste overruns into rivers, and poison bleeds into the air. It is not a proclamation from on high, but rather an admonition from below—not an attempt at supply chain resilience but an opportunity for supply chain reconciliation. But this is not a singular endeavor. The manifest is a social document, and its listing is empowered by the shared understanding of those who read it. The future of DH for the critical study of logistics requires precisely such common cause. We must foster a community of open data and open-data standards, contribute and extend open-source frameworks for common visualization tasks, and imagine both the present moment and a collaborative afterlife for digital projects that are capable of contending with the logistical software systems of supply chain capitalism.

NOTES

1. See https://sourcemap.com/.

2. For the dataset of the Port Botany Visualisation, see Neilson, Ang, and Rossiter. For a discussion of the project, see Nguyen, Zhang, and Simoff.

3. See https://www.trase.earth/.

4. This brings with it, of course, all the limitations of the dashboard. As Shannon Mattern writes: "The acknowledged partiality of the dashboard's rendering might make us wonder what is bracketed out. Why, all the mud of course! All the dirty (un-'cleaned') data, the variables that have nothing to do with key performance (however it's defined), the parts that don't lend themselves to quantification and visualization."

5. See Panopticon, https://altair.com/panopticon; InetSoft, https://www.inetsoft.com; Log-hub, https://log-hub.com.

6. See http://followthethings.com.

7. End points are server locations, usually Universal Resource Locators (URLs), where an application programming interface (API) receives requests about a specific resource on its server. These are usually used to provide programmatic access to databases by external developers or internal developers working on a different software application.

8. See Open Supply Hub, https://opensupplyhub.org/; Wikirate, https://wikirate.org/.

9. See Panjiva Supply Chain Intelligence, https://panjiva.com/; ImportGenius, https://www.importgenius.com/; MarineTraffic, https://www.marinetraffic.com/en/ais/home/; ImportYeti, https://www.importyeti.com; Port Examiner, https://portexaminer.com/.

10. See https://digitalchicagohistory.org/exhibits/show/charnley-persky-house/archaeology-at-the-charnley-pe.

11. See https://www.empirelogistics.org/about/.

12. See Abandoned Seafarer Map, https://abandonedseafarermap.cargo.site; ROUTES Dashboard, http://www.routesdashboard.org/; Feral Trade, https://feraltrade.org/.

13. See Surfacing, https://surfacing.in/; Anatomy of an AI System, https://anatomyof.ai/; Feral Atlas, https://feralatlas.org/.

14. See http://www.landesgeschichte.uni-goettingen.de/handelsstrassen/map.php.

15. See Old Weather, https://www.oldweather.org/; Trading Consequences, https://web.archive.org/web/20220929115427/https://tradingconsequences.blogs.edina.ac.uk/. On the Trading Consequences project, see also Hinrichs et al.

16. See Familiars, https://www.wesleygoatley.com/familiars/; Cargonauts, https://cargonauts.net/; Landscapes of Fulfillment, https://landscapes-of-fulfillment.org/.

17. See https://manifest.supplystudies.com/.
18. See https://sourcemap.com/. On Sourcemap, see also Bonanni et al.
19. See https://manifest.supplystudies.com/data/.
20. See https://manifest.supplystudies.com/; https://github.com/hock/Manifest.

BIBLIOGRAPHY

Ader, Eliza Jacob Bolton, and Miriam Matthiessen. Abandoned Seafarer Map. 2022. https://abandonedseafarermap.cargo.site.

Bernes, Jasper. "Logistics, Counterlogistics and the Communist Project." *Endnotes* 3 (2013): 172–201.

Bernstein, William Z., Devarajan Ramanujan, Niklas Elmqvist, et al. "ViSER: Visualizing Supply Chains for Eco-Conscious Redesign." In *Proceedings of the ASME 2014 International Design Engineering Technical Conferences and Computers and Information in Engineering Conference. Volume 4: 19th Design for Manufacturing and the Life Cycle Conference; 8th International Conference on Micro- and Nanosystems*, Buffalo, New York, August 17–20, 2014.

Berry, David M. "The Computational Turn: Thinking About the Digital Humanities." *Culture Machine* 12 (2011): 1–22.

Bhandar, Brenna. *Colonial Lives of Property: Law, Land, and Racial Regimes of Ownership*. Duke University Press, 2018.

Bonanni, Leonardo, Matthew Hockenberry, David Zwarg, et al. "Small Business Applications of Sourcemap: A Web Tool for Sustainable Design and Supply Chain Transparency." In *Proceedings of the SIGCHI Conference on Human Factors in Computing Systems*, 937–46. ACM, 2010. https://doi.org/10.1145/1753326.1753465.

Brown, Jackie. "Source Material." *Real Life*. March 8, 2021. https://reallifemag.com/source-material/.

Bruns, Friedrich, and Hugo Weczerka. *Hansische Handelsstraßen*. Böhlau, 1962.

Cargonauts. Home page. n.d. https://cargonauts.net/.

Charnley-Persky House Archaeological Project. "Archaeology at the Charnley-Persky House." Lake Forest College Digital Chicago Project, 2016. https://digitalchicagohistory.org/exhibits/show/charnley-persky-house/archaeology-at-the-charnley-pe.

Commodities of Empire. Home page. n.d. British Academy. https://commoditiesofempire.org.uk/.

Cook, Ian, et al. "followthethings.com: Analysing Relations Between the Making, Reception and Impact of Commodity Activism in a Transmedia World." In *Innovations Sociales: Comment les Sciences Sociales contribuent à transformer la Société*, edited by Ola Söderström, Laure Kloetzer, and Hugues Jeannerat, 50–61. Université de Neuchâtel, 2017.

Cowen, Deborah. *The Deadly Life of Logistics: Mapping Violence in Global Trade*. University of Minnesota Press, 2014.

Cowen, Deborah. "Logistics' Liabilities." *Limn* 1, no. 1 (2011). https://limn.it/articles/logistics-liabilities/.

"A Dictionary, or Alphabetical Explanation of Most Difficult Terms Commonly Used in Merchandise and Trade . . ." *Edward Hatton's Merchant's Magazine*, xiv, 234/1 (1697).

Drucker, Johanna, David Kim, Iman Salehian, et al. "Introduction to Digital Humanities Course Book." 2014. https://searchworks.stanford.edu/view/11649226.

Empire Logistics. "About." 2020. https://www.empirelogistics.org/about/.

Empire Logistics. "Supply Chain Infrastructure" (map). 2020. https://www.empirelogistics.org/sci-map/.

Farris, Martin T. "Evolution of Academic Concerns with Transportation and Logistics." *Transportation Journal* 37, no. 1 (Fall 1997): 42–50.

Feral Trade. Home page. n.d. Edited by Kate Rich. https://feraltrade.org/.

Helena Kennedy Centre. Forced Labour Lab. 2023. https://www.shu.ac.uk/helena-kennedy-centre-international-justice/research-and-projects/all-projects/forced-labour-lab.

Goatley, Wesley. "Familiars." n.d. Wesley Goatley. https://www.wesleygoatley.com/familiars/.

Hinrichs, Uta, Beatrice Alex, Jim Clifford, et al. "Trading Consequences: A Case Study of Combining Text Mining and Visualization to Facilitate Document Exploration." *Digital Scholarship in the Humanities* 30, no. suppl_1 (2015): i50–i75. https://doi.org/10.1093/llc/fqv046.

Hockenberry, Matthew, Nicole Starosielski, and Susan Ziegler. *Assembly Codes: The Logistics of Media*. Duke University Press, 2021.

Hockenberry, Matthew, and Colette Perold. "Manifest: Digital Humanities Platform for the Critical Study of Logistics." Digital Humanities: Digital Humanities Advancement Grants. HAA-290391-23. 2023–2024.

Klinger, Julie, Dawn Fallik, Xi Peng, et al. "Characterizing the Global Illicit Trade in Energy-Critical Materials Using Machine Learning, Remote Sensing, and Qualitative Research." National Science Foundation Award 2039857. 2021. https://www.nsf.gov/awardsearch/showAward?AWD_ID=2039857.

Landscapes of Fulfillment. Home page. n.d. https://landscapes-of-fulfillment.org/.

Laseter, Tim, and Keith Oliver. "When Will Supply Chain Management Grow Up?" *Strategy and Business* 32 (2003). https://www.strategy-business.com/article/03304.

Manifest. "About Manifest." n.d. https://manifest.supplystudies.com/about/.

Manifest. GitHub repository. 2023. https://github.com/hock/Manifest.

Manifest. Home page. n.d. https://manifest.supplystudies.com/.

Mattern, Shannon. "Mission Control: A History of the Urban Dashboard." *Places Journal* (March 2015). https://placesjournal.org/article/mission-control-a-history-of-the-urban-dashboard/.

Matthiessen, Miriam, and Anne Lee Steele. "Rendering Supply Chains Research and Its (Dis)contents: An Antipaper on Open Knowledge and Maintenance as a Research Ethos." *APRJA* 11, no. 1 (2022): 11–27.

Meneses, Luis, and Richard Furuta. "Shelf Life: Identifying the Abandonment of Online Digital Humanities Projects." *Digital Scholarship in the Humanities* 34 (2019): i129–i134.

Neilson, Brett, Ien Ang, and Ned Rossiter. "Port Botany Visualisation." Research Data Australia. Accessed August 27, 2023. https://researchdata.edu.au/port-botany-visualisation/2368203.

Nguyen, Quang Vinh, Kang Zhang, and Simeon Simoff. "Unlocking the Complexity of Port Data with Visualization." *IEEE Transactions on Human-Machine Systems* 45, no. 2 (April 2015): 272–79. https://doi.org/10.1109/THMS.2014.2369375.

Old Weather. Home page. n.d. https://www.oldweather.org/.

Panopticon. Home page. n.d. https://altair.com/panopticon.

Parikka, Jussi. "A Care Worthy of Its Time." In *Debates in the Digital Humanities 2019*, edited by Matthew K. Gold and Lauren F. Klein, 449–52. University of Minnesota Press, 2019.

Pasternak, Shiri, and Tia Dafnos. "How Does a Settler State Secure the Circuitry of Capital?" *Environment and Planning D: Society and Space* 36, no. 4 (2018): 739–57.

Port Botany Visualisation. Brett Neilson, Ian Ang, and Ned Rossiter, and Ien Ang. Transit Labour. Java. 2012. https://researchdata.edu.au/port-botany-visualisation/2368203.

Posner, Miriam. "Breakpoints and Black Boxes: Information in Global Supply Chains." *Postmodern Culture* 31, no. 3 (2021).

Posner, Miriam. "The Software That Shapes Workers' Lives." *New Yorker.* March 12, 2019.

Reducing Opportunities for Unlawful Transport of Endangered Species (ROUTES). "ROUTES Dashboard." n.d. http://www.routesdashboard.org/.

Rossiter, Ned. "Materialities of Software." In *Advancing Digital Humanities,* edited by Paul Longley Arthur and Katherine Bode, 221–40. Palgrave Macmillan, 2014.

Rossiter, Ned. *Software, Infrastructure, Labor: A Media Theory of Logistical Nightmares.* Routledge, 2016.

Schmidt, Ben. "Reading Digital Sources: A Case Study in Ship's Logs." *Sapping Attention* (blog), November 15, 2012. http://sappingattention.blogspot.com/2012/11/reading-digital-sources-case-study-in.html.

Schwartz, Elizabeth, Lucian Li, David Satten-Lopez, and Eleanor Colbert. "Scanning at Scale: Digitization Labor in the Internet Archive." In *Cultures of Scale: Disciplines, Data, and Labor,* edited by Joshua Ortiz Baco, Jim Casey, Benjamin Charles Germain Lee, and Sarah H. Salter. University of Minnesota Press, forthcoming.

Sharpe, Christina. *In the Wake: On Blackness and Being.* Duke University Press, 2016.

Sourcemap. Home page. 2023. https://sourcemap.com/.

Starosielski, Nicole. "Resource Operations of the Ecological Digital Humanities." *PMLA* 131, no. 2 (2016): 401–409.

Steinberg, Marc. "From Automobile Capitalism to Platform Capitalism: Toyotism as a Prehistory of Digital Platforms." *Organization Studies* 43, no. 7 (2022): 1069–1090.

Surfacing. Nicole Starosielski, Erik Loyer, and Shane Brennan. Home page. n.d. https://surfacing.in/.

Trading Consequences. Home page. n.d. https://web.archive.org/web/20220929115427/https://tradingconsequences.blogs.edina.ac.uk/.

Trase. Home page. n.d. Stockholm Environment Institute and Global Canopy. https://www.trase.earth/.

Tsing, Anna. "Supply Chains and the Human Condition." *Rethinking Marxism: A Journal of Economics, Culture & Society* 21, no. 2 (2009): 148–76.

Viabundus Pre-modern Street Map. Bart Holterman, et al. V. 1.2. 2022. http://www.landesgeschichte.uni-goettingen.de/handelsstrassen/map.php.

Wright, Spencer. "Dong Xi—A Study of Stuff." *The Prepared,* November 24, 2019. https://web.archive.org/web/20211117022357/https://theprepared.org/features-feed/dong-xi-a-study-of-stuff.

PART I][Chapter 5

Critical Studies of Tech Stacks
What Can Technologies Tell Us About a Lab Culture?

URSZULA PAWLICKA-DEGER, ARIANNA CIULA,
AND MIGUEL VIEIRA

This chapter proposes a methodological approach that integrates laboratory ethnography with network analysis to understand the role of technologies in shaping lab cultures. By *lab cultures,* we mean research environments whose technological milieu and social-philosophical character influence each other. Our particular focus is on digital humanities (DH) labs, and our case study is being carried out at King's Digital Lab (KDL) at King's College London, where we have been embedded in compound participant-observer positions that allow us to use observation, interviews, and access to lab documents to conduct "polyvocal" ethnography (Nelson, 65–66).[1] We draw on ethnographical methods in the mode of science and technology studies (STS), laboratory studies (Latour and Woolgar), and workplace and organization studies (Gellner and Hirsch). To further investigate interconnections between infrastructures, organizations, and their cultures, we apply STS and social-science network visualization methods that position human and technological agents as nodes in networks of relations (Venturini, Munk, and Jacomy). We also apply related network analysis methods (van Geenen et al.) that have been used by others to represent sociotechnical assemblages such as infrastructure (Edwards), data (Bounegru and Gray), and online communities (Grandjean and Mauro). In general, our interest is the entanglement of infrastructures and ethnography: how infrastructures can be approached ethnographically (Star), and reciprocally, how ethnographies of digital technologies can be approached infrastructurally (Knox). Combining network analysis with infrastructural ethnography allows us to analyze the way that technologies and cultural practices in a DH lab mutually shape each other.

We investigate in particular what we will call the *technology stack* (or *tech stack*) of KDL—a set of technologies for developing projects (software, standards, systems, programming languages, and equipment) that is also a network of social actors. These actors include the university, for-profit-company, nonprofit-organization, and community sectors that develop technologies and—engaging with them through the technologies that it uses—KDL's research software engineer (RSE) team. "RSEs," or "the RSE team," as the KDL members refer to themselves, is a designation for staff researchers who originated in science fields, but it is increasingly recognized in DH and other areas in the United Kingdom, Europe, and Australasia as a way to bring more professional recognition to such positions by providing better-defined job roles and support for career progression.[2] The picture of relationality that emerges in studying the implicit network of KDL's tech stack—following Star's precept that "infrastructure is a fundamentally relational concept" (380)—is one of a number of complex interactions of closed and open, corporate and nonproprietary, and institutional and local lab values. Ultimately, we aim to showcase how such a network of relations configures the culture of DH labs.

Contributing to the growing field of research on how technology influences epistemic and cultural processes,[3] this chapter is an example of what we call *critical studies of tech stacks*. In a DH lab, technology makes the culture that shapes knowledge creation.

Setting the Scene of the Case Study

While scientific laboratories have been analyzed through ethnographic methods (complemented more recently by STS approaches), DH labs have been less so; and the emerging scholarship on such labs has tended to foreground other aspects of their culture than those related to their technological stacks.[4] To bring attention to the culture of DH lab tech stacks, we offer a case study of KDL.

Like other DH laboratories, KDL can be characterized generally as a site that enables "research through the provision of infrastructure, tools, and methods" (Smithies and Ciula, 157). But KDL is also one of the largest, most developed, and most professionalized DH lab environments, with an especially extensive tech stack that affords unique opportunities for closely examining the relation of technology and culture. KDL is rooted deeply in the history of DH from the 1970s on. At King's College London, it emerged in a lineage that included the creation of the Research Unit in Humanities Computing in 1992 (which evolved into its own center in 1995), the Department of Digital Humanities in 2009–2010 (already de facto a department from 2002; Short et al.), and then the KDL itself in 2015. The KDL thus arose sufficiently far in the growth of DH for its founding director, James Smithies, and others to reflect on the lab's legacy and scope (Smithies et al.) and respond to new concerns about the professional careers of RSEs as part of complex relations of technology, workflows, and labor in DH (Smithies and Ciula; Ciula and Smithies; Smithies, ffrench, and Ciula).[5] Taking inspiration from established technology industry practices in project management and product development, the KDL RSE team collaboratively created and refined for the lab a Software Development Lifecycle (SDLC) for Research Software Engineering[6] schema as a generalized, staged process for developing projects that can help manage those complex relations.[7] In particular, the team adapted the Agile dynamic systems development method (DSDM) framework (Wikipedia) and released for public use related project templates and guidance.[8] But while these SDLC documents summarize how the team develops and stages project life cycles, they are merely the scaffolding rather than the substance, which consists of the expertise of the team members in their day-to-day interactions with technologies, project partners, cross-functional teams in the faculty and university, and external stakeholders (including funders and vendors). The relationality of KDL's research and development work thus stems not only from its processual nature (one iteration of product development informing others incrementally in a nonlinear, creative way) but from its socially negotiated nature: how the team's work interconnects with the expectations of others (first of all, funders and project partners) in the adoption, adaptation, refinement, and limiting of technologies. This matrix of social relations fused with its tech stack is the scene of KDL's lab culture.[9]

A Technology Network Perspective

To study KDL's tech stack and the social relations that it mediates, we created an interactive network visualization in a data notebook on the Observable visualization platform (Pawlicka-Deger et al.).[10] This visualization can be explored online, and it is also represented in selected views in Figures 5.1, 5.2, and 5.4. Based on ethnographic data gathered manually by Pawlicka-Deger during February to December 2021 from KDL's internal documents and public resources (including GitHub repositories for projects and a StackShare page of tools, services, and packages), the visualization shows nodes for 220 software and hardware technologies at the time of our study.[11] For example, ArcGIS, Gephi, JavaScript, XML, Zotero, and others are software tools and protocols familiar to digital humanists. Each technology is associated through edges with nodes representing the following attributes, which can be switched on or off to filter the visualization:

- *developer* (e.g., Adobe, Apache Software Foundation, King's College London, University of Hamburg)
- *function* (e.g., databases, graphic editors, text annotation, storytelling platforms)
- *permission* (proprietary or open source)
- *sector* (communities, for-profit companies, universities, nonprofit organizations).[12]

To get a sense of how the KDL tech stack evolves, we also tracked which technologies have been considered for adoption, which are actively used, and which have become legacy tools.

Figure 5.1 shows a simplified view of the visualization, with just the nodes for sector type and permission type activated to allow an observer to focus on these attributes of KDL technologies. And Figure 5.2 shows additional attribute nodes that have been turned on to reveal the full complexity of the KDL tech stack, a crowded and tangled engine room of technologies associated with 136 developer nodes (with 113 distinct developers), 79 function nodes (with 45 distinct functions), 221 permission nodes (165 open-source and 56 proprietary), and 221 sector nodes (61 community, 88 for-profit company, 32 university, and 40 nonprofit organization) in various states of being considered, actively used, or deprecated to legacy status.

That tangle of these technologies is not just a puzzle for running a DH lab's organizational and operational workflow. It is also about how technological practice is shaped by, and in turn modifies, the social ethos of the people who work with those

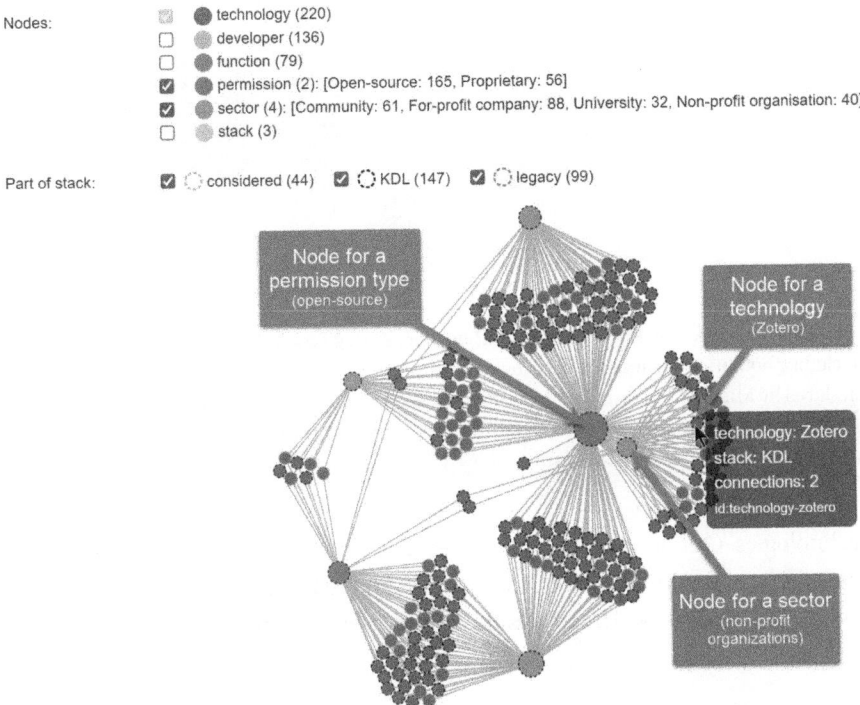

Figure 5.1. Network visualization of KDL's technology stack during February to December 2021, with just the sector type and permission type nodes turned on as attributes. Here, a specific technology node (Zotero) appears in a part of the graph associated with the nonprofit organization sector and the open-source permission type. (Clicking on the Zotero node highlights the edges connecting it to the nonprofit and open-source nodes; hovering on the Zotero node shows data about it.) Graph produced by Miguel Vieira using images published via the Observablehq.com platform at https://observablehq.com/@jmiguelv/dhlab-kdl-technology-network under a CC BY 4.0 license.

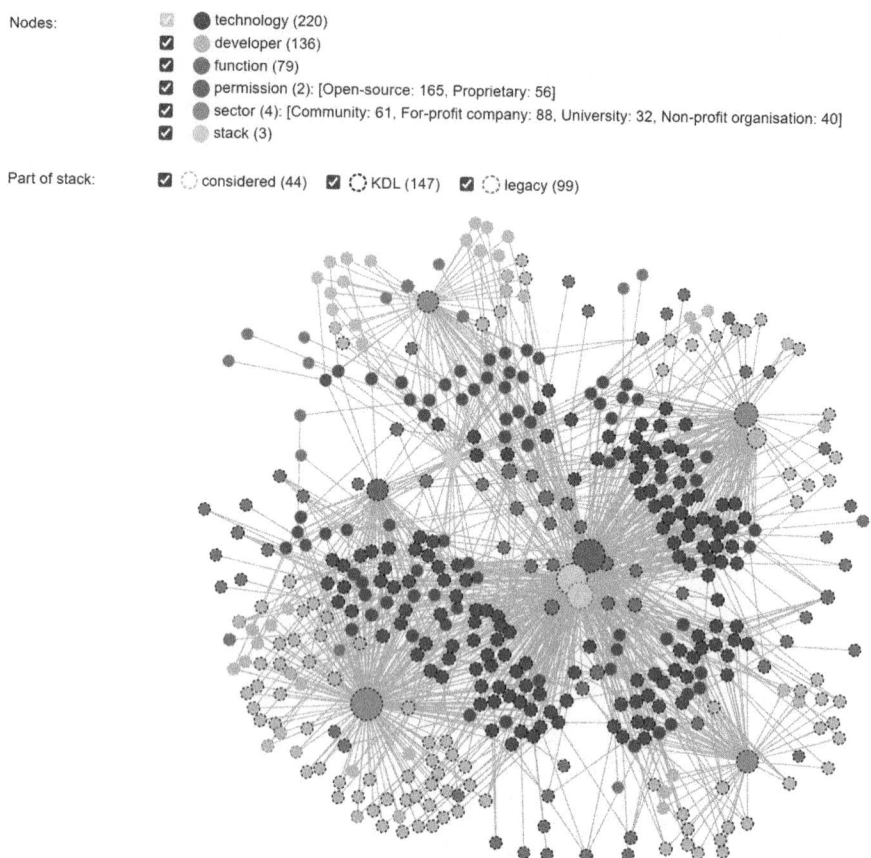

Figure 5.2. Network visualization of KDL technologies during February to December 2021 (including both used and legacy technologies), with all attribute nodes activated. Graph produced by Miguel Vieira using images published via the Observablehq.com platform at https://observablehq.com/@jmiguelv/dhlab-kdl-technology-network under a CC BY 4.0 license.

technologies. Our network visualization allows us to observe the constantly adjusting balance—one involving both alignments and tensions—between technological and social domains, which creates the culture of a DH lab.

Interconnections Between Tech Stacks and Lab Culture

By observing the combination of everyday technological and human practices at KDL and analyzing the lab's tech stack with the aid of our network visualization, we can gain insights into what we call KDL's *open-source pragmatism* culture. On the one hand, this is a culture characterized by an ethos of transparency, sustainability, and experimentation consistent with an open-source model and good research software engineering methods, including Agile DSDM for project management (Wikipedia). Much like companies that operate in a manner influenced by open-source models (Labourey), the lab embodies the idea of open-source values at a cultural level. But on the other hand, as the mention of "companies" hints, the lab's culture is also shaped by its need to work in conjunction with management and development models native to other sectors of society and other units of the lab's own university. For-profit, nonprofit, and university organizational cultures thus pragmatically conjoin with KDL's lab culture—a complex process of alignment and tension mediated, as mentioned previously (and as we will return to in this discussion), by the fact that open-source values now to some degree move among all these sectors. KDL is a DH lab that is distinctive in part because its team not only uses but reflectively considers technologies in this larger cultural picture.

Deferring for the moment the open-source side of KDL's open-source pragmatism, we can begin first on the most pragmatic side of the lab's culture by observing that the greatest share of technologies at the lab—88 in number, representing 40 percent of the total—comes from the for-profit sector (e.g., companies like Adobe, Google, and Microsoft).[13] (See Figure 5.3.) The dominance of the business sector in the software market from which KDL draws tools raises important issues regarding licenses, dependencies, and design biases. As an example, consider Figshare, a platform from the Digital Science company for storing, sharing, and managing data. Moving a university's repositories to Figshare, as King's College London did in 2021, is often justified as a solution for "re-integrating a presently splintered scholarly infrastructure" (Plantin, Lagoze, and Edwards). However, such a move

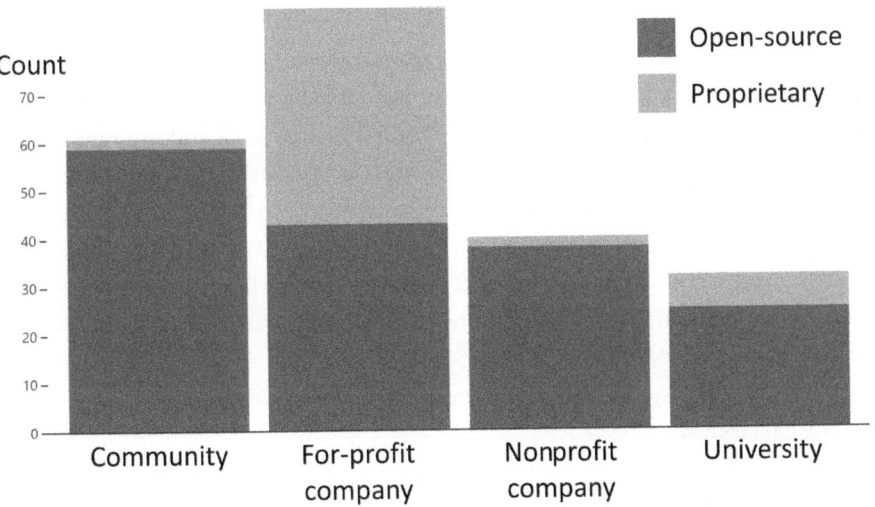

Figure 5.3. Distribution of technologies in the KDL during February to December 2021 by sector of origin and permission type. The tallest bar in the graph shows technologies from the for-profit company sector, which contributes both proprietary and open-source software used by the lab. Graph produced by Miguel Vieira using images published via the Observablehq.com platform at https://observablehq.com/@jmiguelv/dhlab-kdl-technology-network under a CC BY 4.0 license.

risks overcentralizing research data flow if it is not complemented by robust expertise and policies for research data management, data life-cycle support, and assessment of the implications of technical architectures. (As a cautionary tale, when YouTube temporarily terminated Cornell University Library's entire account due to "sensitive content" in videos of lectures [Bright; Butler], many on social media criticized the university's reliance on outsourced technology infrastructure and advocated for hosting resources in institutional repositories.) In general, the platformization of university resources can be a form of "technological dystopia" (Farnell) because "the reorganisation of cultural practices and imaginations around these platforms" (Poell, Nieborg, and van Dijck) poses risks related to security, surveillance, privacy, and datafication.

Why does KDL rely on so many technologies from the for-profit sector? One pragmatic answer, of course, is that the lab is not a stand-alone facility but rather part of its university, where institution-level policies negotiate and integrate technologies to meet multiple needs across King's College London as a whole. Technologies from the for-profit sector enter strongly into the mix because, though not a perfect fit, their design as enterprise-level technologies are often suited to the whole-organization integration that universities need and cannot fully develop or support themselves. Thus during the Covid epidemic, for example, the need for technology integration led King's College London to opt for Microsoft Teams as its main remote communication platform instead of the Zoom product chosen by most other universities. The decision was driven by the integration that Teams provides with other Microsoft services like SharePoint, which the university uses for data storage and sharing, and also by considerations of security and compliance with regulations.[14] Large institutions such as major universities adopt technical solutions based on a mix of criteria (e.g., security, cost efficiency, unification of platforms) that are assessed, steered, and funded at an enterprise level and are rarely determined through shared governance with local units. Such institutional decisions then have a cascading effect on the choice of systems at such local units as KDL. For instance, in 2020, KDL transferred project management data from Google Drive to SharePoint to align with its university's information technology (IT) professional service department's recommendations and with universitywide security policies. Whatever the reservations of the KDL team about either Google or Microsoft data management products, institutional compliance was the decisive factor for change in the lab's tech stack.

Another pragmatic reason for reliance on technologies from the for-profit sector that is internal to the KDL lab itself is integration at a different level: workflow management. For example, the lab coordinates its day-to-day operational workflow and communications through Slack, a for-profit chat platform popular in industry and in many project teams in the RSE and DH communities. But Slack is not supported by the King's College London IT service department, which provides

Teams (and the Microsoft social networking services named Yammer) for similar functions. KDL and the Department of Digital Humanities at King's had adopted Slack before the university offered these Microsoft products. That fact, combined with Slack's wide usage elsewhere along its efficiency and reliability, made Slack KDL's continued choice for internal workflow integration contra universitywide integration. Slack is one instance of the for-profit technologies that (along with ClickUp and Microsoft project management software, for example) the lab uses for operational management, storage, and communication.

Pragmatic needs thus account for much of the preponderance of technologies from the for-profit sector seen in our network visualization of the KDL tech stack. But having said that, KDL indeed has a culture not just of pragmatism but what we term *open-source pragmatism*. This is a culture in which a critical stance toward proprietary commercial tools motivates exploring alternatives that align with the lab's baseline open-source ethos. For instance, when in 2022 the European Data Protection Supervisor called for banning Google Analytics in European countries (see noyb), the KDL team sought more ethical open-source options, such as AWStats, Cloudflare Web Analytics, and GoAccess. The challenge was to find solutions that enhance workflow efficiency, provide relevant traffic statistics for assessment and reporting, and adhere to sustainable and open-source principles. Ultimately, KDL chose GoAccess, an open-source web analytics tool respectful of user privacy that comes from the sector labeled "community" in our network visualization. This is an example of how KDL explores a vast range of tools from the community, nonprofit, and university sectors, whose solutions—many of them open source—collectively make up 60 percent of the lab's tech stack. Solutions developed by the community, usually shared through GitHub, are typically open source and free. Examples in addition to GoAccess that KDL uses include LeafletJS for mapping, TextBlob for natural language processing, and Debian as a Linux-based operating system. The university sector also plays a significant role in the development of open-source and free tools created by DH centers or labs, as well as other campus units staffed by RSEs, alternative-academic (alt-ac) staff researchers, and other research technical professionals. Examples of university-created open-source software include tools for annotation (e.g., Archetype, produced by King's College London), timeline visualization (e.g., TimelineJS, produced by researchers at Northwestern University), and machine learning (e.g., GloVe, produced by researchers at Stanford University). Altogether, as revealed by our network visualization, open-source technologies for design, development, testing, and production of digital projects make up 75 percent of the KDL tech stack.

The bias toward open-source where possible reflects the lab's philosophy of transparency at its foundation (King's Digital Lab),[15] which accords not just with universitywide policies and the requirements of funding organizations for open publication of data, code, and software, but also with general calls to reject "Big

Edu-Tech" systems in favor of standardized, "Libre open-source" software or software developed within institutions (Farnell). Transparency is crucial to the lab's practices, characterizing its approach to providing access to its code, web application programming interfaces (APIs), project templates, and more.[16] Transparency involves exposing practices, tasks, and costs to stakeholders to foster intellectual rigor and to create more credible funding applications. Transparency is also a lived value in KDL's research software engineering operations, which are steered collectively through documented regular meetings (held via platform-supported gatherings such as Project Pipeline and Timebox) and daily updates on progress and obstacles shared through the lab's "Standup" Slack channel. Access by all lab members and project partners to communication and management channels, as well as opportunities for everyone to join in collaborative critiques and reviews, provide better traction on what happens in otherwise obscure digital processes.

KDL's preference for open-source technologies also reflects the need for tools that can be adapted for project requirements by allowing flexible and creative customization. RSEs such as those making up the KDL team are a group that is increasingly professionalized, advanced, and creative in the DH community in adapting or further developing tools. At KDL, RSE positions are defined in such a way that the research staff can use 10 percent of their time for independent experimental work, including hacking, tinkering with, and iterating open-source technologies to tailor them to individual projects and infrastructures. Examples include working with rapidly evolving technologies such as augmented reality, artificial intelligence (AI; Hall), and machine learning methods (Noël; Bode et al.) to push them beyond the constraints of proprietary tools. RSEs in KDL are thus like developers generally in the open-source community, who play a crucial role in adapting and professionalizing open-source products (Volpi). Such developers act as the main customers, but also coproducers, of open-source products—discovering, downloading, reengineering, and integrating them into their projects.

But, of course, open-source technologies are also challenged by problems of reliability and security. Thus, the lab complements its open-source ethos with another ethos that is equally important: sustainability. Sustainability is a crucial aspect of digital research production that the lab takes seriously to keep projects accessible and retrievable for several years (even decades, in some cases). Sustaining open-source software in working form can be challenging and labor-intensive when software is considered as what Marisa Leavitt Cohn calls "lived temporalities of code"—that is, "an object subject to continuous change and lived with over time as it evolves" (Cohn, 423). This is why the lab sometimes favors some kinds of open-source software over others. Analysis of the data behind our network visualization thus shows that while KDL considers many open-source software tools developed in the university sector, ultimately it tends not to adopt those solutions, for reasons that come down to problems with sustainability. For instance, in 2017, KDL assessed adopting Omeka, an open-source digital asset management system originating from the university sector that is well known in DH for curating and publishing online collections and exhibits. At that time, the lab's assessment indicated limitations in Omeka's ability to support complex metadata schemas, manage version control, and handle risks associated with dependencies on other tools and software components. The lab thought that for its specific purposes, its ability to customize the system and make projects based on it more sustainable was too limited.[17] As an alternative, the lab opted for another open-source solution for content management and modular-component web publishing that was developed outside the university sector in the nonprofit organization sector: the Django Python web framework. Since that time, KDL developers have tailored Django for greater, sustainable control over lab projects. Currently, the lab is undergoing a further consolidation stage, where an archival and static-first approach leading to static sites development from the outset is preferred and implemented.[18] In short, KDL prioritizes solutions that are tested, flexible, and integrated enough with the lab's tech stack and workflow to dovetail the two ideals of open source and sustainability. Rather than pursuing an idealistic notion of permanent sustainability, the lab prioritizes pragmatic solutions that create a best-fit alignment between research software engineering best

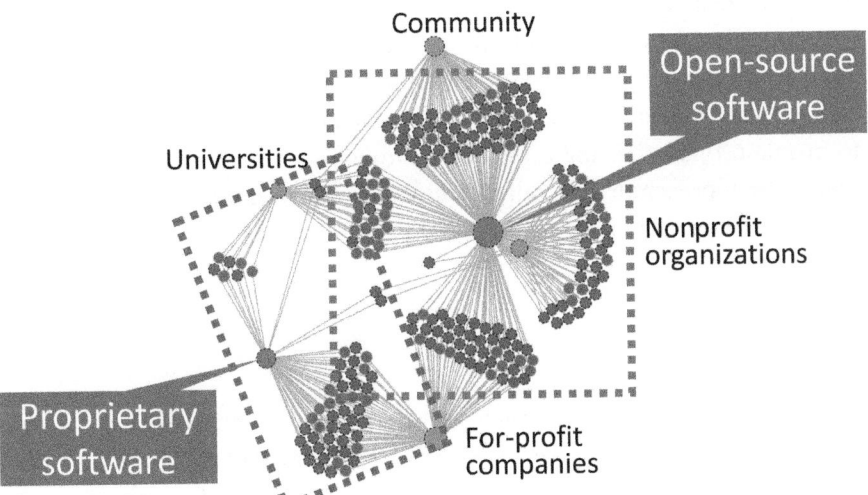

Figure. 5.4. Open-source technologies in the KDL tech stack and their sources in the community, universities, nonprofit, and for-profit sectors (by contrast with proprietary technologies). Graph produced by Miguel Vieira using images published via the Observablehq.com platform at https://observablehq.com/@jmiguelv/dhlab-kdl-technology-network under a CC BY 4.0 license.

practices and the lab's ethos, all aimed at ensuring the sustainability of relevant components of digital projects.[19]

Indeed, in the antithetical spirit of open-source pragmatism, such an alignment is epitomized by the open-source software developed by commercial companies that KDL uses in its tech stack. After all, there is no longer a necessary division between open-source products, associated with free and collaborative solutions, and software from the for-profit sector. The open-source market has undergone significant changes due to the monetization of open software and the fact that big tech companies offer proprietary cloud-service platforms, such as Amazon Web Services, on which others build open-source products. In KDL's tech stack, 27 percent of open-source tools are developed or owned by companies like Google, Facebook, and Microsoft. (See Figure 5.4 for a view of the network visualization showing open-source technologies in the KDL stack developed by various sectors.)

While not exhaustive, our network visualization of the KDL tech stack has allowed us to observe and offer critical reflections on the relations between a DH lab and the for-profit, nonprofit, community, and university sectors of digital technology. The visualization adds nuance beyond any simplistic open- versus closed-systems analysis, showcasing how the freedom of a DH lab to define its technical strategy is inevitably counterbalanced by organizational policies and management practices. The emergence of a culture of RSE teams in DH with a self-aware ethos of openness, creativity, transparency, and sustainability has the potential to tune that balance so that the values of RSE, alt-ac, and other researchers—values shared by the DH community at large—factor meaningfully into what would otherwise be overly centralized enterprise cultures. However, critical scrutiny will be needed to maintain that balancing act enacted through tech stacks and their management. The approach that we call *critical studies of tech stacks* contributes to such scrutiny of how DH labs are embedded in institutional, technical, operational, and sociocultural contexts.

NOTES

1. Urszula Pawlicka-Deger conducted an ethnographic study of KDL from February to December 2021 while she was a Marie Curie Research Fellow at the lab during 2020–2023; Arianna Ciula was cosupervisor of the fellowship and is currently director and senior research software analyst of the lab; and Miguel Vieira is the principal research software engineer at the lab. Pawlicka-Deger conducted twenty-five interviews with KDL members and others involved in the lab's work. Some of the interviews have been transcribed and published in the Zenodo repository (Pawlicka-Deger, "DH Lab: Interview Transcripts").

2. On RSEs in the arts and humanities, see Sichani et al.; Smithies, "Research Software (RS) Careers." See also DHTech, the movement recently recognized as an ADHO Special Interest Group around the development and use of software in DH and its activities (https://dh-tech.github.io/). Except in the sciences, the term and concept of RSEs is less well known in North America DH, where attention has focused instead on the only partly similar notion of alternative-academic (alt-ac) researchers in staff positions and where the early "hack versus yack" and "Do you have to know how to code?" controversies in DH discouraged notions of DH that could seem to be exclusionary of participants who are not programmers. On alt-ac, see Nowviskie, "The #alt-Ac Track." On "hack versus yack," see Nowviskie, "On the Origin of 'Hack' and 'Yack.'" On the "Do you have to know how to code?" controversy, see Stephen Ramsay's reflections in the aftermath of this question that he asked at the MLA conference in 2011 (Ramsay). It is noteworthy that *The RSE Turn in DH* was featured as a topic of an international panel at the Digital Research Infrastructure for the Arts and Humanities (DARIAH) annual conference in June 2024: https://www.conftool.net/dariah2024/index.php?page=browseSessions&form_session=67.

3. For examples of such research, see Berry and Fagerjord; Smithies, *Digital Humanities and the Digital Modern*; Plantin, Lagoze, and Edwards; van Geenen.

4. For examples of studies of DH labs, see Ciula and Smithies; Foka et al.; Kemman; Lang; Oiva and Pawlicka-Deger; Pawlicka-Deger, "The Laboratory Turn"; Pawlicka-Deger, "Feasibility Documents as Critical Structuring Objects"; Pawlicka-Deger and Thomson; Smithies and Ciula; and Wershler, Emerson, and Parikka.

5. The situation of RSEs and their relations to technology, workflows, and labor mentioned here refer specifically to the higher-education institutional context and other structures developed in the United Kingdom. While these tend to resonate with comparable developments in the wider European context, they might not be applicable or even relevant across different nations. (See also n. iii in this chapter on differences with the North American context, where the term "RSE" has not been as broadly adopted in fields outside the sciences.)

6. https://kdl.kcl.ac.uk/blog/sdlc-for-rse/.

7. https://kdl.kcl.ac.uk/blog/sdlc-for-rse/.

8. https://github.com/kingsdigitallab/sdlc-for-rse/wiki.

9. Of course, the KDL technology stack evolves dynamically both on a project-by-project basis and over the long term. Seen in overview, KDL's technology has undergone several major stages of change spanning from the lab's origin in 2015, when it hosted over 100 projects inherited from the King's Department of Digital Humanities, to the present, when it is currently migrating much of its software from on-premise infrastructure to the OpenStack infrastructure of King's College London's centralized e-Research service unit. At the start-up in 2015, KDL's tech stack (based on that of the preexisting Department of Digital Humanities) was already significant by DH standards. It included rack servers supporting 400 gigabytes of RAM, over 180 virtual machines, and 27 terabytes of data.

The back end and front end of the lab's projects used application frameworks comprising a mix of technologies: Java (with custom-based frameworks such as rdb2java and DJFacet), Python (particularly the web framework Django), the bespoke XML-based publishing solutions XMod and Kiln, and PHP-based frameworks such as WordPress, Omeka, and Typo3 (Ciula and Smithies). By 2018, KDL's infrastructure included over 200 virtual machines running Linux and was hosted in physical cabinets located at the University of London Computing Centre. KDL systems managers then completed a complex process of remediation and upgrading in December 2021 to improve the sustainability of infrastructure and ensure its alignment with the university's e-Research and IT service units. In combination with KDL's archiving and sustainability program (which requires regularly assessing hosted environments and communicating with faculty, partners, and project owners about maintenance options), this process reduced the number of virtual machines to 110, which KDL standardized by limiting each one to specific principal components (e.g., Django/Python, JavaScript, and Docker). In addition, KDL dedicates eleven servers to centralized services such as image and other data storage, email, and user authentication. The lab uses a range of network, security, and website analytics tools for monitoring and upgrade processes. As described elsewhere (Smithies and Ciula; Ciula and Smithies; and De Roure, Moore, Page et al.), KDL's infrastructure is managed through distributed accountability and regular cycles of upgrades, assessment, and remediation. Going forward, KDL will maintain a smaller on-premises footprint for projects with specialized needs, such as software for testing machine learning methods and digital creativity applications (e.g., for the production of immersive experiences enhanced by AI and Internet of Things applications such as Room Is Sad). Integration with the university's e-Research services unit will provide KDL with a flexible long-term infrastructure road map that is aligned to its systems in terms of architecture and security and supported by centralized university research infrastructure expertise—a step that will make possible a major change in systems management, as well as environmental sustainability. In the last leg of migration, KDL will host all its servers on the university's CREATE OpenStack system, with two dedicated hypervisor platforms for virtual machines. Upgrading KDL's core production infrastructure from on-premises infrastructure to the e-Research unit's infrastructure will contribute to the university's net zero carbon institutional target by 2025.

10. https://observablehq.com/@jmiguelv/dhlab-kdl-technology-network.

11. For the dataset (in .csv format) underlying the visualization, see Pawlick-Deger, "King's Digital Lab Technology Network" (dataset).

12. For technologies developed by individuals, we assigned the "community" sector.

13. Note that the percentages mentioned in this chapter should be interpreted as approximate because technologies included in the data collection could not always be completely and separately enumerated. Open-source libraries, for example, proliferate in the KDL software development life cycle, but they were excluded from the data collection process as it would have been impractical to compile an exhaustive list.

14. This information was requested to central ITS and obtained via email in April 2021.

15. https://2015.kdl.kcl.ac.uk/how-we-work/philosophy/.

16. See, for example, the way that KDL makes its project templates available in its GitHub repository for "Document Templates," https://github.com/kingsdigitallab/sdlc-for-rse/wiki/Document-templates.

17. Omeka, of course, has since further evolved. KDL's technology assessments are constrained by deadlines and the requirements of specific projects; they are thus necessarily not comprehensive or general in their conclusions.

18. https://kdl.kcl.ac.uk/slides/2024-london-static-first/.

19. KDL has implemented an archiving approach and framework that include comprehensive sustainability assessment overseen by a dedicated committee. The framework offers maintenance and archiving options such as converting web resources to static sites or packaging them for migration.

BIBLIOGRAPHY

Agile Business Consortium. *Agile PM: Agile Project Management Handbook V2*, 2014.

Berry, David M., and Anders Fagerjord. *Digital Humanities: Knowledge and Critique in a Digital Age.* Polity Press, 2017.

Bode Katherine, Arianna Ciula, Galen Cuthbertson, et al. "Critical Modelling of Extensive Literary Data: An Experiment." *King's Digital Lab* (blog), May 9, 2023. https://kdl.kcl.ac.uk/blog/cmeld-ae/.

Bounegru, Liliana, and Jonathan Gray. *The Data Journalism Handbook: Towards a Critical Data Practice.* Amsterdam University Press, 2021.

Bright, Susie. "Terminated." *Susie Bright's Journal* (blog), June 17, 2022. https://susiebright.substack.com/p/terminated.

Butler, Matt. "Cornell Library YouTube Page Restored After Termination Last Week Over Nudity Content." *The Ithaca Voice*. June 24, 2022, https://ithacavoice.org/2022/06/cornell-library-youtube-page-restored-after-termination-last-week-over-nudity-content/.

Ciula, Arianna, and James Smithies. "Sustainability and Modelling at King's Digital Lab: Between Tradition and Innovation." In *On Making in the Digital Humanities: The Scholarship of Digital Humanities Development in Honour of John Bradley,* edited by Julianne Nyhan, Geoffrey Rockwell, Stéfan Sinclair, and Alexandra Ortolja-Baird, 78–104. University College London Press, 2023. https://doi.org/10.14324/111.9781800084209.

Cohn, Marisa Leavitt. "Keeping Software Present: Software as a Timely Object for STS Studies of the Digital." In *digitalSTS: A Field Guide for Science & Technology Studies,* edited by Janet Vertesi and David Ribes, 423–46. Princeton University Press, 2019.

De Roure David, John Moore, Kevin Page, et al. "DigiSpec: Scoping Future Born-Digital Data Services for the Arts and Humanities: Case Reports." Zenodo, 2022. https://doi.org/10.5281/zenodo.4716148.

DHTech, an ADHO Special Interest Group. Home page. 2023. https://dh-tech.github.io.

Eby, Michael. "Agile Workplace." *New Left Review,* August 20, 2021. https://newleftreview.org/sidecar/posts/agile-workplace.

Edwards, Paul N. *A Vast Machine. Computer Models, Climate Data, and the Politics of Global Warming.* MIT Press, 2013.

Farnell, Andy. "We Can't Teach in a Technological Dystopia." *Times Higher Education*, March 4, 2021. https://www.timeshighereducation.com/features/we-cant-teach-technological-dystopia.

Foka, Anna, Anna Misharina, Viktor Arvidsson, et al. "Beyond Humanities Qua Digital: Spatial and Material Development for Digital Research Infrastructures in HumlabX." *Digital Scholarship in the Humanities 33,* no. 2 (2018): 264–78.

Gellner, David, and Eric Hirsch, eds. *Inside Organizations: Anthropologists at Work.* Berg, 2001.

Gold, Matthew K. "An Open Opportunity: Free Software, Community-Supported Infrastructure, and the People's University." Cistudies.org Initiative channel, YouTube, September 26, 2021, https://www.youtube.com/watch?v=XTpdfmEneYg.

Grandjean, Martin, and Aaron Mauro. "A Social Network Analysis of Twitter: Mapping the Digital Humanities Community." *Cogent Arts & Humanities 3,* no. 1 (2016). https://doi.org/10.1080/23311983.2016.1171458.

Hall, Elliott. "Ghosts in the Machine." *King's Digital Lab* (blog), July 31, 2017. https://kdl.kcl.ac.uk/blog/ghosts-machine/.

Hall, Elliott. "Room Is Sad." *King's Digital Lab* (blog), June 23, 2023. https://kdl.kcl.ac.uk/blog/room-sad/.

Jakeman, Neil. "Software Development Lifecycle for Research Software Engineering." *King's Digital Lab* (blog), August 19, 2020. https://kdl.kcl.ac.uk/blog/sdlc-for-rse/.

Kemman, Max. *Trading Zones of Digital History.* De Gruyter Oldenbourg, 2021.

King's Digital Lab (KDL). "Document Templates." King's Digital Lab, 2023. https://github.com/kingsdigitallab/sdlc-for-rse/wiki/Document-templates.

King's Digital Lab (KDL). "Frequently Asked Questions: What Project Partners Might Want to Know About KDL." King's Digital Lab, 2023. https://kdl.kcl.ac.uk/how-we-work/faq-partners/.

King's Digital Lab (KDL). "How We Work: Our Philosophy." King's Digital Lab, 2022. https://2015.kdl.kcl.ac.uk/how-we-work/philosophy/.

Knox, Hannah. "An Infrastructural Approach to Digital Ethnography: Lessons from the Manchester Infrastructures of Social Change Project." In *The Routledge Companion to Digital Ethnography,* edited by Larissa Hjorth, Heather Horst, Anne Galloway, Genevieve Bell, 354–62. Routledge, 2016.

Labourey, Sacha. "Why Companies Should Model Their Culture After Open Source." *Forbes*, June 29, 2021. https://www.forbes.com/sites/forbestechcouncil/2021/06/29/why-companies-should-model-their-culture-after-open-source/.

Lang, Sarah. "Experiments in the Digital Laboratory." In *Fabrikation von Erkenntnis. Experimente in den Digital Humanities,* edited by Manuel Burghardt, Lisa Dieckmann, Timo Steyer, et al. Melusina Press, 2021. https://www.melusinapress.lu/read/melusina-8f8w-y749-eitd/section/ea3e7e69-ea18-4378-901b-fc1b8352a86a.

Latour, Bruno, and Steve Woolgar. *Laboratory Life: The Construction of Scientific Facts.* Princeton University Press, 1979.

Nelson, Katie. "Doing Fieldwork: Methods in Cultural Anthropology." In *Perspectives: An Open Invitation to Cultural Anthropology,* 2nd ed., edited by Nina Brown, Laura Tubelle de González, and Thomas McIlwraith, 45–69. American Anthropological Association, 2018. https://perspectives.americananthro.org/Chapters/Fieldwork.pdf.

Noël, Geoffrey. "How KDL Applies Machine Learning to Research Projects: A Subjective Retrospective." *King's Digital Lab* (blog), April 19, 2022. https://kdl.kcl.ac.uk/blog/how-kdl-applies-machine-learning-research-projects/.

Nowviskie, Bethany. "The #alt-Ac Track: Negotiating Your 'Alternative Academic' Appointment." *Chronicle of Higher Education,* 2010. https://www.chronicle.com/blogs/profhacker/the-alt-ac-track-negotiating-your-alternative-academic-appointment-2.

Nowviskie, Bethany. "On the Origin of 'Hack' and 'Yack.'" In *Debates in the Digital Humanities 2016,* edited by Matthew K. Gold and Lauren F. Klein, 66–70. University of Minnesota Press, 2016. https://dhdebates.gc.cuny.edu/read/untitled/section/a5a2c3f4-65ca-4257-a8bb-6618d635c49f.

noyb. "EDPS Sanctions Parliament over EU-US Data Transfers to Google and Stripe." January 11, 2022, https://noyb.eu/en/edps-sanctions-parliament-over-eu-us-data-transfers-google-and-stripe.

Oiva, Mila, and Urszula Pawlicka-Deger. "Lab and Slack. Situated Research Practices in Digital Humanities—Introduction to the DHQ Special Issue." *Digital Humanities Quarterly* 14, no. 3. (2020). http://www.digitalhumanities.org/dhq/vol/14/3/000485/000485.html.

Pawlicka-Deger, Urszula. "DH Lab: Interview Transcripts." Zenodo, 2023. https://doi.org/10.5281/zenodo.7880810.

Pawlicka-Deger, Urszula. "Feasibility Documents as Critical Structuring Objects: An Approach to the Study of Documents in Digital Research Production." *Convergence: The International Journal of Research into New Media Technologies* 29, no. 3 (2022): 746–65. https://doi.org/10.1177/13548565221111073.

Pawlicka-Deger, Urszula. "King's Digital Lab Technology Network" (dataset). King's College London. December 23, 2021. https://doi.org/10.18742/17372021.v1.

Pawlicka-Deger, Urszula. "The Laboratory Turn: Exploring Discourses, Landscapes, and Models of Humanities Labs." *Digital Humanities Quarterly* 14, no. 3. (2020). http://www.digitalhumanities.org/dhq/vol/14/3/000466/000466.html.

Pawlicka-Deger, Urszula, and Thomson Christopher, eds. *Digital Humanities and Laboratories: Perspectives on Knowledge, Infrastructure and Culture.* Routledge, 2024.

Pawlicka-Deger, Urszula, and Miguel Vieira, with Arianna Ciula and Tiffany Ong. "King's Digital Lab Technology Network" (visualization). *Observable,* December 10, 2021. https://observablehq.com/@jmiguelv/dhlab-kdl-technology-network.

Plantin, Jean-Christophe, Carl Lagoze, and Paul N. Edwards. "Re-integrating Scholarly Infrastructure: The Ambiguous Role of Data Sharing Platforms." *Big Data & Society* 5, no. 1 (2018): 1–14. https://doi.org/10.1177/2053951718756683.

Poell, Thomas, David Nieborg, and José van Dijck. "Platformisation." *Internet Policy Review* 8, no. 4 (2019). https://doi.org/10.14763/2019.4.1425.

Ramsay, Stephen. "On Building." Stephen Ramsay (blog), January 11, 2011. https://web.archive.org/web/20150318002543/http://stephenramsay.us/text/2011/01/11/on-building/.

Short, Harold, Julianne Nyhan, Anne Welsh, and Jessica Salmon. "Collaboration Must Be Fundamental or It's Not Going to Work: An Oral History Conversation Between Harold Short and Julianne Nyhan." *Digital Humanities Quarterly* 6, no. 3 (2012). http://www.digitalhumanities.org/dhq/vol/6/3/000133/000133.html.

Sichani, Anna-Maria, Ruth Ahnert, James Baker, et al. "iDAH Research Software Engineering (RSE) Steering Group Working Paper." Zenodo, 2023. https://doi.org/10.5281/zenodo.8177926.

Smithies, James. *The Digital Humanities and the Digital Modern.* Palgrave Macmillan, 2017.

Smithies, James. "Research Software (RS) Careers: Generic Learnings from King's Digital Lab, King's College London (v. 6.2)." Zenodo, 2019. https://doi.org/10.5281/zenodo.2564790.

Smithies, James, and Arianna Ciula. "Humans in the Loop: Epistemology & Method in King's Digital Lab." In *Routledge International Handbook of Research Methods in Digital Humanities,* edited by Kristen Schuster and Stuart Dunn, 155–72. Routledge, 2020.

Smithies, James, Patrick ffrench, and Arianna Ciula. "*Droit de cité:* The Digital Lab as Digital Milieu." In *Digital Humanities and Laboratories: Perspectives on Knowledge, Infrastructure and Culture,* edited by Urszula Pawlicka-Deger and Christopher Thomson, 52–66. Routledge, 2023. https://doi.org/10.4324/9781003185932.

Smithies, James, Carina Westling, Anna-Maria Sichani, et al. "Managing 100 Digital Humanities Projects: Digital Scholarship & Archiving in King's Digital Lab." *Digital Humanities Quarterly* 13, no. 1 (2019). http://www.digitalhumanities.org/dhq/vol/13/1/000411/000411.html.

Spiro, Lisa. "'This Is Why We Fight': Defining the Values of the Digital Humanities." In *Debates in the Digital Humanities,* edited by Matthew K. Gold, 16–34. University of Minnesota Press, 2012.

Star, Susan Leigh. "The Ethnography of Infrastructure." *American Behavioral Scientist* 43, no. 3 (1999): 377–91.

van Geenen, Daniela. "Critical Affordance Analysis for Digital Methods: The Case of Gephi." In *Explorations in Digital Cultures,* edited by Marcus Burkhardt, Mary Shnayien, and Katja Grashöfer, 1–21. meson press, 2020. https://doi.org/10.25969/mediarep/14855.

van Geenen, Daniela, Jonathan W. Y. Gray, Liliana Bounegru, et al. "Staying with the Trouble of Networks." *Frontiers in Big Data* 5 (2023). https://doi.org/10.3389/fdata.2022.510310.

Venturini, Tommaso, Anders Kristian Munk, and Mathieu Jacomy. "Actor-Network Versus Network Analysis Versus Digital Networks: Are We Talking About the Same Networks?" In *digitalSTS: A Field Guide for Science & Technology Studies,* edited by Janet Vertesi and David Ribes, 510–24. Princeton University Press, 2019.

Volpi, Mike. "How Open-Source Software Took over the World." *TechCrunch,* January 12, 2019. https://techcrunch.com/2019/01/12/how-open-source-software-took-over-the-world/.

Wershler, Darren, Lori Emerson, and Jussi Parikka. *The Lab Book: Situated Practices in Media Studies.* University of Minnesota Press, 2021.

Wikipedia. "Dynamic Systems Development Method." February 13, 2022, at 02:56. Accessed August 1, 2023. https://en.wikipedia.org/wiki/Dynamic_systems_development_method.

PART I][Chapter 6

Shadow Libraries and Pirate Infrastructures

MARTIN PAUL EVE

How broken is our current infrastructure of scholarly communications? And to what extent can the decentralized system of academic journal, book, software, and multimedia publishing even actually be termed an "infrastructure"? In her analysis of infrastructure, Susan Leigh Star writes that infrastructures are so routinized as to be "boring things" that "by definition [are] invisible" (377, 380). She adds that infrastructure appears just to work unnoticed in the background except "when it breaks" (382). When people begin talking and writing about an infrastructure, it is because something has broken.

Scholarly communications systems (*scholcomms*) fulfill many of Star's conditions for the definition of "infrastructure." Certainly, most academics do not inquire about their functionality and expect the system "just to work." Scholarly communications also tick off all the checkboxes for defining infrastructures. For instance, they are embedded, transparent, and built on the installed base of the university. In fact, it has become routine, both in academic circles and in the wider popular higher education press, to refer to scholcomms as part of the infrastructure of research (Plantin, Lagoze, and Edwards; Guldi; Mintz). There are even a set of principles around scholarly infrastructure that spell out good behavior for technology providers working in this space (Bilder, Lin, and Neylon). In the humanities disciplines, such a focus on scholcomms as infrastructure has been a central tenet of the Andrew W. Mellon Foundation's Scholarly Communications Program, which sought, under the directorship of Donald J. Waters, to create an open and accessible end-to-end *infrastructure* for humanities research (Waters).

That said, the past two decades have seen unprecedented levels of metadiscursive analysis of scholarly publishing infrastructure, indicating that something has gone awry. Most of this discussion has centered on the changes wrought by the advent of digital technologies, including questions of publisher labor, digital availability, and the continued relevance, or otherwise, of print (Bhaskar). Perhaps the most strident commentary has been the demand for open access (OA) to scholarly and research publications (Suber; Eve, *Open Access and the Humanities*; Eve and

Gray). *Open access (OA)* refers to removing price and permission barriers to access to peer-reviewed scholarship. It means that users can download such works without charge and that readers may reuse the work in more ways than is permitted under fair use or fair dealings provisions in copyright law (usually via a Creative Commons license). One of the broken things in the infrastructure of scholarly communications, such a discourse implicitly claims, is that readers have to pay for access to scholarship and research.

How did "paying for scholarship" become seen as a breakdown of our research dissemination infrastructure? The first answer to this question comes from a long-term trend in corporate consolidation. Over the past thirty years, ever-greater portions of the research publication infrastructure have become owned by an ever-smaller number of companies. Most scientific research publishing, for instance, is routed through just five large corporate entities (Lawson, Gray, and Mauri; Fyfe et al.; Lawson). Furthermore, these corporations operate on extreme profit margins that exceed 35 percent in some cases. This is more than Big Pharma or oil companies.

Consequently, the system has been branded an "oligopoly" in academic studies and the popular press (Monbiot; Larivière, Haustein, and Mongeon). Hence, at least one reason why some advocates pursue OA is to see such profiteering edged out of research publication. In its purest forms, this argument critiques big publishing and for-profit practices, arguing that eleemosynary forms are better suited for this space. There is a flip side to this argument, however. In using technological capacity as an argument for lowering costs, such arguments risk rehearsing the tenets of technosolutionism. At its worst, this logic becomes a neoliberal exercise, where such advocates devalue all publisher labor (claiming that "you can just put a PDF on a website for no cost"). In doing so, they then enforce merely a new regimen of austerity (Adema and Moore; Moore; Eve, "Open Access and Neoliberalism").

However, there is a more profound rationale that underpins the demand for OA, which brings us to this chapter's topic: shadow libraries. The underwriting logic for why scholarly communications should be open and accessible in the digital era is that scholarship, particularly in scientific, medical, and other areas critical to people's well-being, is different from other forms of publishing. Medical research saves lives, for example. As with criticisms of private health-care systems, it is easy for detractors of corporations to note (correctly) that corporations making 35 percent profit margins are withholding potentially lifesaving research materials from those who cannot pay. The moral and ethical standpoint here—and the primary claim as to why scholarly communications infrastructures are now broken in the digital era—is that people are dying in the service of profit. Such an argument is sharpened by the fact that such unequal outcomes are unequally distributed: it is those from the Global South who suffer the most at the hands of seemingly neocolonial structures of research publication (Mboa Nkoudou, "The (Unconscious?) Neocolonial Face of Open Access"; Babini; Mboa Nkoudou, "Epistemic Alienation in African Scholarly Communications"; Packer; Roh, Inefuku, and Drabinski). In this light, some operatives have begun to take action on their own. Aaron Swartz, for instance, famously tried to liberate the collection of articles provided through JSTOR, with tragic consequences (Swartz; Hockenberry). Others, as we shall see, have been even more ambitious.

What We Do in the Shadow Libraries

Two decades of calls for OA have made a difference. Much research work is now available openly. New business models have emerged to support the publisher labor that underwrites such publications (although not all of these have met with academic favor) (Digital Science et al.). Nonetheless, substantial portions of the research literature remain behind paywalls or in paid-for books, with no open, free digital correlate. Even if the OA movement succeeded in its goals tomorrow, vast historical reservoirs of scholarship would remain locked away for the duration of the international copyright term (seventy years after the author's death in many cases). OA gets us partially toward a dream of free accessibility of the research literature, but it is not the whole story.

Frustrated by the slow growth of OA and the realization that even OA will not achieve 100 percent access, various groups and individuals worldwide have built a series of digital infrastructures, called *shadow libraries*, that circumvent publisher paywalls. *Shadow* is an apt term here. In Star's definition, we recall, one of the crucial features of infrastructure is that, when they are working, they are *invisible*. The emergence of their "shadows" implies that these infrastructures, with all their flaws, are no longer invisible; they now occlude the light. The most well known and significant of these libraries are Sci-Hub, Library Genesis, Anna's Archive, and Memory of the World, each hosting substantial volumes of material.[1] In addition, UbuWeb, which explicitly describes itself as a "pirate shadow library," hosts hundreds of thousands of specifically avant-garde digital artifacts. Other shadow libraries include Monoskop ("an independent web-based educational resource and research platform for arts, culture and humanities founded in 2004") and Aaaaarg (originally AAARG, an acronym of Artists, Architects, and Activists Reading Group), both of which bill themselves specifically in educational terms.

These libraries exist on a spectrum of legal antagonism. UbuWeb, for instance, traffics in specifically avant-garde material. These artifacts sit outside the conventional sphere of commercial exchange, quite simply because, although they may hold historical and cultural significance, nobody is willing to pay for them. As such, despite the fact that "by the letter of the law, the site is questionable," UbuWeb has "never been sued—never even come close" (UbuWeb). Sites like Monoskop tend also to function more as aggregators than actual hosts of pirated material. Again, the legal status of such linking is unclear.[2] Memory of the World, by contrast, veers much more deeply into clearly illicit territory in many jurisdictions, hosting copies of in-copyright works that have commercial value. Yet the site situates itself within a specific lineage of free and public libraries as a public good that exists for the education of the people (Doran). At the very extreme end of the potentially illegal spectrum sit Library Genesis, Sci-Hub, and Anna's Archive (the latter of which only came to prominence in the later stages of writing this chapter).

Library Genesis holds 33 terabytes of books and more than 60 terabytes of scientific journal articles on behalf of Sci-Hub (Eve, "Lessons from the Library").[3] These pirate libraries are extensive, providing access to almost all published academic journal output (Banks; Himmelstein et al.). However, it is worth noting that national deposit libraries, backed by government mandates and legal structures, are substantially more extensive than the shadow archives. For instance, the British Library has a central collection of approximately 170 million items total and 13.5 million books, with a digital collection over a petabyte in size (British Library). By contrast, Library Genesis stores fewer than 3 million books, although it constantly grows. While shadow libraries are large, they struggle to compete on a scale with formal, resourced institutions.

The terminology of shadows that surrounds these archives is itself of interest (as is the designation of "pirate"). As historians studying color have shown, the idea of the "shadow" has changed throughout history. Until the eighteenth century, shadows were not defined as colorless opposites of objects. Instead, before then, shadows were partial transparencies and outlines, primarily in art and heraldry. Perhaps most important, though, in heraldic terms, shadows were used to reveal a hidden part of the family tree, not a mirror of it as might be expected (Pastoureau 26–27). Shadows, in heraldry, show illegitimate family links—children born out of wedlock—rather than formal reflections of the main subjects. Hence, as Nanna Bonde Thylstrup observes, it is worth reflecting on the "inherently unstable form of shadow libraries as a cultural construct" (Thylstrup 98). While we can think of shadow libraries as "mirror libraries" that ape mainstream, legitimate bibliographic venues, the terminology of shadows allows us to consider them within a more complex lineage of subversion and counternorms. Shadow libraries, the bastards of archival worlds, upset our relationships with traditional information retrieval.

Shadow libraries define themselves against a standard of morality that attempts to transcend local lawmaking and national-level jurisprudence; copyright laws, they say, are wrong. "Within decades," writes one operator of Library Genesis, "generations of people everywhere in the world will grow up with access to the best scientific texts of all time. [. . .] the quality and accessibility of education to the poor will grow dramatically too. Frankly, I see this as the only way to naturally improve mankind: we need to make all the information available to them at any time" (Bodó 25). The amateur "archivists" who have taken it upon themselves to seed Library Genesis on BitTorrent believe, indeed, that their "initiative fulfills United Nations/ UNESCO world development goals that mandate the removal of restrictions on access to science" and "limiting and delaying humanity's access to science isn't a business, it's a crime, one with an untold number of victims and preventable deaths" (u/shrine, "Library Genesis Project Update"). But, as before, the ethical foundations on which these initiatives rest are rooted in the discourse of profit against lives.

Shadow libraries straightforwardly appear, from the perspective of the Global North, to be illegal. They are premised, after all, on a total disregard for copyright law and the contracts that academics have signed. Regardless of whether they do so from the standpoint of any moral superiority, courts in the United States have continually found against Alexandra Elbakyan, the founder of Sci-Hub ("The Library of Alexandra," as a recent podcast quipped [WNYC Studios]), awarding millions of dollars in compensation to publishing companies such as Elsevier (Schiermeier; Harris and Barrett). However, an uncomplicated designation of illegality for these shadow libraries does not hold. In many jurisdictions worldwide, substantially more liberal laws permit the distribution of educational material, even when it is under copyright.

A good example is India, where Elbakyan is currently defending a lawsuit filed by Elsevier, Wiley India Pvt. Ltd and the American Chemical Society at the Delhi High Court (Reddy and Mishra). It is currently unclear whether she will succeed in defending Sci-Hub in this jurisdiction, but the court did not simply throw the case out as a clear-cut verdict.

Several complex legitimation discourses circle projects such as Sci-Hub and chime with longer-term economic rationalities. Even the terminologies of *archive*—so contested already—and *library*, let alone *pirate*, confer various statuses of veridiction upon these infrastructures.[4] However, one of the perpetually resurfacing accusations is that Sci-Hub is run by the Russian state and is designed to harvest university credentials from unwitting users for malicious foreign agents (Harris and Barrett; Erkal; Masnick). The argument here goes that users who wish to support the (they believe) laudable goals of Sci-Hub donate their library credentials, but Russian government agents are then using these for other purposes. Such rumors—while remaining unverified—are fueled in part by the fact that Sci-Hub grew, according to Balázs Bodó and Elbakyan, from the sharing cultures of Soviet communism. This stands in stark contrast to other piracy cultures, such as the warez scene, which thrives on cutthroat competition that looks more akin to a capitalist economy (Bodó; Elbakyan, "A Robin Hood in the World of Science"; Eve, *Warez*). As such, elements of Sci-Hub's ideology are economic anathema to U.S., U.K., and European governments operating under the political rationalities of neoliberalism (defined as the disenchantment of politics by economics, as per Davies). In this political climate, and particularly since the Russian invasion of Ukraine in 2022, it then becomes difficult to distinguish whether these allegations have merit or if they are mere smear campaigns made by powerful economic actors. Elbakyan denies these accusations (Elbakyan, "How the Chronicle Is Trying to Malign Sci-Hub").

Disciplines and Hierarchy

There is, of course, a specific disciplinary resonance to the rationale for shadow libraries. Although Library Genesis is a nonspecific library/archive, Sci-Hub explicitly encodes *science* in its name (although it is notable that *science,* in the European context, includes the humanities disciplines). Research that "saves lives" implies the fields of biomedicine, medical physics, virology, and other disciplines that contribute toward the extension and preservation of human vitality, rather than "merely" its enrichment. Research in other spaces cannot posit the same urgency. That "nobody will die" if they cannot read the latest literary criticism seems somewhat self-evident. Hence, an argumentative terrain emerges from shadow libraries in which it is harder to justify the pirating of literary criticism than, say, cancer research. In the case of the latter, if a for-profit publisher refuses to make this research available, it is easy to argue that people may die as a result, and this would be, in some moral sense, a criminal outcome. The same cannot be said of the humanities disciplines (and much social-scientific research).

Thus, according to the arguments of these pirate sites, there comes a specific hierarchy of disciplinary values that places the humanities far lower than many of their scientific counterparts. Indeed, the arguments for access and preservation, whether they come from the formal open-access movement to scholarship or from pirate archives, are usually centered around scholarship's impact on the world. When the value of that scholarship is more centered on cultural understanding, it becomes harder to justify a copyright-violating crusade because the external moral and legal blame is significantly less. That is to say that arguments for the ethical necessity of OA to research in the medical space often center on the fact that it can be seen as morally unacceptable to profit from the restriction of medical scholarship, which could lead to people dying. This argument has more or less traction depending on the stance toward health care in the nation in question. Economic *advancement*—a concept that embeds a specific telos of progress—is undoubtedly, if the United States is anything to go by, no sign of a move toward universal health care. Nevertheless, universal health care is a sign of accepting a "right to health," included in several declarations, such as the Universal Declaration of Human Rights, the International Covenant on Economic, Social, and Cultural Rights, and the Convention on the Rights of Persons with Disabilities. In cultures where people have a right to health or health care—and where it is morally unacceptable to deny such things to people based on their wealth—blocking access to the medical literature with paywalls becomes a social ill.

Perhaps it is simply true that preserving and improving life and health are fundamentally more critical tasks than understanding culture. Suppose that we return to taxonomies such as Abraham Maslow's well-known hierarchy of needs. In that case, it seems clear that without health and security, one is unlikely to profit significantly from cultural and humanistic endeavors.[5] However, this is also not straightforwardly true. It is not possible to advance medically without comprehending medical history and medical ethics, for example. Indeed, basic medical ethics training is part of any medical degree, which includes the infamous history of human experimentation during World War II and the Tuskegee Syphilis Study. Science fiction has also often served as the primary moral compass by which the efforts of science are hypothesized; warnings against the tide of unmitigated progress or imaginary explorations of human interactions with technological advancement (e.g., McLane). To hierarchize the humanities at the top of a pyramid of needs—as though they are merely a luxury—is to neglect their embeddedness in how we understand and measure the natural sciences.

Nevertheless, shadow libraries predicate their ethical existence on access to the scientific and medical literature. They justify their international presence by gesturing toward the vast global inequalities in current scholarly communications practices. Their missions are couched in discourses of scientific enrichment and civil rights–era formulations. As the authors of one letter of solidarity put it: "This is the time to recognize that the very existence of our massive knowledge commons is an act of collective civil disobedience" (Barok et al.). However, while the rhetoric of civil disobedience here at once invokes freedom rides and other racially liberating protests in the U.S. context, the primary objection that stands in the way of shadow libraries is the civil contract/tort law against individuals. While civil rights protests were primarily against statutes, the primary alleged wrong of shadow libraries is to violate the copyright of individuals. Indeed, it may be true that Library Genesis and Sci-Hub operate at such a scale that they count as conspiracies to violate copyright (a criminal offense) rather than merely infringements on individual rights. But the fundamental logic to which this civil disobedience objects is the individual ownership of published research in many jurisdictions: tort copyright law. The protest is, to some extent, against the very idea of intellectual property (see Vaidhyanathan; Johns; Stallman).

Technical Capacity and Decentralization: Shadow Libraries as Open Scholarly Infrastructure

As with many pirate cultures, shadow libraries play a game of hide-and-seek (or whack-a-mole) with law enforcement. They also face a fundamental paradox. To be most useful, a shadow library must be well known and prominent. There is little point in establishing a pirate archive designed to liberate all human knowledge if nobody can use it. On the other hand, all increases in exposure lead to additional attention from law enforcement and the ever-prevalent risk of shutdown. To counter this threat, shadow libraries adopt principles of technological and, to some extent, organizational decentralization.

These resilience principles have also been well articulated in the formal research space. Geoffrey Bilder et al., for instance, present a set of Principles of Open Scholarly

Infrastructure (POSI) designed to promote the durability of research publishers. Some of these principles are directly applicable to (and applied by) shadow libraries, whereas others are relevant only to legitimate organizations. For instance, the idea that shadow libraries "cannot lobby" would be strongly resisted by those who believe that such archives have a political function. By contrast, the idea that there are formal incentives to "fulfill mission and wind down" is a gray area. This clause of POSI is meant to ensure that organizations do not prioritize their own ongoing existence as a fundamental necessity, if the mission for which they were established has been completed. Yet if we achieved 100 percent OA tomorrow (including access to the back archive of in-copyright scholarship), would shadow libraries continue to exist? This is not clear.

However, in the context of the principle that POSI calls "insurance," which means that the infrastructure and its data can be reconstructed by a third party, shadow libraries most viably demonstrate adherence to some of the tenets of POSI where they most strongly converge with durability requirements. Library Genesis and Anna's Archive, for instance, offer torrent archives of all the material that they cache. Boldly, in 2019, several "seedbox" firms stepped up to the challenge of preserving this large and distributed digital archive of books and journal articles (u/shrine, "Charitable Seeding Update"). As of 2022, the entire archive of Library Genesis files and its associated metadata database is available in distributed downloadable form. The source code for the platform is also openly available. That said, this is not universally the case of different shadow libraries. For example, while POSI stipulates that "all software required to run the infrastructure should be available under an open-source license," Sci-Hub does not currently offer its source code. In a move that seems to go somewhat against the open principle of the site, Elbakyan launched a fundraising drive in 2021 with the promise of open-sourcing the platform, alongside the development of various neural network–based text and data-mining facilities (Maxwell). The slightly problematic logic here resides in demanding payment to open the platform, even while stipulating that it is unethical to pay publishers that do not publish openly. This fundraising drive does, also and ironically, place Sci-Hub's activities back within POSI's frame of sustainable infrastructure. The basic fact is that, without a revenue stream, people cannot eat. Sci-Hub's need for fundraising, to pay for the time spent working on the platform, counts, to some extent, as a sustainability plan.

That said, Sci-Hub's functionality rests on a certain level of security by obscurity. The universal proxy system that Sci-Hub uses can retrieve any scholarly article given just a DOI. However, this comes with a particular download "signature." To remain undetected by publishers that seek to block it, Sci-Hub's background fetching mechanism must log in to publisher websites and appear to mimic a legitimate user request. It should not appear as a rogue agent. Security by obscurity is not a recommended practice in the information security industry. However, in Sci-Hub's case, keeping the source code secret is one way to protect the signature of its download mechanism.

Other, more radical proposals have been mooted to distribute Library Genesis and Sci-Hub's content. The most extreme is a move to a system called the InterPlanetary File System (IPFS) protocol (Rahalkar and Gujar). IPFS yields a distributed, addressable system that assigns a hash to every object. This hashing—a one-way system to reduce a message to a unique key—means that looking up an object's location in the system can be done swiftly. Crucially, as the designers put it, "IPFS has no single point of failure, and nodes do not need to trust each other" (Benet).

Importantly, IPFS has various characteristics of permanence that inhere in its system of *pinning*. In this structure, objects are deemed immutable (pinned) and pulled down permanently to mirroring clients, thereby ensuring a type of digital preservation between clients. As the designer notes, "this also makes IPFS a Web where links are permanent, and Objects can ensure the survival of others they point to" (Benet). While various other systems have been proposed—energy-inefficient blockchains are the recurrent terrible idea beloved of overly keen technophiles (Wang and Zhao)—IPFS offers the most realistic chance for shadow libraries to implement widely distributed, geographically redundant, and disaster-resilient digital preservation.

IPFS solves many, but not all, problems for shadow libraries. It is not possible, for instance, to participate anonymously, meaning that so long as shadow libraries remain outlawed, it will still be possible to identify a potentially infringing user by Internet Protocol address. It is also unknown whether IPFS can realistically scale to the level that these shadow libraries require (for an explanation of the technical challenges, see Anna). Having to spread 33 terabytes over 2.5 million unique objects is a significant hurdle to clear. However, it will be challenging for formal publishers to shut down the shadow libraries if these difficulties can be overcome. The principles of open scholarly infrastructure would protect these shadow organizations as much as they might protect a legitimate counterpart.

One of the reasons why I believe that we should consider shadow libraries to be part of the infrastructure of contemporary scholarly communications is that they resiliently adhere to a set of formally articulated principles of such infrastructure. While they fall on several criteria—as do their legal counterparts (Bilder)—the undercover architectures of shadow libraries exhibit many of the characteristics of permanence that we expect from scholcomms infrastructure. These heretical infrastructures are also becoming embedded as a silent norm in the academy, with an entire cohort of students learning that the easiest way to access paywalled research is via piracy (Owens). The distribution and technical hardiness of these infrastructures mean that they are here to stay.

Where Next?

Shadow libraries are the symptom of ills in scholcomms. However, there are also an entire set of disagreements around their purpose. While many would argue that

Sci-Hub, Library Genesis, and Anna's Archive are potent motivators to "change the system" of scholarly communications to open access, the founder of the former disagrees. As Elbakyan puts it: "My view is completely different. For me, Sci-Hub has a value by itself, as a website where users can access knowledge. . . . The system has to be changed so that websites like Sci-Hub can work without running into problems. Sci-Hub is a goal, changing the system is one of the methods to achieve it. . . . Sci-Hub is not going to die when research articles will be free" (Elbakyan, "Sci-Hub Is a Goal").

The digital humanities (DH) disciplines have often entirely sidestepped this problem of poor access to scholarship. By beginning from a principle of OA to new forms of research output—such as interactive digital websites—it can appear as though the challenges that shadow libraries address are irrelevant in this born-open space. This is not strictly true. We know, for instance, that one of the ways in which DH disciplines make themselves commensurate (and comeasurable) with conventional scholarship is by aping the forms of their peers (Eve, "Violins in the Subway"). DH spaces have journal articles and books, and when they adopt these forms, they are not always open by default. Further, DH scholarship has to cite existing work. It is not enough to set one's new work aside as though it were a totally new field, with no connection to the past. No matter how hard it tries, DH finds itself enmeshed in the issues of access to which shadow libraries are often illegally devoted.

Is the future a world where centralized sites, such as Sci-Hub, amalgamate all research findings and anybody can read such work without paying? Time and court cases will tell whether they will be legal. However, in reality, that future is already here. Whether publishers like it or not, and whether legal or not, we already have a system where almost all newly published knowledge is freely available to read. The question becomes, instead, about the future of publishing labor. If this situation eventually leads to libraries and universities no longer paying for scholarship—and even open research—what is the future of research publishing? Who will perform the professionalized functions of platform development, peer review oversight, typesetting, copyediting, proofreading, digital preservation, and identifier assignment and maintenance? Are we happy to outsource these labor functions to amateur pirates funded by cryptocurrencies? Furthermore, if such outfits were professionalized, would they be so different from being publishers in their own right? It is to these questions—of the logic of extinction or exhaustion of scholarly communications—that shadow libraries apocalyptically gesture.

NOTES

1. In this chapter, I do not hyperlink directly to shadow libraries for multiple reasons. The first is that such links are highly unstable, as the libraries change their locations on a frequent basis to avoid detection. The second is that I do not wish to give any impression of support for such libraries.

2. The official webpage of Monoskop notes, "Many of the titles in the bibliographies are linked to electronic versions of publications made available on Monoskop or other free/libre libraries."

3. Library Genesis includes not just books but scientific articles that it caches for SciHub.

4. For some background on archives, see Derrida; and Caswel.

5. For a critique of Maslow, though, see Tay and Diener.

BIBLIOGRAPHY

Adema, Janneke, and Samuel A. Moore. "Collectivity and Collaboration: Imagining New Forms of Communality to Create Resilience in Scholar-Led Publishing." *Insights: The UKSG Journal* 31 (2018). https://doi.org/10.1629/uksg.399.

Anna. "Putting 5,998,794 Books on IPFS." *Anna's Blog*. November 19, 2022. http://annas-blog.org/putting-5,998,794-books-on-ipfs.html.

Babini, Dominique. "Toward a Global Open-Access Scholarly Communications System: A Developing Region Perspective." In *Reassembling Scholarly Communications: Histories, Infrastructures, and Global Politics of Open Access*, edited by Martin Paul Eve and Jonathan Gray, 331–41. MIT Press, 2020.

Banks, Marcus. 2016. "What Sci-Hub Is and Why It Matters." *American Libraries* 47, no. 6 (2016): 46–49.

Barok, Dušan, Josephine Berry, Bodó Balázs, et al. "In Solidarity with Library Genesis and Sci-Hub." *Custodians.Online* (blog). November 30, 2015. http://custodians.online.

Benet, Juan. "IPFS—Content Addressed, Versioned, P2P File System." *ArXiv:1407.3561 [Cs]*, July 14, 2014. http://arxiv.org/abs/1407.3561.

Bhaskar, Michael. *The Content Machine: Towards a Theory of Publishing From the Printing Press to the Digital Network*. Anthem, 2013.

Bilder, Geoffrey. "POSI Fan Tutte." *Crossref* (blog), March 8, 2022. https://www.crossref.org/blog/posi-fan-tutte/.

Bilder, Geoffrey, Jennifer Lin, and Cameron Neylon. "Principles for Open Scholarly Infrastructures-V1." 2015. https://doi.org/10.6084/m9.figshare.1314859.v1.

Bodó, Balázs. "The Genesis of Library Genesis: The Birth of a Global Scholarly Shadow Library." In *Shadow Libraries: Access to Educational Materials in Global Higher Education*, edited by Joe Karaganis, 25–52. MIT Press, 2018.

British Library. "Facts and Figures of the British Library." British Library. 2020. https://www.bl.uk/about-us/our-story/facts-and-figures-of-the-british-library.

Caswell, Michelle. "'The Archive' Is Not an Archives: On Acknowledging the Intellectual Contributions of Archival Studies." *Reconstruction: Studies in Contemporary Culture* 16, no. 1 (2016). https://escholarship.org/uc/item/7bn4v1fk.

Davies, William. *The Limits of Neoliberalism: Authority, Sovereignty and the Logic of Competition*. SAGE, 2014.

Derrida, Jacques. *Archive Fever: A Freudian Impression.* Translated by Eric Prenowitz. University of Chicago Press, 1995.

Digital Science, Daniel Hook, Mark Hahnel, and Ian Calvert. "The Ascent of Open Access." *Digital Science.* 2019. https://doi.org/10.6084/m9.figshare.7618751.v2.

Doran, Aideen. "Free Libraries for Every Soul: Dreaming of the Online Library." Memory of the World. 2014. https://www.memoryoftheworld.org/blog/2019/10/25/free-libraries-for-every-soul/.

Elbakyan, Alexandra. "How The Chronicle Is Trying to Malign Sci-Hub." *Engineuring* (blog). July 9, 2021. https://engineuring.wordpress.com/2021/07/09/how-the-chronicle-is-trying-to-malign-sci-hub/.

Elbakyan, Alexandra. "A Robin Hood in the World of Science: Alexandra Elbakyan." İnterview by N. Ezgi Altınışık, Alp Öztarhan, and Emel Güneş. 2021. http://bilimveaydinlanma.org/a-robin-hood-in-the-world-of-science-alexandra-elbakyan/.

Elbakyan, Alexandra. "Sci-Hub Is a Goal, Changing the System Is a Method." *Engineuring* (blog), March 11, 2016. https://engineuring.wordpress.com/2016/03/11/sci-hub-is-a-goal-changing-the-system-is-a-method/.

Erkal, Esra. "Allegations Linking Sci-Hub with Russian Intelligence." Elsevier Connect. December 20, 2019. https://www.elsevier.com/connect/allegations-linking-sci-hub-with-russian-intelligence.

Eve, Martin Paul. "Lessons from the Library: Extreme Minimalist Scaling at Pirate Ebook Platforms." *Digital Humanities Quarterly* 16, no. 2 (2022). http://www.digitalhumanities.org/dhq/vol/16/2/000587/000587.html.

Eve, Martin Paul. *Open Access and the Humanities: Contexts, Controversies and the Future.* Cambridge University Press, 2014. https://doi.org/10.1017/CBO9781316161012.

Eve, Martin Paul. "Open Access and Neoliberalism: A Response to Holmwood and Marcuello-Servós." In *The Social Production of Knowledge in a Neoliberal Age: Debating the Challenges Facing Higher Education,* edited by Justin Cruickshank and Ross Abbinnett, 205–10. Rowman and Littlefield International, 2022. https://eprints.bbk.ac.uk/id/eprint/44173/.

Eve, Martin Paul. "Violins in the Subway: Scarcity Correlations, Evaluative Cultures, and Disciplinary Authority in the Digital Humanities." In *Digital Technology and the Practices of Humanities Research,* edited by Jennifer Edmonds, 105–22, 2020. Open Book Publishers.

Eve, Martin Paul. *Warez: The Infrastructure and Aesthetics of Piracy.* punctum, 2021.

Eve, Martin Paul, and Jonathan Gray, eds. *Reassembling Scholarly Communications: Histories, Infrastructures, and Global Politics of Open Access.* MIT Press, 2020.

Fyfe, Aileen, Kelly Coate, Stephen Curry, et al. "Untangling Academic Publishing: A History of the Relationship Between Commercial Interests, Academic Prestige and the Circulation of Research." Zenodo, 2017. https://doi.org/10.5281/zenodo.546100.

Guldi, Jo. "Scholarly Infrastructure as Critical Argument: Nine Principles in a Preliminary Survey of the Bibliographic and Critical Values Expressed by Scholarly Web-Portals for Visualizing Data." *Digital Humanities Quarterly* 14, no. 3 (2020). http://www.digitalhumanities.org/dhq/vol/14/3/000463/000463.html.

Harris, Shane, and Devlin Barrett. 2019. "Justice Department Investigates Sci-Hub Founder on Suspicion of Working for Russian Intelligence." *Washington Post.* December 19, 2019. https://www.washingtonpost.com/national-security/justice-department-investigates-sci-hub-founder-on-suspicion-of-working-for-russian-intelligence/2019/12/19/9dbcb6e6-2277-11ea-a153-dce4b94e4249_story.html.

Himmelstein, Daniel S., Ariel Rodriguez Romero, Jacob G Levernier, et al. 2018. "Sci-Hub Provides Access to Nearly All Scholarly Literature." *eLife.* February 9, 2018. https://doi.org/10.7554/eLife.32822.

Hockenberry, Benjamin. "The Guerilla Open Access Manifesto: Aaron Swartz, Open Access and the Sharing Imperative." *Lavery Library Faculty/Staff Publications* (November 2013): 1–7.

Johns, Adrian. *Piracy: The Intellectual Property Wars from Gutenberg to Gates.* University of Chicago Press, 2011.

Larivière, Vincent, Stefanie Haustein, and Philippe Mongeon. 2015. "The Oligopoly of Academic Publishers in the Digital Era." *PLOS One* 10 (6): e0127502. https://doi.org/10.1371/journal.pone.0127502.

Lawson, Stuart. "Open Access Policy in the UK: From Neoliberalism to the Commons." Doctoral dissertation, Birkbeck, University of London, 2019. https://hcommons.org/deposits/item/hc:23661/.

Lawson, Stuart, Jonathan Gray, and Michele Mauri. "Opening the Black Box of Scholarly Communication Funding: A Public Data Infrastructure for Financial Flows in Academic Publishing." *Open Library of Humanities* 2, no. 1 (2016). https://doi.org/10.16995/olh.72.

Masnick, Mike. 2020. "Academic Publishers Get Their Wish: DOJ Investigating Sci-Hub Founder for Alleged Ties to Russian Intelligence." *Techdirt.* January 3, 2020. https://www.techdirt.com/2020/01/03/academic-publishers-get-their-wish-doj-investigating-sci-hub-founder-alleged-ties-to-russian-intelligence/.

Maxwell, Andy. 2021. "Sci-Hub Pledges Open Source & AI Alongside Crypto Donation Drive." *TorrentFreak.* August 9, 2021. https://torrentfreak.com/sci-hub-pledges-open-source-ai-alongside-crypto-donation-drive-210809/.

Mboa Nkoudou, Thomas Hervé. "Epistemic Alienation in African Scholarly Communications: Open Access as a *Pharmakon.*" In *Reassembling Scholarly Communications: Histories, Infrastructures, and Global Politics of Open Access,* edited by Martin Paul Eve and Jonathan Gray, 25–40. MIT Press, 2020.

Mboa Nkoudou, Thomas Hervé. "The (Unconscious?) Neocolonial Face of Open Access." YouTube, November 29, 2017. https://web.archive.org/web/20220627151827/https://www.youtube.com/watch?v=-HSOzoSLHL0.

McLane, Maureen N. *Romanticism and the Human Sciences: Poetry, Population, and the Discourse of the Species.* Cambridge Studies in Romanticism 41. Cambridge University Press, 2000.

Memory of the World. Home page. n.d. https://www.memoryoftheworld.org/blog/tag/library/.

Mintz, Steven. 2022. "The Humanities' Scholarly Infrastructure Is in Utter Disarray." *Inside Higher Ed.* July 18, 2022. https://www.insidehighered.com/blogs/higher-ed-gamma/humanities%E2%80%99-scholarly-infrastructure-utter-disarray.

Monbiot, George. 2011. "Academic Publishers Make Murdoch Look Like a Socialist." *The Guardian.* August 29, 2011. http://www.guardian.co.uk/commentisfree/2011/aug/29/academic-publishers-murdoch-socialist.

Moore, Samuel. "Common Struggles: Policy-Based vs. Scholar-Led Approaches to Open Access in the Humanities." Doctoral dissertation, King's College London, 2019. https://hcommons.org/deposits/item/hc:24135/.

Owens, Brian. "Sci-Hub Downloads Show Countries Where Pirate Paper Site Is Most Used." *Nature.* February 25, 2022. https://doi.org/10.1038/d41586-022-00556-y.

Packer, Abel L. "The Pasts, Presents, and Futures of SciELO." In *Reassembling Scholarly Communications: Histories, Infrastructures, and Global Politics of Open Access*, edited by Martin Paul Eve and Jonathan Gray, 297–313. MIT Press, 2020.

Pastoureau, Michel. "L'incolore n'existe Pas." In *Points de Vue: Pour Philippe Junod*, edited by Danielle Chaperon and Philippe Kaenel, 21–36. Champs Visuels. L'Harmattan, 2003.

Plantin, Jean-Christophe, Carl Lagoze, and Paul N Edwards. "Re-integrating Scholarly Infrastructure: The Ambiguous Role of Data Sharing Platforms." *Big Data & Society* 5, no. 1 (2018). https://doi.org/10.1177/2053951718756683.

Rahalkar, Chaitanya, and Dhaval Gujar. "Content Addressed P2P File System for the Web with Blockchain-Based Meta-Data Integrity." *ArXiv:1912.10298 [Cs],* January 2020. http://arxiv.org/abs/1912.10298.

Reddy, Hrishikesh, and Shivang Mishra. "Sci-Hub Case: Legally Removing the Barriers in the Way of Science." *NLUJ Law Review* (blog), April 29, 2021. http://www.nlujlawreview.in/sci-hub-case-legally-removing-the-barriers-in-the-way-of-science/.

Roh, Charlotte, Harrison W. Inefuku, and Emily Drabinski. "Scholarly Communications and Social Justice." In *Reassembling Scholarly Communications: Histories, Infrastructures, and Global Politics of Open Access*, edited by Martin Paul Eve and Jonathan Gray, 41–52. MIT Press 2020.

Schiermeier, Quirin. "US Court Grants Elsevier Millions in Damages from Sci-Hub." *Nature News.* June 22, 2017. https://doi.org/10.1038/nature.2017.22196.

Sci-Hub. Home page. n.d. https://sci-hub.ru/.

Stallman, Richard M. 2015. "Did You Say 'Intellectual Property'? It's a Seductive Mirage." Gnu.Org. April 20, 2015. https://www.gnu.org/philosophy/not-ipr.en.html.

Star, Susan Leigh. "The Ethnography of Infrastructure." *American Behavioral Scientist* 43, no. 3 (1999): 377–91. https://doi.org/10.1177/00027649921955326.

Suber, Peter. *Open Access.* Essential Knowledge Series. MIT Press, 2012. http://bit.ly/oa-book.

Swartz, Aaron. "Guerilla Open Access Manifesto." In *The Boy Who Could Change the World.*, 26–27. Verso, 2015.

Tay, Louis, and Ed Diener. "Needs and Subjective Well-Being Around the World." *Journal of Personality and Social Psychology* 101, no. 2 (2011): 354–65. https://doi.org/10.1037/a0023779.

Thylstrup, Nanna Bonde. *The Politics of Mass Digitization*. MIT Press, 2018.

u/shrine. "Charitable Seeding Update: 10 Terabytes and 900,000 Scientific Books in a Week with Seedbox.Io and UltraSeedbox." Reddit—r/Seedboxes, 2019. https://www.reddit.com/r/seedboxes/comments/e3yl23/charitable_seeding_update_10_terabytes_and_900000/.

u/shrine. "Library Genesis Project Update: 2.5 Million Books Seeded with the World, 80 Million Scientific Articles Next." Reddit—r/DataHoarder. 2020. https://www.reddit.com/r/DataHoarder/comments/ed9byj/library_genesis_project_update_25_million_books/.

Vaidhyanathan, Siva. *Copyrights and Copywrongs: The Rise of Intellectual Property and How It Threatens Creativity*. NYU Press, 2003.

Wang, Yu, and Liangbin Zhao. "Blockchain for Scholarly Journal Evaluation: Potential and Prospects." *Learned Publishing* 34, no. 4 (2021): 682–87. https://doi.org/10.1002/leap.1408.

Waters, Donald J. "The Emerging Digital Infrastructure for Research in the Humanities." *International Journal on Digital Libraries,* October 7, 2022. https://doi.org/10.1007/s00799-022-00332-3.

WNYC Studios. "The Library of Alexandra." Radiolab Podcasts. April 7, 2023. https://www.radiolab.org/podcast/library-alexandra.

PART II

DIGITAL HUMANITIES (AND CRITICAL INFRASTRUCTURE STUDIES)

PART II][Chapter 7

Digital Humanities and the Energetics of Big Data

JAVIER CHA AND IAN M. MILLER

In 2008, the National Library of Korea prepared to open its Digital Library annex, a key milestone in realizing its vision of becoming a "ubiquitous library" (Yu). However, this transition was met with an unanticipated obstacle: energy. The deployment of an online database with 100 million electronic items, radio frequency identification (RFID) tagging, and computer vision–based tracking systems resulted in a marked escalation in electrical demand (Yu). This increase was further intensified by the installation of user terminals, backup systems, and creative media studios (Yu). Consequently, the National Library of Korea had to seek an additional 400 million Korean won (approximately US$350,000) to ensure the provision of adequate power for these new services.[1]

The National Library of Korea's experience serves as a reminder of the substantial energy required to operate digital infrastructure and foreshadows similar challenges now becoming more apparent in digital humanities (DH). As DH moves beyond personal computing into the realm of big data, the ecological impact of the systems that researchers utilize regularly has become a pressing concern. As per Doug Laney's 3Vs and the *Oxford English Dictionary*'s formal definition of big data, its dynamic and distributed nature, along with its vast scale and complexity, require an extensive, energy-intensive infrastructure.[2] The DH community has promoted minimal computing as a strategy to reduce the environmental costs associated with creating, operating, maintaining, and preserving DH projects (Risam and Gil).[3] However, minimal computing did not anticipate the recent surge in large-scale artificial intelligence (AI) systems trained on big data, and the radical shift toward practices reliant on centralized computing clusters.

Our investigation into the intersection of DH and the energetics of big data is structured as two sections, each aimed at inviting debates, disagreements, and further discussions. First, we argue that big data marks a radical departure from predigital media in terms of resource and energy usage. While recording information on epigraphs, paper, woodblocks, and movable types requires raw materials and human labor, no external energy is necessary to read it. Conversely, hard disk drives (HDDs), solid-state drives (SSDs), and optical discs depend on electricity for both data storage and retrieval. Although the energy demand for a single personal computer can be met with a solar panel the size of a backpack, hundreds of workstations already begin to cause logistical problems, as the National Library of Korea discovered. Corporate data centers, which form the backbone of big data, and their attendant information communication technology (ICT) infrastructures have much greater power requirements. Kak, the South Korean tech giant Naver's flagship data center, handles a data volume equivalent to 1 million National Libraries of Korea (Lee). This facility consumes an immense 156,875 MWh of electricity annually (Naver), sourced from six nearby hydroelectric power stations (Naver Business Platform, 60), and the 120,000 servers (26) hosted in the facility generate a tremendous amount of heat as a by-product of data processing. Today's age of big data prompts DH to address the direct links among data flows, power generation, grid networks, carbon footprint, and cooling.

The second section turns to paradoxes and trade-offs. The big data turn in DH is yet to grapple with the massive capital and energy flows required for handling exabyte-scale data streams and beyond. The National Library of Korea's new Data Preservation Center in P'yŏngch'ang, with the capacity to store 14 million items of cultural significance, including archival-grade optical discs in a climate-controlled setting, is currently under construction with a budget of 61 billion Korean won (US$46 million) (Pak). In comparison, Naver has invested a staggering 1.9 trillion Korean won (US$1.5 billion) in ICT infrastructure from 2019 to 2022 (Pak). Meanwhile, the global expenditure on cloud services reached a remarkable US$178 billion in 2021 (Synergy Research Group). These figures raise critical questions about the relationship between DH and big data. Given that only a fraction of today's big data is likely to be preserved for posterity, it remains uncertain whether future humanities scholars will have the financial and electrical resources to access the hypothetical mega-archives of tomorrow.

From an environmental standpoint, leading cloud service providers, including Amazon, Microsoft, Alphabet (Google), Meta (Facebook), Naver, and Alibaba, are at the forefront of innovations and investments in data center efficiency and renewable energy sources. However, their business models inadvertently contribute to a modern Jevons paradox, where increased efficiency leads to higher overall consumption, output, and a greater reliance on centralized infrastructure. Despite ongoing initiatives to decentralize the web and reduce the dominance of Big Tech, practical alternatives remain elusive. In fact, to date, the rise of large-scale AI has accelerated the concentration of computing power and energy use in Big Tech infrastructure to an unprecedented level. Our research suggests that a viable approach to addressing these challenges lies in drawing insights from historical perspectives, reliable statistics, technical analysis, regional differences, and the coexistence of older and newer information regimes.

The Energetics of Data Production and Preservation

To understand how the energy consumption of big data stacks up against that of nondigital media, we build upon the insights of Gilbert Shapiro, John Markoff, and Silvio R. Duncan Baretta (115–17). Their framework, which assesses the likelihood of a historical document's survival, encompasses various phases: recording, reproduction, preservation, cataloging, and publication.[4] Our back-of-the-envelope calculations estimate the energy costs associated with each juncture, factoring in both the embodied energy inherent in the materials used for crafting these records and the cumulative human and machine labor for their reproduction and preservation (see Table 7.1).[5]

When considering the energy overheads of nondigital and digital media, a dichotomy emerges. The creation and replication of nondigital artifacts, while more energy-intensive initially, starkly contrast with that of digital materials, which demand consistent energy input for their sustenance and access. The history of information media uncovers a trend toward the diminishing of energy needed to record a given amount of content. The process of inscribing text onto bronze, such as those found in early China's ritual vessels, required about 3 GJ, and an equal amount of energy was needed for its replication. A substantial part of this energy was invested in the creation of the bronze itself. In contrast, transcribing manuscripts onto paper or parchment drastically slashed this energy demand to between 0.5 to 2.5 MJ per page, even though the same amount of energy was still needed to make copies of the original.

The introduction of print marked a leap in reducing energy usage, particularly for reproduction. While engraving a page onto a woodblock demanded about six to sixty times more energy than manuscript writing, the subsequent printing process drastically cut the energy costs to about 200 kJ per page, a mere tenth of the energy needed for manual transcription. Advancements in printing technology continued to drive down these costs. The energy required for producing the first copy of a page dropped to about 2 MJ for hand letterpress and under 1 MJ for rotary print. Digital technology has decreased energy consumption, eclipsing all prior media

Table 7.1. Estimated Energy Consumption of Various Recording Media

Media	Production	Reproduction
1. Bronze	3 GJ/kg	3 GJ/kg
2. Manuscript	0.5–2.5 MJ/page	0.5–2.5 MJ/page
3. Woodblock	15–30 MJ/page	< 200 kJ/page
4. Hand letterpress	1.45–2.1 MJ/page	< 200 kJ/page
5. Rotary print	600–950 kJ/page	< 200 kJ/page
6. Digital media	30–100 kJ/page	< 100 kJ/page

forms in volume and reach. Digital text production outpaces the finest print technologies by a factor of 10 in energy efficiency, and digital replication surpasses a two-thousandfold increase in efficiency.

The remarkably low production and reproduction costs of digital media come with caveats: reduced longevity and higher upkeep demands. Volatile memories, such as L1, L2, and L3 caches and dynamic random access memory (DRAM), immediately lose the loaded data in the event of power interruption. Nonvolatile storage, including HDDs, SSDs, optical discs, and magnetic tape drives, require electricity, albeit for different reasons. To avert data deterioration, these devices require climate control, with optimal conditions for magnetic tapes and optical discs being the temperature range of 15–25°C and humidity levels between 30 percent and 50 percent (CLIR). HDDs, particularly helium-sealed drives, demonstrate greater resilience and tolerate a broader temperature range, from –40–70°C (AKCP), and they need to be shielded from direct sunlight. In addition, consider how modern cloud computing categorizes data as hot, warm, or cold according to usage patterns and energy use. SSDs designed for handling high-demand or "hot" data, averages between seven and ten years, while SSDs without access to a power source may risk data loss over time due to the degradation of their NAND flash components.[6] A cold-storage facility like Meta's Prineville location stores infrequently accessed data on low-energy archival-grade HDDs and Blu-ray discs, with shelf lives of thirty and one hundred years, respectively (Bandaru and Patiejunas; Miller; Hogan). Millenniata's M-DISC and Microsoft's Project Silica exceptionally provide data preservation on media designed to last more than 1,000 years, but their adoption remains limited (see Hachman; Salter). In the ICT sector, linear tape open (LTO) tape drives, with about thirty years of service life, are a favored backup solution (Hewlett Packard Enterprise). Most of these digital memory and storage systems rely on continuous power for data longevity.

In contrast, preserving paper primarily involves maintaining a dry environment and avoiding temperature fluctuations, while bronze requires minimal upkeep (see Figure 7.1). The lifespan of digital media, usually ranging several years to a couple of decades, is significantly less than the centuries that paper can last or the millennia for bronze artifacts. The ICT industry's strategies for preserving big data via redundancy and duplication mirrors the way that historical records have been preserved by repeatedly creating copies. For example, updates to early modern Chinese genealogies occurred approximately every sixty years (Pieke, 113), while reproductions of classical texts took place about once per century. One could argue that digital media's need for frequent duplication diverges from the slower duplication processes seen with other media. For our purposes, the aggregate energy needed to preserve and maintain the integrity of big data in cloud facilities is the ultimate dilemma. Even though each act of digital replication consumes negligible amount of power, the aggregate effect of these repeated operations leads to substantial energy use over time.

Media	Production Cost	Preservation Cost	Lifetime	Frequency of Reproduction
Bronze	very high	practically zero	centuries to millennia	centuries
Paper	medium	very low	decades to centuries	decades
Digital	low	low	years to decades	nearly instant and on demand

Figure 7.1. Production and preservation of different recording materials.

To sum up, the power needed for digital publication is considerably less than that for a printed book (which in turn is more energy-efficient than manuscript production), and far less than the energy needed for creating epigraphic materials. Media with higher energy demand at the moment of creation typically boast longer lifespans, while those requiring less energy facilitate a greater volume of work. This dynamic results in digital materials outnumbering printed media, with printed works surpassing both manuscripts and epigraphic artifacts in quantity. Scholars specializing in early periods tend to depend on materials preserved on more enduring media or those that have been replicated over time, typically missing out on ephemera. In contrast, researchers focused on more contemporary periods have access to a plethora of ephemera, but face uncertainties regarding their long-term preservation.

Paradoxes and Trade-offs

The ways that a modern DH researcher's work is dependent on Big Tech infrastructure come at the cost of energy consumption and environmental footprints, serving as yet another reminder of the paramount importance of recognizing the materiality of digital technologies. To reference a few well-known relevant studies on this topic, Friedrich Kittler's "There Is No Software" has significantly contributed to our understanding of the physical foundations of digital technologies. Matthew Kirschenbaum's *Mechanisms* and *Bitstreams* have demonstrated the application of forensics in the study of electronic literature and digital collections. Kate Crawford's *Atlas of AI*, in turn, has taken a critical look at the AI industry's dependence on precious minerals, water, and electricity (23–51), extending beyond Kittler's perspective, which was limited to silicon. Thomas Mullaney's introduction to *Your Computer Is on Fire* also stresses that "nothing is virtual" (5). Further, Nathan Ensmenger urges a reconsideration of "the Cloud as a factory, and not as a disembodied computational device" (43). Mullaney's advocacy for a critical stance toward the tech industry, described as a "call to arms" with "unapologetically direct and bold arguments" (8), and Crawford's examination of the AI industry merit careful contemplation. At the same time, it is also important to be specific and precise.

Contrary to the centrality of lithium in Crawford's discussion of AI, it is a critical mineral used in the production of batteries for mobile devices and electric vehicles, not the semiconductors, data storage units, and telecommunication networks that underpin cloud computing and AI.

Adding to this challenge is the Jevons paradox, which observes how efficiency gains encourage more utilization. "According to a principle recognized in many parallel instances," William S. Jevons astutely noted, "new modes of economy will generally lead to an increase in consumption" (103). Originally articulated in the context of coal use, the Jevons paradox offers a compelling lens for understanding the burgeoning of Web 2.0 services, in which massive capital investments and the development of centralized infrastructures drove efficiency gains. The technological strides in recent years have resulted in the seamless weaving of digital technologies into various facets of daily life, including academic research.

Recent market data confirms this trend. From 2010 to 2020, global server shipments increased from 8.9 million to 12.15 million units (Alsop), and about 1.3 billion personal computers continued to be operational as of 2016 despite a marginal downturn in sales (Statista Research Department, "Installed Base of Personal Computers"). Meanwhile, smartphone subscriptions surged from 3.6 billion in 2016 to 6.4 billion in 2022 and an anticipated 7.7 billion by 2028 (Statista Research Department, "Number of Smartphone Subscriptions"). The Web 2.0 trend of end users as both consumers and producers of digital content has consequentially fueled an exponential increase in global data volumes. Total data volume is projected to increase from approximately 41 (Taylor) to 59 ZB (IDC) in 2019 to 175 (Reinsel, Grantz, and Rydning) or 181 ZB (Taylor) by 2025.

However, a closer examination of the Jevons paradox in the Web 2.0 transition reveals a more intricate picture than what meets the eye. ICT efficiency gains and market demands manifest in surprising ways. The disparity between the proposed Kryder's Law and actual outcomes illustrates this point. In 2005, Mark Kryder projected that digital storage capacity increases would outpace Moore's Law (Walter) and anticipated the availability of 40-TB disks at US$40 by 2020 (Kryder and Kim, 3406). Contrary to these forecasts, the ICT sector focused less on storage volume and more on performance enhancement and power conservation, driven by the proliferation of mobile devices, cloud computing, and social media. This mismatch between predicted and actual trajectories underlines the unpredictability and complex nature of technological advancements. Theoretical models provide a framework, but real-world tech development and its impact on resources and consumption tend to diverge from expected outcomes.

From 2008 to 2023, high-end consumer and enterprise-grade SSDs experienced more than a hundredfold performance growth in both sequential and 4K random input/output operations per second (IOPS). In 2008, a standard Seagate HDD required 11,772 seconds (3.27 hours) for sequential 1-TB reads and 1,048,576 seconds (more than twelve days) for the same amount at 4K random blocks. The

2023 SSD completes these tasks in under two minutes and approximately two hours, respectively. (See the relative time measures in Figure 7.2, calculated based on the performance measures shown in Table 7.2.)[7]

In addition, the shift from mechanical disks to flash memory has brought considerable power efficiency gains. In 2009, Intel's premium MLC and SLC SSDs already showcased a 27- to 213-fold decrease in energy per terabyte compared to HDDs. In 2023, Crucial's flagship model achieved a further 6- to 10-fold reduction in electricity consumption per terabyte for sequential reads relative to Intel's 2009 models. (See the relative energy consumption measures in Figure 7.3, calculated on the basis of energy efficiency measures shown in Table 7.3.[8])

The environmental and energy costs of global ICT infrastructure similarly present a multifaceted picture. Phrases such as "more than 2 percent of global energy use" (Mullaney, 5) and "70 billion kilowatt-hours of electricity in 2016 in the United States alone" (Ensmenger, 34) demand a more nuanced interpretation. Crawford's assertion that "the tech sector will contribute 14 percent of global greenhouse emissions by 2040" (42) carries an important caveat. This is a worst-case scenario based

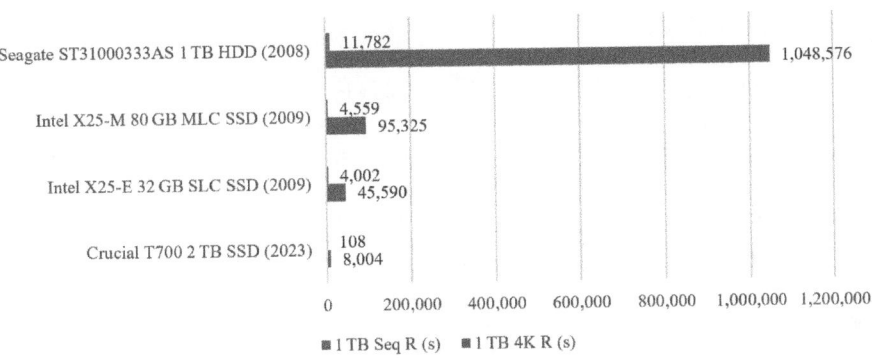

Figure 7.2. Time measures (in seconds) for hard drives reading 1 TB of data from 2008 to 2023.

Table 7.2. Performance Gains in HDDs and SSDs from 2008 to 2023

Storage Unit	Sequential Read (MB/s)	Sequential Write (MB/s)	32K IOPS (MB/s)	4K IOPS (MB/s)
Seagate ST31000333AS 1 TB HDD (2008)	89	82	7	1
Intel X25-M 80 GB MLC SSD (2009)	230	77	89	11
Intel X25-E 32 GB SLC SSD (2009)	262	173	138	23
Crucial T700 2 TB SSD (2023)	9,731	9,780	4,742	131

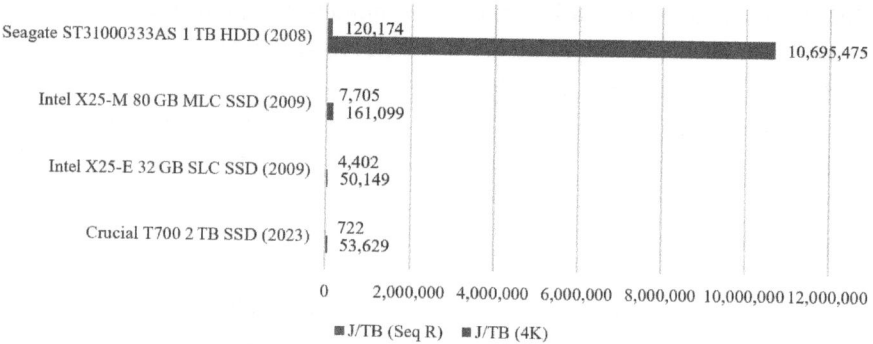

Figure 7.3. Energy consumption (in joules) of hard drives per terabyte of data from 2008 to 2023, based on efficiency measures given in Table 7.3.

Table 7.3. Efficiency Gains in SSDs vs. HDDs from 2008 to 2023

Storage Unit	Time for Seq Read of 1 TB (s)	Time for 4K Read of 1 TB (s)	Energy per TB Seq Read (J)	Energy per TB 4K Read (J)
Seagate ST31000333AS 1 TB HDD (2008)	11,782	1,048,576	120,174	10,695,475
Intel X25-M 80 GB MLC SSD (2009)	4,559	95,325	7,705	161,099
Intel X25-E 32 GB SLC SSD (2009)	4,002	45,590	4,402	50,149
Crucial T700 2 TB SSD (2023)	108	8,004	722	53,629

on an exponential fit and contingent on current trends continuing "if unchecked" (Belkhir and Elmeligi, 448), and with "large uncertainty about the lifecycle annual footprint computers . . . and displays" (458). Relative data is partially measures of ICT's outpaced growth vis-à-vis other industries, while projections are scenario-based ranges with minimum and maximum values. The analysis also varies depending on whether the focus is solely on data center operations or also encompasses the manufacturing, installation, and usage of optical fibers, wireless networks, personal computers, mobile devices, televisions, and gaming consoles. In 2005, data centers accounted for about 1 percent of the world's electricity, half of which was dedicated to cooling and power distribution (Koomey, 1). By 2020, data centers were responsible for 1.4 to 1.6 percent of the global carbon footprint, with communication networks and personal digital devices adding another 1.7–2 percent, while communication networks, desktops, notebooks, displays, tablets, and smartphones emitted an additional 1.7 to 2 percent.[9]

Eric Masanet and his team present a compelling counternarrative to the dire forecasts about data center energy consumption. Their research indicates that, as of 2020, global electricity usage by data centers remained steady, at about 1.1 to 1.5 percent, due to a concomitant rise in overall power generation to meet the lifestyle demands of the growing middle-class population, particularly in emerging economies (Masanet et al., 984; Statista Research Department, "Electricity Generation Worldwide"; Tweed). Masanet et al. remind us to be careful with "oft-cited yet simplistic analyses" (984) that overlook simultaneous trends in energy efficiency. From 2010 and 2018, there was a remarkable twenty-five-fold increase in data center storage capacity (Masanet et al., 985), but this expansion was counterbalanced by a significant ninefold reduction in the amount of energy required per unit of storage (984). While the ICT sector is a major energy consumer and contributor to emissions, it will not consume one-fifth of the world's electricity by 2025 (Vidal), nor will Japan's data centers gobble up the entire nation's electricity supply by 2030 (Bawden). A balanced perspective is vital for crafting effective strategies, policies, and collective responses.

What if the objective is to drastically reduce greenhouse gas emissions in absolute terms and increase the use of renewable energy to 100 percent? The conundrum is that the ICT sector, as one of the largest carbon polluters, has responded with commitments to the transition to renewables. Take US-based Big Tech for example. Google uses a credit-matching strategy in which it purchases the same amount of renewable energy as its facilities consume. Despite Crawford's criticism of this practice (43), it is still a positive development and a step in the right direction. Meta's global operations have been powered entirely by renewable energy since 2018, and the company aims to achieve net zero emissions across its entire supply chain by 2030 (Parekh). In 2021, Amazon Web Services reported that 85 percent of the electricity used in Amazon's businesses was derived from renewable sources, with the goal of reaching 100 percent by 2025 (Amazon). Recently, the cloud giant added 2.7 GW of low-carbon energy capacity in South America, India, and Poland, as well as investing in 379 renewable projects worldwide (Robinson). Despite these efforts, however, Amazon's overall carbon footprint increased by 19 percent due to the company's growth outperforming its green initiatives (Richardson).

The world outside the United States merits more than a passing mention. In northern Europe, Meta operates data centers in Odense and Luleå that use 100 percent renewables according to the abovementioned global commitment (Edelman; Meta), and Google powers its Hamina location with electricity from a wind farm in northern Sweden (Alley). However, one should be cautious about generalizing Nordic facilities as uniformly green. As Julia Velkova notes (672), renewable sources accounted for less than 10 percent of the electricity used at the Yandex data center in Mäntsälä, Finland, in 2018 and 2019. By contrast, Latin America demonstrates a latecomer's advantage in building new ICT infrastructure powered primarily by renewables. Between 2010 and 2022, the renewable capacity in Latin America and the Caribbean has nearly doubled from 168 to 314 GW (Fernández, "Renewable Energy Capacity in Latin America"), with Brazil leading the charge (Fernández, "Leading Countries"). According to Boston Consulting Group, the region's cloud market is worth US$10 billion as of 2022 and will grow at a 30 percent annual rate (Boston Consulting Group). To meet this demand, Microsoft's Chile, Ascenty, Scala Data Centers have already built or are in the process of building carbon-neutral data centers (News Center Microsoft Latinoamérica; *DF SUD*; Bnamericas).

China also deserves special attention. The People's Republic is home to one-quarter of the world's data centers and the world's largest emitter of greenhouse gases (BBC News). Despite its continued reliance on coal, China has reduced its coal use for electricity generation from 72.4 percent in 2005 to 56.8 percent in 2020 (Cheng) and has emerged as the leading producer of renewable energy in the world. As of 2024, China's total renewable capacity was 1,828 GW, more than four times that of the United States (Fernández, "Leading Countries"). The issue is that renewable energy appears to be slow to reach the ICT demand concentrated on the east coast. In 2021, BloombergNEF ranked the e-commerce giant Alibaba as the largest buyer of renewable energy among Chinese companies (Alibaba Group), and Alibaba pledged to achieve carbon neutrality in all its operations by 2030 (Greenpeace East Asia). During the same year, however, Alibaba Cloud reported that only 21.6 percent of its electricity usage came from clean energy sources, which was the national average in China in 2018 (Greenpeace). Aware of this issue, the Ministry of Industry and Information Technology and National Development and Reform Commission have mandated that all new hyperscale data centers attain a power usage effectiveness (PUE) ratio of 1.25 by 2025 (Xue). China's challenge is that that remote solar and wind generators in Gansu, Ningxia, Qinghai, Xinjiang, and Inner Mongolia must be connected to digital infrastructure in populous regions via ultra-high-voltage electricity transmission lines. This centrally planned, supply-driven strategy is met with resistance from various interest groups and will take time to gain local acceptance (Chen).

The foregoing discussion is written from the perspective of an intellectual and an environmental historian striving for a balance of ideational and materialist viewpoints. Our interest in this topic originates from many hours of musing in graduate school over a narrative of world history from the perspective of information and energy regimes. As area specialists of medieval Korea and early modern China, we approach global and comparative scholarship with an emphasis on the necessity of engaging with sources from multiple continents and in various languages. Our background in premodern history informs our tendency to stress the importance of problematizing historical temporalities in the Annales sense of short-term events, medium-term conjunctures, and enduring *longue durée* trends.

This chapter's genesis is also marked by a confluence of unforeseen events. In 2021, when the editors of this *Debates in the Digital Humanities* volume announced the call for papers, Bitcoin's value reached its all-time-high and the web3 community was in the midst of buoyant optimism about the imminent coming of a decentralized web. By the time we completed this chapter, however, the ICT discourse radically shifted, with an overwhelming attention given to large-scale AI and centralized infrastructure, spearheaded by the advent of OpenAI's ChatGPT. Concurrently, Javier Cha, one of the authors of this chapter, conducted interviews and field research exploring data centers and digital archive solutions, notably at Google's Hamina location, Microsoft Research's Project Silica, and the Arctic World Archive in Svalbard.

Cha's journey from Hong Kong to Svalbard required more than twenty hours of air travel. During the descent at Longyearbyen, the sobering spectacle of barren landscapes, receding glaciers, and snowless peaks revealed the stark realities of climate change in a region warming at quadruple the global average. The surreal depositing of a symbolic piqlFilm in a decommissioned coal mine within the Arctic Circle fostered a mix of humility and cautious optimism.[10] The striking sight of air conditioners installed inside the Global Seed Vault, a response to the permafrost failing to maintain the required coolness during summer, raised questions about the longevity of the digital data stored in the Arctic World Archive. This poignant experience profoundly echoed what Bethany Nowviskie means by "graceful degradation, preservation, memorialization, apocalypse, ephemerality, and minimal computing" (i12) with respect to DH practice in the Anthropocene.

The path forward for DH requires an in-depth understanding of the energy dynamics in digital information processing. Advanced cold-storage solutions such as Project Silica, M-Disc, and piqlFilm prompt new inquiries about the ultra-long-term preservation of digital data, spanning thousands to millions of years. Yet, in the immediate term, the transformation of the humanities into a substantial energy consumer is a pressing concern. Minimal computing principles, while valuable, are inadequate for addressing this surge in energy demand. Unexpectedly, the recent shift to AI-optimized servers in data centers has led to a marked increase in electricity and water usage, far exceeding that of conventional Web 2.0 setups (Langley). Renewable energy sources power many of these centers, and district heating systems in Nordic and Baltic regions show how excess heat can be effectively reused (Velkova). Our own experiences with LLM training on high-end NVIDIA guzzlers, however, lead us to question the necessity of such tasks and think about future implications of these trends.

Our hope is that the DH community uses its expertise in digital technologies to bridge the gap between the research of activist groups and corporate green claims and come up with some sound solutions and responses for mitigating big data's environmental impact. Addressing the environmental impact of digital technology will require working together with scientists, engineers, educators, and policy-makers, especially in the communities most affected by climate change. The ability of DH scholars to add our deep understanding of digital media and computational methods to the traditional humanities' strengths in critical thinking, cultural understanding, and linguistic skills positions us in a unique way to contribute meaningfully to this dialogue.

NOTES

Javier Cha would like to acknowledge the support of the Innovative and Pioneering Young Researchers Scheme at Seoul National University, as well as the Seed for Basic Research for New Staff and the Dean's Development Fund from the Faculty of Arts at the University of Hong Kong.

1. This account is based on personal communication between Chinho Pak and Javier Cha, one of the authors of this chapter, on November 14, 2022 as well as an email sent by Pak to Cha on December 7, 2022.

2. 3Vs (volume, velocity, and variety) is a widely accepted framework for defining the key characteristics of big data. The concept is credited to Doug Laney, who originally formulated it as 3Ds: data volume, data velocity, and data variety. For a discussion on the implications of big data for DH, see Cha.

3. *Digital Humanities Quarterly* published fourteen articles on minimal computing in its summer 2022 issue. The guest editors, Roopika Risam and Alex Gil, provide an excellent overview of the lofty goals and complexities of pursuing minimal computing in the DH (Risam and Gil).

4. Our modification has added the word "reproduction" to highlight the important differences in the energy costs of different media.

5. A1: According to Smil (209, box 4.20), the approximate amount of fuel required to smelt copper is 1 kg charcoal and 1 kg slag, and the specific energy content of charcoal is 28–32 MJ/kg (12, box 1.4). Ores typically contain 0.6–1 percent copper. Using the higher purity estimate, about 100 kg of charcoal, or about 3 GJ, is needed to produce a 1-kg copper object, which is a reasonable estimate of the energy needed to produce bronze. Notably, this is significantly higher than the 90–100 MJ/kg required to produce copper using modern techniques. The high energy cost of materials completely outweighs the energy cost of human labor, which ranges from 6 to 10 MJ/person/day, depending on the size of the person and the difficulty of the labor (Smil, 19, box 1.10). A2: This figure is adapted from Smil, 19, box 1.10, which estimates the energy cost of human labor as about 7.5 MJ/person/day and accounts for the relatively light labor of writing. Assuming a median productivity of 16 folio pages/day at 7.5 MJ/person/day, the labor cost per page is approximately 0.47 MJ. A lower productivity, perhaps 2–4 pages per day, would result in a much lower efficiency at 0.94–1.88 MJ per page. In contrast, the higher scribal efficiencies observed in the late Middle Ages would have yielded a much higher efficiency of perhaps 0.2–0.3 MJ/page. Assuming a higher cost of paper or vellum than in early modern times, the paper's embodied energy was likely between 0.3 and 0.6 MJ/page. This results in a wide range of potential energy costs for manuscripts, ranging from as low as 0.5 MJ/page to as high as 2.5 MJ/page, with human labor accounting for 60–75 percent. A3: This figure follows Kai-wing Chow's estimates of 200–400 characters per folio side and 100 characters/person/day for carving

woodblock (35–38). Smil (16, box 1.8) calculates embodied energy to be between 1 and 3 MJ/kg for lumber and 23 and 35 MJ/kg for paper. Accounting for the lighter labor and adapting from Smil, 19, box 1.10, the energy cost of human labor turns out to be approximately 7.5 MJ/person/day. Using these numbers, the energy cost of one folio page is 15–30 MJ/page for labor (2–4 days of work at 7.5 MJ/day), 0.75 MJ/page for woodblock (0.5 kg/page at 1.5 MJ/kg), and 0.15 MJ/page for paper (5 g/page at 30 MJ/kg). B3: Assuming that a worker could produce 1,500 pages/day (Chow, 70), the energy expended works out to 0.005 MJ/page. Lucien Febvre and Henri-Jean Martin (132) provide a higher estimate of up to 3,350 impressions/day for printing with movable type, resulting in an energy cost of slightly more than 0.002MJ/page. With several orders of magnitude greater production rates, industrial printing would have truly marginal energy costs for human and machine output. In every instance, the materials cost dominates at 0.15 MJ/page for paper and a somewhat lower cost for ink (perhaps 0.05 MJ/page as a high estimate). The total is about 0.200 MJ/page or 200 KJ/page. A4: Compositors could complete 1 to 3 sets per day, which equates to approximately 1 to 6 pages (i.e. 1 set of 4 pages or 2–3 sets of 2 pages) (Febvre and Martin, 131–32). This results in labor costs of 1.25 to 1.91 MJ/page for setting forms (excluding checking) and 0.002 to 0.004 MJ per page for printing. Paper and ink cost 0.2 per page in terms of materials. Because type could be reused, it is much more difficult to estimate the cost of fixed materials, but they were probably marginal compared to the cost of the labor and paper. The total comes to somewhere between 1.45 and 2.10 MJ/page. A5: This figure is from Burr. A6: Assuming a computer operating at 75 w, five minutes of typing would require 22.5 kilojoules of energy. This represents a high-end estimate of power consumption. In actuality, while Ian Miller was typing this page, his 60-w laptop consumed less than 5 w on average, or 1.5 kilojoules of energy consumed in five minutes. If this is performed on cloud-hosted documents, the cost could be multiplied by two to four, or 3 to 6 kJ for five minutes. A human typist would consume an astoundingly similar amount of energy as a computer running full tilt. 7.5 MJ/day translates to 5.2 kJ/min, or approximately 26 kJ for five minutes. A slower typist taking ten minutes would increase these figures by twofold. On Miller's laptop, it takes between 29 and 64 kJ to type a page of text, depending on my typing speed and power consumption. An extremely high-energy use case could reach 100 kJ/page. B6: For instance, it took approximately two seconds to copy and load a one-page document on Google Drive, which caused the laptop's power consumption to spike to approximately 20 w. Assuming that this resulted in a similar power surge on the server side, it took approximately 80 J to generate the document. Longer documents do not increase the amount of time required in a linear fashion, so their per-page energy consumption decreases to probably just a few joules.

6. The minimal charge requirement of SSDs for long-term data storage is poorly understood. In 2015, ExtremeTech reported that SSDs removed from a power source could begin to lose data after only 20 to 105 weeks (Hruska). This claim turned out to be a misinterpretation of the data presented by Alvin Cox of JEDEC, a semiconductor trade organization and standardization body (Ung). In addition, SSDs have finite read/write cycles, which can shorten its lifespan, but this characteristic is unrelated to energy use.

7. The performance measures for each HDD or SSD in this table are from PassMark Software's Hard Drive Benchmarks—Seagate ST31000333AS 1TB HDD, https://www.harddrivebenchmark.net/hdd.php?id=112; Intel X25-M 80GB MLC SSD, https://www.harddrivebenchmark.net/hdd.php?id=11053; Intel X25-E 32GB SLC SSD, https://www.harddrivebenchmark.net/hdd.php?id=1651; Crucial T700 2TB SSD, https://www.harddrivebenchmark.net/hdd.php?id=34572. Among various metrics for measuring the performance of HDDs and SSDs, the distinction between sequential reads and 4K random IOPS is crucial. Sequential reads measure the speed at which data is read from the storage device when the data is arranged contiguously or in order, such as when transferring a large movie file from one location to another. On the other hand, 4K random IOPS is a performance measure for tasks dealing with small (i.e., 4K) data chunks located randomly across HDDs and SSDs, which is useful for understanding scenarios where the workload involves a lot of small, scattered day-to-day tasks, such as web browsing, multitasking environments, and server applications where access patterns involve numerous small files.

8. The estimated energy consumed per terabyte is calculated by multiplying the average wattage by the time in seconds required to read 1 TB. For example, Crucial T700 2TB SSD, with an average power consumption of 6.7 w, requires 107.75 seconds to read 1 TB sequentially, returning 722 J per TB. The individual power requirement figures have been obtained as follows: Seagate ST31000333AS 1TB, average power requirement 10.2 w (source: Schmid and Roos, "Tom's Winter 2008," 15); Intel X25-M 80GB MLC SSD, average power requirement 1.69 w (source: Schmid and Roos, "Intel X25-E," 7; Intel X25-E 32GB SLC SSD, average power requirement 1.1 w); Crucial T700 2TB SSD, average power requirement 6.7 w (source: "Crucial T700 2 TB").

9. This figure is obtained by multiplying 3.1 to 3.6 percent of total relative carbon footprint by 45 percent and 55 percent, respectively (see Belkhir and Elmeligi, 457).

10. piqlFilm is a long-term digital preservation medium developed by the Norway-based company Piql. It records textual and visual information by printing directly onto film and binary data as high-definition QR codes on film coated with silver halides, engineered to resist degradation. Piql also operates the Arctic World Archive in Svalbard, where reels of piqlFilm are sealed in containers and stored inside a decommissioned coal mine with the intention of being maintained at optimal conditions by the surrounding permafrost. GitHub is archived in the Arctic World Archive.

BIBLIOGRAPHY

AKCP. "The Impact of Temperature on IT Storage: Enhancing Performance and Longevity." August 2023. https://web.archive.org/web/20250312000644/https://www.akcp.com/blog/how-temperature-affects-it-data-storage/.

Alibaba Group. "Environmental, Social, and Governance Report 2022." 2022, https://data.alibabagroup.com/ecms-files/1452422558/5feb0e46-f04b-4d9c-9568-e4a5912db37e.pdf.

Alley, Alex. "Swedish Wind Energy Project Lifts off, Will Power Finnish Google Data Center." *Data Center Dynamics,* November 8, 2019. https://www.datacenterdynamics.com/en/news/swedish-wind-energy-project-lifts-will-power-finnish-google-data-center/.

Alsop, Thomas. "Server Shipments Worldwide from 2010 to 2020." *Statista,* November 28, 2022. https://www.statista.com/statistics/219596/worldwide-server-shipments-by-vendor/.

Amazon. "Delivering Progress Every Day. Amazon's 2021 Sustainability Report." 2021. https://sustainability.aboutamazon.com/2021-sustainability-report.pdf.

Bandaru, Krish, and Kestutis Patiejunas. "Under the Hood: Facebook's Cold Storage System." *Facebook Engineering,* May 4, 2015. https://engineering.fb.com/core-data/under-the-hood-facebook-s-cold-storage-system/.

Bawden, Tom. "Global Warming: Data Centres to Consume Three Times as Much Energy in Next Decade, Experts Warn." *The Independent,* January 23, 2016. https://www.independent.co.uk/climate-change/news/global-warming-data-centres-to-consume-three-times-as-much-energy-in-next-decade-experts-warn-a6830086.html.

BBC News. "Report: China Emissions Exceed All Developed Nations Combined." May 7, 2021, https://www.bbc.com/news/world-asia-57018837.

Belkhir, Lotfi, and Ahmed Elmeligi. "Assessing ICT Global Emissions Footprint: Trends to 2040 & Recommendations." *Journal of Cleaner Production* 177 (2018): 448–63. https://doi.org/10.1016/j.jclepro.2017.12.239.

Bnamericas. "Scala data centers el el primero en la industria en ser 100% neutral en Alcance 3, cubriendo las emisiones indirectas de carbono." December 23, 2022. https://www.bnamericas.com/es/noticias/scala-data-centers-es-el-primero-en-la-industria-en-ser-100-neutral-en-alcance-3-cubriendo-las-emisiones-indirectas-de-carbono.

Boston Consulting Group. "Los servicios en la nube crecerán un 30% al año en Latinoamérica." September 21, 2022. https://www.bcg.com/press/21september2022-los-servicios-en-la-nube-creceran-un-30-al-ano-en-latinoamerica.

Burr, Christina. "Defending 'The Art Preservative': Class and Gender Relations in the Printing Trades Unions, 1850–1914." *Labour/Le Travail* 31 (1993): 47–73.

Cha, Javier. "Big Data Studies: The Humanities in Uncharted Waters." *Korean Studies* 47 (2023): 274–99. https://doi.org/10.1353/ks.2023.a908625.

Chen, Gang. *The Politics of Renewable Energy in China.* Edward Elgar Publishing, 2019.

Cheng, Evelyn. "China Has 'No Other Choice' But to Rely on Coal Power for Now, Official Says." CNBC, April 29, 2021. https://www.cnbc.com/2021/04/29/climate-china-has-no-other-choice-but-to-rely-on-coal-power-for-now.html.

Chow, Kai-wing. *Publishing, Culture, and Power in Early Modern China.* Stanford University Press, 2004.

Council on Library and Information Resources (CLIR). "5. How Can You Prevent Magnetic Tape from Degrading Prematurely?" https://www.clir.org/pubs/reports/pub54/5premature_degrade/.

Crawford, Kate. *Atlas of AI: Power, Politics, and the Planetary Costs of Artificial Intelligence.* Yale University Press, 2021.

"Crucial T700 2 TB," *TechPowerUp*. https://www.techpowerup.com/ssd-specs/crucial-t700-2-tb.d1464.

DF SUD. "Ascenty invertirá US$ 290 millones para la construcción de cinco data centers en Brasil, Chile y Colombia." November 9, 2022. https://dfsud.com/tecnologia-y-startup/ascenty-invertira-us-290-millones-para-la-construccion-de-cinco-data.

Edelman, Lauren. "Facebook's Hyperscale Data Center Warms Odense." *Tech at Meta*. July 7, 2020. https://tech.facebook.com/engineering/2020/7/odense-data-center-2/.

Ensmenger, Nathan. "The Cloud Is a Factory." In *Your Computer Is on Fire*, edited by Thomas S. Mullaney et al., 29–49. MIT Press, 2021.

Febvre, Lucien, and Henri-Jean Martin. *The Coming of the Book: The Impact of Printing, 1450–1800*. Verso Books, 2010.

Fernández, Lucía. "Leading Countries in Installed Renewable Energy Capacity Worldwide in 2024 (in Gigawatts)." *Statista*, March 2025. https://www.statista.com/statistics/267233/renewable-energy-capacity-worldwide-by-country/.

Fernández, Lucía. "Renewable Energy Capacity in Latin America and the Caribbean from 2010 to 2022 (in Gigawatts)." *Statista*, April 27, 2023. https://www.statista.com/statistics/665458/renewable-energy-capacity-latin-america-caribbean/.

Greenpeace. "Powering the Cloud: How China's Internet Industry Can Shift to Renewable Energy." Greenpeace, 2019. https://www.greenpeace.org/static/planet4-eastasia-stateless/2019/11/7bfe9069-7bfe9069-powering-the-cloud-_-english-briefing.pdf.

Greenpeace East Asia. "Alibaba Pledges Carbon Neutrality in Its Operations by 2030: Greenpeace Response." Greenpeace, December 20, 2021. https://www.greenpeace.org/eastasia/press/7123/alibaba-pledges-carbon-neutrality-in-its-operations-by-2030-greenpeace-response/.

Hachman, Mark. "New 1,000-Year DVD Disc Writes Data in Stone, Literally." *PCMag*, August, 15. 2011. https://www.pcmag.com/archive/new-1000-year-dvd-disc-writes-data-in-stone-literally-286353.

Hewlett Packard Enterprise. "QuickSpecs: HPE LTO Ultrium Storage Supplies." October 4, 2021, p. 9, https://www.hpe.com/psnow/doc/c04154430.pdf.

Hogan, Mél. "Data Flows and Water Woes: The Utah Data Center." *Big Data & Society* 2 (2015): 1–12, https://doi.org/10.1177/2053951715592429.

Hruska, Joel. "SSDs Can Lose Data in as Little as 7 Days Without Power." *ExtremeTech*, May 11, 2015. https://www.extremetech.com/computing/205382-ssds-can-lose-data-in-as-little-as-7-days-without-power.

International Data Corporation (IDC). "IDC's Global DataSphere Forecast Shows Continued Steady Growth in the Creation and Consumption of Data." May 8, 2020. https://web.archive.org/web/20200604065157/https://www.idc.com/getdoc.jsp?containerId=prUS46286020.

Jevons, William Stanley. *The Coal Question: An Enquiry Concerning the Progress of the Nation, and the Probable Exhaustion of Our Coal-mines*. Macmillan, 1865.

Kirschenbaum, Matthew G. *Bitstreams: The Future of Digital Literary Heritage*. University of Pennsylvania Press, 2021.

Kirschenbaum, Matthew G. *Mechanisms: New Media and the Forensic Imagination.* MIT Press, 2008.

Kittler, Friedrich A. "There Is No Software." In *The Truth of the Technological World: Essays on the Genealogy of Presence,* 219–29. Stanford University Press, 2014.

Koomey, Jonathan G. "Worldwide Electricity Used in Data Centers." *Environmental Research Letters* 3, no. 3 (2008): 1. https://doi.org/10.1088/1748-9326/3/3/034008.

Kryder, Mark H., and Chang Soo Kim. "After Hard Drives—What Comes Next?" *IEEE Transactions on Magnetics,* 45, no. 10 (2009): 3406–13. https://doi.org/10.1109/TMAG.2009.2024163.

Laney, Doug. "3D Data Management: Controlling Data Volume, Velocity, and Variety." META Group, February 6, 2001, https://web.archive.org/web/20120304154148/https://blogs.gartner.com/doug-laney/files/2012/01/ad949-3D-Data-Management-Controlling-Data-Volume-Velocity-and-Variety.pdf.

Langley, Hugh. "Google's Water Use Is Soaring. AI Is Only Going to Make It Worse." *Business Insider,* July 25, 2023. https://www.businessinsider.com/google-water-use-soaring-ai-make-it-worse-data-centers-2023-7.

Lee, Philip. "Naver Opens Massive New Data Center in Sejong City." *The Pickool,* November 6, 2023, https://www.thepickool.com/naver-opens-massive-new-data-center-in-sejong-city/.

Masanet, Eric, Arman Shehabi, Nuoa Lei, et al. "Recalibrating Global Data Center Energy-Use Estimates." *Science* 367, no. 6481 (2020): 984–86. https://doi.org/10.1126/science.aba3758.

Meta. "LULEÅ: Meta Data Centers." https://web.archive.org/web/20220127144603/https://datacenters.fb.com/wp-content/uploads/2021/12/Lulea.pdf.

Miller, Rich. "Inside Facebook's Blu-Ray Cold Storage Data Center." Data Center Frontier, July 1, 2015. https://datacenterfrontier.com/inside-facebooks-blu-ray-cold-storage-data-center/.

Mullaney, Thomas S. "Your Computer Is on Fire." In *Your Computer Is on Fire,* edited by Thomas S. Mullaney et al., 3–9. MIT Press, 2021.

Naver. "Data Center Gak." Home page. 2023. https://datacenter.navercorp.com/green/green-energy.

Naver Business Platform. *Teitŏ Sentŏ Kak.* Iro, 2015.

News Center Microsoft Latinoamérica. "Microsoft Chile anuncia que su dataemovableupará energía 100% renovable de AES Andes." Microsoft, April 6, 2022. https://news.microsoft.com/es-xl/microsoft-chile-anuncia-que-su-datacenter-ocupara-energia-100-renovable-de-aes-andes/.

Nowviskie, Bethany. "Digital Humanities in the Anthropocene." *Digital Scholarship in the Humanities* 30, suppl. 1 (2015): i4–i15. https://doi.org/10.1093/llc/fqv015.

Oxford English Dictionary. "Big Data." 2023. https://www.oed.com/dictionary/big-data_n.

Pak, Sanghyŏn. "Kungnip Chungang Tosŏgwan, P'yŏngch'ang Munhŏn Pojon'gwan sŏlgyean 'Muhan ŭi kil' sŏnjŏng." *Yŏnhap nyusŭ,* August 5, 2021, https://www.yna.co.kr/view/AKR20210805077000005.

Pak, Sŏngu. "Tijitŏl Syŏttaun Taeung Tallattŏn Iyu Nŭn . . . Teitŏ Sentŏ 1 Cho Ssŭn Neibŏ, Munŏbal Hwakchangman K'ak'ao." *Chosun Biz,* October 19, 2022. https://biz.chosun.com/it-science/ict/2022/10/19/6NCZQGNEE5AAHKSE753JXEAMHE/.

Parekh, Urvi. "Achieving Our Goal: 100% Renewable Energy for Our Global Operations." *Tech at Meta,* April 14, 2021. https://tech.facebook.com/engineering/2021/4/renewable-energy/.

PassMark Software. *Hard Drive Benchmarks.* 2024. https://www.harddrivebenchmark.net/.

Pieke, Frank N. "The Genealogical Mentality in Modern China." *The Journal of Asian Studies* 62, no. 1 (February 2003): 113, https://doi.org/10.2307/3096137.

Reinsel, David, John Gantz, and John Rydning. *Data Age 2025: The Digitization of the World from Edge to Core.* An IDC White Paper. International Data Corporation, Seagate, 2018. https://www.seagate.com/files/www-content/our-story/trends/files/idc-seagate-dataage-whitepaper.pdf.

Richardson, Tim. "Amazon: Our Carbon Footprint Went up 19% Last Year but We Grew Even More Than That, So 'Carbon Intensity' Is Down." *The Register,* July 1, 2021. https://www.theregister.com/2021/07/01/amazon_carbon_footprint/.

Risam, Roopika, and Alex Gil. "Introduction: The Questions of Minimal Computing." *Digital Humanities Quarterly* 16, no. 2 (2022). https://www.digitalhumanities.org/dhq/vol/16/2/000646/000646.html.

Robinson, Dan. "Amazon Adds 2.7 Gigawatts of Renewable Energy to Its Operations." *The Register,* September 21, 2022. https://www.theregister.com/2022/09/21/amazon_27gw_renewable_energy/.

Salter, Jim. "Microsoft's Project Silica Offers Robust Thousand-Year Storage." *Ars Technica,* November 7, 2019. https://arstechnica.com/gadgets/2019/11/microsofts-project-silica-offers-robust-thousand-year-storage/.

Schmid, Patrick, and Achim Roos. "Intel X25-E Walks All over the Competition." *Tom's Hardware,* February 27, 2009. https://www.tomshardware.com/reviews/intel-x25-e-ssd,2158.html.

Schmid, Patrick, and Achim Roos. "Tom's Hardware 2008 Hard Drive Guide." *Tom's Hardware,* November 24, 2008. https://www.tomshardware.com/reviews/hdd-terabyte-1tb,2077.html.

Shapiro, Gilbert, John Markoff, and Silvio R. Duncan Baretta. "The Selective Transmission of Historical Documents: The Case of the Parish Cahiers of 1789." *Histoire & Mesure* 2, no. 3/4 (1987): 115–72, https://doi.org/10.3406/HISM.1987.1328.

Smil, Vaclav. *Energy and Civilization: A History.* MIT Press, 2018.

Statista Research Department. "Electricity Generation Worldwide from 1990 to 2022." *Statista,* January 16, 2024, https://www.statista.com/statistics/270281/electricity-generation-worldwide/.

Statista Research Department. "Installed Base of Personal Computers (PCs) Worldwide from 2013 to 2019." *Statista,* September 15, 2016, https://www.statista.com/statistics/610271/worldwide-personal-computers-installed-base/.

Statista Research Department. "Number of Smartphone Subscriptions Worldwide from 2016 to 2021, with Forecasts from 2023 to 2028." *Statista*, December 4, 2023, https://www.statista.com/statistics/330695/number-of-smartphone-users-worldwide/.

Synergy Research Group. "As Quarterly Cloud Spending Jumps to over $50B, Microsoft Looms Larger in Amazon's Rear Mirror." February 3, 2022. https://www.srgresearch.com/articles/as-quarterly-cloud-spending-jumps-to-over-50b-microsoft-looms-larger-in-amazons-rear-mirror.

Taylor, Petroc. "Volume of Data/Information Created, Captured, Copied, and Consumed Worldwide from 2010 to 2020, with Forecasts from 2021 to 2025." *Statista*, November 16, 2023. https://www.statista.com/statistics/871513/worldwide-data-created/.

Tweed, Katherine. "Electricity Use Could Soar as Global Middle Class Embraces Air Conditioning." *IEEE Spectrum*, May 4, 2015. https://spectrum.ieee.org/electricity-consumption-could-soar-as-global-middle-class-embraces-ac.

Ung, Gordon Mah. "Debunked: Your SSD Won't Lose Data If Left Unplugged After All." *PCWorld*, May 21, 2015. https://www.pcworld.com/article/427602/debunked-your-ssd-wont-lose-data-if-left-unplugged-after-all.html.

Velkova, Julia. "Thermopolitics of Data: Cloud Infrastructures and Energy Futures." *Cultural Studies* 35, no. 4–5 (2021): 663–83. https://doi.org/10.1080/09502386.2021.1895243.

Vidal, John. "'Tsunami of Data' Could Consume One Fifth of Global Electricity by 2025." *The Guardian*, December 11, 2017. https://www.theguardian.com/environment/2017/dec/11/tsunami-of-data-could-consume-fifth-global-electricity-by-2025.

Walter, Chip. "Kryder's Law." *Scientific American*, August 1, 2005, https://www.scientificamerican.com/article/kryders-law/.

Xue, Yujie. "Climate Change: China's Data Centres and Telecoms Networks in Beijing's Sights as Key Targets for Decarbonisation." *South China Morning Post*, October 13, 2022. https://www.scmp.com/business/article/3195779/climate-change-chinas-data-centres-and-telecoms-networks-beijings-sights.

Yu Sŏkchae. "Tijit'ŏl K'ont'aench'ŭ 1 Ŏk Kŏn Kŏmsae Kungnip Chungang Tosŏgwan Yubikwŏt'ŏsŭ Pyŏnsin." *Chosun Ilbo*, December 29, 2008. https://www.chosun.com/site/data/html_dir/2008/12/28/2008122800686.html.

PART II][Chapter 8

Alternative Infrastructures for Digital Equity
Community-Based Internet Access

ALEX WERMER-COLAN, GRANT WYTHOFF, ALLAN GOMEZ, AND DEVREN WASHINGTON

Infrastructure as a Commons

Crowding Philadelphia's rooftops and building facades are row after row of south-facing satellite dishes, comically placed in questionable locations and serving equally questionable purposes. These receivers exemplify the atomization of residential housing for media consumption, with each household paying for its own exclusive gateway to a satellite orbiting the Earth. Satellite dishes aren't the only things here that are built on principles of exclusion and separation.

The highly entangled backsides of Philly rowhouses tell the cabled version of this story: each residence receives its own connection to a wired telephone or Internet service provider (ISP), while the rights (and tools) needed to cut through sidewalks and fences to install copper or fiber-optic cables remain prohibitively expensive. Only the wealthiest telecom companies and ISPs can afford to build this vital infrastructure. The tortured history behind the current telecommunications landscape in Philadelphia suggests that most of these for-profit companies never got the memo about sharing economies in the "City of Brotherly Love."

These dominant models of telecommunication infrastructure create needlessly exclusive realms of information access. But in the neighborhoods of Kensington and Fairhill in North Philadelphia, a growing group of over one hundred volunteers and five staff members are experimenting with a new model. These community organizers, technologists, librarians, scholars, students, and local residents work for a project called Philly Community Wireless (PCW), a collective social impact initiative consisting of allied community organizations, nonprofits, libraries, and universities in Philadelphia.[1] PCW focuses on expanding access to free, net-neutral broadband across many of the most redlined and underserved districts of the city, starting in predominantly Latinx North Philadelphia communities (Philly Community Wireless, "About").

Born of the pandemic and the need that it exacerbated for broadband access and digital equity, PCW has sought to adapt alternative models for bridging the digital divide, specifically community Wi-Fi initiatives in regional cities similar to Philadelphia, the poorest big city in the United States (Jones and Duchneskie).[2] The authors of this chapter—with backgrounds in community technology and digital humanities (DH)—are cofounders, staff, and Board members of PCW. In this chapter, we step back from our on-the-ground experiences over the last five years to explore in a broader context the potential of community-controlled broadband networks and the underlying wireless mesh network technology. In doing so, we offer a framework for critical praxis in DH, academic scholarship, and social justice work.

Philly Community Wireless and the Stakes of Community Broadband

For the past five years, PCW has been building a wireless mesh network that now provides high-speed internet access to tens of thousands of unique client devices and thousands of individuals in North Philadelphia. PCW's network, each node of which serves not one but many, is an infrastructure built on the existing architecture of the city. Wireless mesh networks involve a distributed system of routers and antennas that allow a single source of bandwidth to be shared among a broader group of users, with little cost required for a sustained connection. Every new mesh node becomes an integrated part of the same network. Mesh technology allows a single building, owned and used by residents, businesses, or community organizations, to leverage its real estate for broadcasting Wi-Fi to its nearby neighbors, who can host devices that relay, extend, and fold the signal into spaces (blocks, alleyways, residences, public spaces) where internet would not otherwise be accessible. In mesh networks, every rooftop that houses an access point provides Wi-Fi to the building's residents, as well as to neighboring buildings and passersby on the street. And with medium- to long-range connectivity, rooftop relays can bridge access across neighborhoods, eventually extending the mesh across vast swaths of the city.

PCW's mesh technology enables community members to control the distribution of bandwidth around their neighborhood while the mesh maintains cohesion between Wi-Fi nodes. For users who previously either did not have access to the internet or had to resort to costly cellular and broadband provider plans, the ideal result is continuous connectivity, both indoors and outdoors, to an open-access, public Wi-Fi network across a large geographic area. Besides the ability to move freely through one's neighborhood without a paid internet connection, there are other advantages. Unlike 4G and 5G cellular data networks, for example, fixed wireless and Wi-Fi mesh technologies allow for an alternative model of locally owned and managed access, access that is not tied to any one device (smartphones or hot spots) or service contract with a for-profit company. The primary aim of PCW is the construction of baseline connectivity across the cityscape on a very different model than that of for-profit ISPs parceling out each customer's Wi-Fi access and data usage separately. The goal is for no one to be left behind in the effort to bridge the digital divide, including people without permanent habitation. Collaborating with community organizations of several kinds—including informal neighborhood groups, community development corporations (CDCs), public libraries, city-owned properties and communal garden spaces, rehab institutions and churches, as well as private businesses, wealthy property owners, and real estate companies—PCW adapts emerging internet technologies, especially for fixed wireless broadband and Wi-Fi, to build an alternative model of internet provision in U.S. cities at a time when government and private institutions cannot singlehandedly ensure access to the infrastructures that communities need to live, work, and prosper.

In this discussion, we situate contemporary conversations on community technology within an international history of community-focused approaches to communications infrastructure. That context helps clarify differences between community mesh networks in Philadelphia (the focus of our own work) and other cities like Detroit, Pittsburgh, and New York City, while bringing into relief the broader U.S. political bias toward ensuring internet access for "unserved" rural areas over "underserved" urban localities.[3] Our question is how infrastructure cooperatives that provide community mesh networks for urban environments can facilitate collaboration between neighbors while fostering community consensus on network architecture, organizational structure, and the geographic redistribution of digital resources. Ultimately, our theory of change is that access leads to adoption in a fuller social sense: if one empowers the communities most affected by the biases and harms of the tech industry to own and control last-mile internet infrastructure in their neighborhood, such communities will be equipped not just to use broadband technology, but to advocate for better outcomes in the way that such technologies are utilized, governed, and regulated.

In addition, for the DH scholarly community, our question is: How can PCW and projects like it become platforms for research and teaching in the service of activism? By discussing the methodologies of PCW and related community mesh networks, we offer a vision of DH that adopts the activist orientation of multistakeholder projects like Saving Ukrainian Cultural Heritage Online (SUCHO) and the Nimble Tents Toolkit, while enabling community-owned infrastructures to serve as vehicles for critical digital literacies and community-based research on the changing nature of the digital divide.[4] Our final concern is how academics can support the voices and visions of community organizations neighboring their educational institutions. We explore this question by focusing on the reparative work made possible through collaborative efforts to change and expand existing digital infrastructure in the areas where we work, study, and live.

From Infrastructure to Discourse: Lessons from Latin America

In the early 2000s, Allan Gomez, longtime volunteer and staff member at PCW—as well as one of this chapter's authors—participated in several exchanges with community organizers in Latin America who were eagerly debating the implications of new communications infrastructures for social justice work. These instances of activist engagement during the early days of Web 2.0 provided valuable lessons central to PCW's approach to deploying community technology for accessibility and inclusion in what the U.S. federal and state governments today consider "underserved" (but not "unserved") urban areas in the nation.

For Gomez, a crucial moment occurred in Ecuador during the fall of 2002 in a crowded room where international protesters, Indigenous organizers, and a ragtag crew of camera-wielding activists involved in the early stages of a movement for independent journalism debated matters of strategy a few days before protests against the U.S.-led Free Trade Area of the Americas Summit (FTAA) in Quito. This coalescence of several media movements—environmental, Indigenous rights, antiglobalization, and democratic—provided ample opportunity for lively debate about purposes, goals, and tactics aimed at shutting down the undemocratic FTAA, an international conference that sought to "all but force member nations to allow privatization of vital social services including water, energy, education, healthcare, and postal and financial services, whether or not their electorates support it" (Coen). Of all the lessons learned during those heated hours of organizing resistance, what stood out was that nothing should be taken for granted when it comes to building infrastructure aimed at empowering people to determine their future.

Advocates for the newly minted independent media centers of the time were eager to carry on the gospel that a digitized form of journalism and storytelling, distributed throughout the internet, was inherently democratic. And who could argue to the contrary, especially among so-called summit-hoppers attempting to reach transnational audiences? The internet allowed anyone to share their direct and unfiltered viewpoints with the world, in contrast to the status quo of curated, one-way, inaccurate news produced through traditional journalism—the Fourth Estate. While not wrong per se, this optimistic prediction about the internet was also incomplete.

Indigenous leaders pointed out the absurdity of building a democratic platform for their communities on top of what was at the time a new and hardly prevalent (or accessible) technology. These Indigenous leaders persuasively stated that to be more than a tokenized part of this emerging infrastructure, their communities would need to undergo a cultural change, adopt foreign tools, master other languages, develop new skills, and model themselves after the ambitions of the Global North. The digital dream of independent media stood in stark contrast to the reality that most people lacked computers and internet access. (Cell phones were nonexistent in rural areas, and cybercafés, while cheap for foreigners, were costly for locals.) Perhaps the Indigenous leaders' biggest objection was to yet another technology imposed from the outside, one that they could neither own nor control. As Syed Ishtiaque Ahmed, Nusrat Jahan Mim, and Steven J. Jackson point out in their study of infrastructures in a different postcolonial context, Bangladesh, developmentalist discourses assume that conditions in the Global North are the end goal for everyone: "Basic assumptions around infrastructure and development 'failures' may reflect the perspectives of northern actors and institutions, and generate new rounds of 'development projects' that fail to engage post-colonial conditions and experiences (though they may be good at sustaining the institutional machinery of the development industry)" (Ahmed, Mim, and Jackson, 439).

Unfortunately, the wake-up call from Indigenous leaders at the gatherings in Ecuador went unheeded. Despite the best of intentions, media activists unintentionally neglected inclusivity. Inclusive intentions do not matter if the tools used for implementation result in exclusion. From this incident in Ecuador, and from other global stories of groups organizing control of their communications infrastructures, including, for example, community-controlled radio stations in Oaxaca, c.2006 (Zepeda), PCW learned that developing community broadband networks in segregated urban areas demands a collective, intentional practice to ensure long-standing inclusivity and agency for communities. This is the lesson that PCW brought home to apply in its own local community.

Community Technology and the Potential of Mesh Networks

Despite the seeming ubiquity and availability of the internet in the twenty-first century, exclusions still permeate our society and media. Even in prosperous cities across the United States, many cannot afford internet access even as it has become increasingly essential (Sanchez): finding employment, shopping, getting directions, reading the news, and even receiving an education or health care are now nearly impossible without being online. While there is a growing consensus that internet access is a human right (Kravets), the predominant approaches to bridging the digital divide, especially in the United States, remain privatized.

Even when ISPs advertise the sharing of internet connectivity in their marketing materials, they are driven by profit motives. For instance, Amazon initiated a new program called Sidewalk, which allows smart devices like Ring cameras and Echo speakers, if they lose their main connection, to connect automatically to the internet through a neighbor's Amazon-made device (Nield). Yet another example of for-profit internet sharing can be seen in companies that started in the last decade to sell wireless mesh networking equipment so neighbors can share bandwidth between their Internet of Things devices for the purposes of exchanging cryptocurrency tokens (Roose). When vendor lock-in, crypto mining, or data capture is the goal of getting more people online, ISPs remain in an extractive relationship with the communities that need to use them. The companies and people who manage those resources, rather than the communities that end up using them, inevitably determine their purpose and future.

But in recent years, community technology projects in cities across the country, including in the northeast region such networks as PCW, New York City's NYC Mesh, Pittsburgh's Community Internet Solutions (formerly known as MetaMesh), the Detroit Community Technology Project's Equitable Internet Initiative, and Baltimore's Project Waves, have offered alternative models for building infrastructure to ensure widespread internet access and adoption within the framework of community engagement and agency.[5] As part of its growth, PCW has developed relationships with and learned from regional organizations like NYC Mesh, Community Internet Solutions, the Detroit Community Technology Project, Community Tech NY, and Hunts Point Community Network.[6] These community internet networks take various approaches to the technologies that they use, the physical and social landscapes in which they operate, and the infrastructure available to them politically and technically.[7] Each of the organizations behind these networks approaches the role of volunteers and community involvement differently, attempting to balance the goal of decentralization with the demands of running an organization. But across the board, all these organizations document their processes and create inclusive spaces for discussion, yielding a wealth of resources that other community networks can learn from and expand upon.[8]

The shared conversation among these organizations centers on the *community technology* movement, which attempts to merge vital infrastructures with models of inclusivity. As defined by Diana Nucera, a Detroit-based organizer influential in the field, community technology adopts a "principled approach to technology that is grounded in the struggle for a more just digital ecosystem, placing value on equity, participation, common ownership and sustainability" (Nucera, 15). The Community Technology Collective identifies four core principles of community technology:

- *Access:* Digital justice ensures that all members of our community have equal access to media and technology, as both producers and consumers.
- *Participation:* Digital justice prioritizes the participation of people who have been traditionally excluded from and attacked by media and technology.
- *Common ownership:* Digital justice fuels the creation of knowledge, tools, and technologies that are free and shared openly with the public.
- *Healthy communities:* Digital justice provides spaces through which people can investigate community problems, generate solutions, create media, and organize together.

To apply these principles, many community technology organizations employ emerging technologies for wireless mesh networks. While the network structure of most traditional ISPs is centralized and one-way (the ISP is a monolithic hub that transmits network traffic to all users on the receiving end), the network architecture enabled by mesh technology is one in which every access point can connect seamlessly with every other access point within range. Mesh networks enable most nodes on the network to remain operational even when individual hubs go out of service, as well as for nodes to speak to one another without sending signals through a centralized relay station. The technical structure of mesh infrastructure (interconnected, resilient) reflects the social structures that they seek to amplify (democratic, participatory, decentralized). While mesh networking technology (or any technology) is not inherently democratic, and while internet infrastructure is not inherently more liberating just because elements of its architecture can be decentralized (Galloway), mesh networking technologies are nevertheless well designed (i.e., affordable and lightweight) for the growth of community-based and democratically controlled communication technology. PCW's mechanisms for building its network turn the affordances of the physical hardware into a foundation for organizing communities from passive recipients into active agents participating in the growth of critical city infrastructure.

Wireless mesh networks are particularly useful for communities looking to build networks across large geographic spaces with relatively limited cost and labor. In the past few years, the cost of mesh Wi-Fi hardware, including outdoor-rated equipment, has come down from tens of thousands to hundreds of dollars.[9] Network management software has become more user-friendly, diminishing the training time necessary to onboard community members to manage the network. These increasingly accessible methods are, in turn, overcoming many of the legal and practical challenges to right-of-way access for laying down copper or fiber-optic cable. All of this is done by adapting preexisting infrastructure: mounting wireless radios on derelict satellite dishes, television antennas, defunct chimneys, and drainpipes on private houses and community-based organizations' real estate (see Figure 8.1). The lightweight, portable, affordable, and flexible nature of wireless mesh devices—for example, their ability to mount an access point on a fence, then later relocate it to a nearby, better location with ease—is part of what enables the technology to serve as the basis for community organizing and neighborhood redevelopment.

As has been discussed, the technology behind mesh networks can be deployed to very different ends, from crypto-mining to surveillance; mesh technologies are increasingly becoming a standard of consumer home internet products for residents and businesses. Mesh is not the be-all and end-all, but rather a means to a particular end. A community group will never be able to own a satellite, for example, but community members can set up a generator to transmit FM radio in a park or install a Wi-Fi antenna on the exterior of their homes to share their internet with neighbors. Communities should have as many options as possible, including alternatives to mainstream technologies. We live in a society that demands access to the internet to flourish. For that reason, PCW envisions infrastructure as a commons: we believe that broadband access should be made freely available to all.

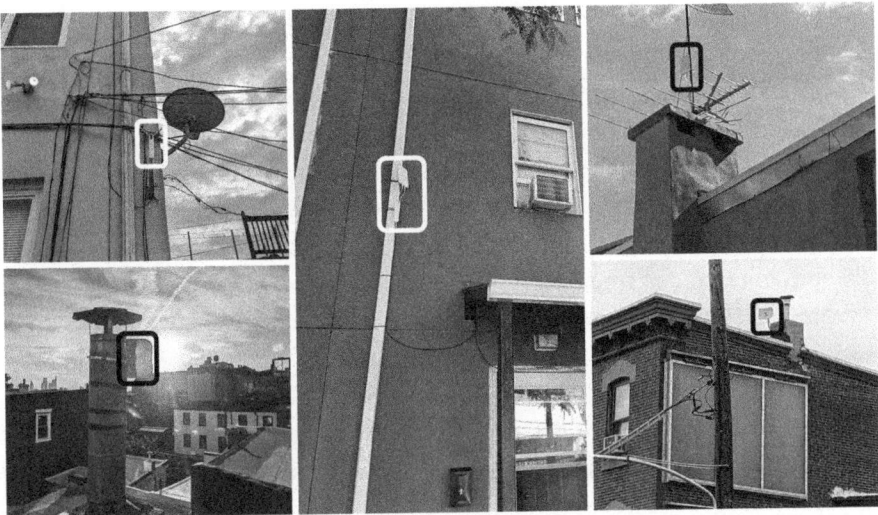

Figure 8.1. Views of PCW antenna installations (outlined in rectangles) in North Philadelphia. Photos by Alex Wermer-Colan.

Mesh Networks in Practice: The Case of Philadelphia

In Philadelphia, PCW found that the primary challenge in constructing alternative infrastructures concerns problems of organizing rather than of physical hardware, or what Lisa Parks, writing about media infrastructure, calls "stuff you can kick." Community organization is the central, yet often overlooked, critical infrastructure. Consider that to connect with PCW, neighborhood residents need to open their doors, allow strangers on their roofs, tolerate holes being drilled in their walls, provide electricity for the antennas, and remain open to recurring visits for tweaks and improvements. Even during the height of the Covid-19 pandemic, during the early years of the PCW project, many residents in North Philadelphia were willing to allow volunteers into their homes to install antennas on the facades of their houses to share internet with their neighbors. For PCW's organizers, one of the earliest measures of success has been the trust and eagerness with which community members provided in-kind resources, such as roof access, electrical power, and a kind welcome to strangers—all to improve the network in a way that benefited their neighbors as much as themselves. In this sense, community networks resonate with what Eduard Arriaga—in his scholarship on how Afro-Latin communities challenge algorithmic determinism—defines as "expanded infrastructure," which includes "how human beings, their cultural assets and knowledge, and their existing social and cultural structures connect with digital tools and digital networks that might already be in place."[10]

By listening to the needs of community neighbors, antenna hosts, network users, and stakeholders, PCW learned that merely focusing on internet access is not enough. Through dialogue, participants in the PCW network provided invaluable insights about how the organization can be more adaptive in designing its network to meet community needs. These participants also promoted engagement with other public resources, like the Free Library of Philadelphia's and the Temple University Digital Equity Center's programs for device distribution and digital literacy training. Central to PCW's approach is collaboration with local community organizations embedded in each neighborhood. Such collaboration enables PCW to provide public hot spots at each community center, using their real estate to broadcast Wi-Fi into the surrounding blocks—establishing a beachhead in the neighborhood that encourages residents and businesses to help spread the network and word of its availability.[11]

Organizations like PCW testify to the resilience of communities in taking control of their own neighborhood infrastructure. But these community projects would not even be necessary in the first place if municipal broadband were legal throughout the United States. As in the case of fifteen other states, Pennsylvania restricts government agencies from providing Wi-Fi for a fee—supposedly because it would interfere with the competitive marketplace. These restrictions have stymied efforts in Philadelphia to address digital inequity (Cooper). Although the City of Philadelphia was one of the earliest adopters of wireless technologies for municipal broadband, developing an expensive plan as early as 2004, private interests quickly complicated those efforts; as a result, free, publicly available wireless internet withered on the vine (Breitbart, Lakshmipathy, and Meinrath).[12] To fill the vacuum left where municipal broadband and public Wi-Fi should be widely available, PCW works with a local, independent wireless Internet service provider (WISP) called PhillyWisper, which donates the bandwidth, as well as labor, training, and hardware, that serves as a foundation for the public mesh network that PCW is building in North Philadelphia.[13]

At the street level, PCW's effort to build a wireless mesh network requires the collaboration of multiple residents and community organizations on every block. PCW's network coverage map shows how the rooftop hubs on various blocks can share Wi-Fi with surrounding areas by broadcasting into parks, public spaces, and neighboring houses (see Figure 8.2). A Wi-Fi access point in one rowhouse can provide Wi-Fi to neighboring rowhouses on each side. Additional mesh access points can extend Wi-Fi networks from these community hubs, allowing the shared network to permeate the neighborhood.

PCW's collaboration with community gardens offers a particularly valuable model for how a community Wi-Fi network can serve as a vehicle for delivering not just internet connectivity, but holistic digital equity services. PCW partnered with organizations like Temple University's Digital Equity Center, the Free Library of Philadelphia, and Holobiont Lab (a climate-change community impact project) to provide clients, such as members and visitors to the Cesar Andreú Iglesias Community Garden, and Norris Square Neighborhood Project's Las Parcelas and Villa Africána Colobó Gardens, with access to free laptops and desktop computers, tablets, office furniture, solar panels, and air quality monitoring sensors.[14] Besides

Figure 8.2. PCW network map showing the network's distribution of Wi-Fi coverage as of February 2025. Underlying map layers copyright OpenStreetMap; map of PCW network created by Felipe Valdez using Mapbox. The most recent version is viewable at https://phillycommunitywireless.org/networkmap/.

working with these community gardens to build physical and digital resources into their public spaces, PCW and its partners included multilingual digital literacy training for community members and organizers at multiple stages of the process, from initial outreach to installation and maintenance.

PCW's installations of outdoor Wi-Fi in community gardens help make the case for the importance of community land ownership and control of public spaces in developing robust community digital equity resources. Cesar Andreú Iglesias Community Garden started in 2012, when the Philly Socialists and the local community claimed unused land to transform it into a collective garden and public park space to fight for their right to use the vacant land during a real estate struggle in North Philadelphia, an area rapidly changing due to rising gentrification. It was in 2022 that PCW (providing Wi-Fi availability) and Holobiont Lab (installing solar panels on a shipping container) brought their support to Iglesias Gardens' ongoing efforts. Iglesias Gardens, shown in the photo in Figure 8.3, includes vegetable and flower gardens, rainwater collection systems, and a public event space called Accessibility Plaza. In the years following the installation of electricity and internet, the garden community continued to grow its infrastructure and organizing at the location. They fought for land rights on neighboring properties and installed more facilities like a performance stage, a compostable toilet, and an outdoor kitchen for hosting gatherings, cookouts, and performances.

PCW's installation at Iglesias Gardens exemplifies what becomes possible through the collective use of technology. By installing a directional Wi-Fi antenna on a neighbor's rooftop, PCW relayed signals from its source tower—operated by PhillyWisper—to the street level, spreading free, open Wi-Fi signals throughout the block (see Figure 8.3). PCW also set up a private Wi-Fi network for the owner of a the new rowhouse, built in the modern style of all new development in a recent wave of gentrification. In this way, community internet projects like PCW can use a single real estate development on a city block to share a public digital utility with the surrounding community, ensuring that public spaces (especially green spaces) have access to affordable "smart" technology and supporting organizations like Iglesias Gardens in their fight for land rights. Recently, Iglesias Gardens has worked with Philly City Council to raise the funds to purchase liens on the land, helping to ensure that more of the land they occupy will not be converted to commercial and residential properties by private developers (Conde).

Organizations like PCW depend on, and facilitate, the success of autonomous, collective organizations like Iglesias Gardens, whose long fight for land ownership in an increasingly gentrified zone of the city has also ensured that sustainable open internet access can be available in a community-controlled public space that would have otherwise been converted to private housing. Iglesias Gardens is also somewhat unique in Philadelphia because unlike most garden spaces, the lack of fencing around the property makes it physically open access. The work by the Iglesias collective to build the technological resources available in their space exemplifies the

Figure 8.3. Drone photo of Cesar Andreú Iglesias Community Gardens, with writing painted on the streets saying "NOT FOR SALE" and "NO ME VOY." Photo by Eli LaBan, June 9, 2023.

way that PCW hopes to contribute to the self-empowerment of local communities and spaces. By leveraging resources from initiatives like PCW, community organizations like Iglesias Gardens can reclaim power and ensure long-term, secondary benefits for the surrounding community. Communities can contribute to, volunteer for, and help maintain social safety nets that include not just community gardens, homeless shelters, and neighborhood cleanup and enrichment programs, but also more technology-oriented projects that grow digital infrastructure where community members live and work. Community organizations and property owners, as well as tenants, can host mesh antennas and help grow the network to share with neighbors. Everybody is welcome to volunteer with PCW to gain fun new experiences, learn about digital technology, and give back to their communities.

Appropriate Technologies and the Digital Humanities

PCW draws participants from many backgrounds, including digital humanists like those among the authors of this chapter who work for universities in the Philadelphia region. Project members help grow both the technical and social infrastructure of PCW by bringing to bear tools and techniques from DH research and teaching—for example, static website generators, digital mapping, project management methodologies, and pedagogical frameworks for training new volunteers and community members. These DH project members (faculty, librarians, and staff), as well as local students in classes at Temple, UPenn, Princeton, and Drexel who have partnered with PCW, are inspired by the activist orientation of such previously mentioned academic projects as SUCHO and the Nimble Tents Toolkit, sharing their dedication to networks of academic collaboration for addressing crises in real time.

However, DH members of PCW also recognize the limitations of some of the tools, techniques, and theories that they use in academic work when applied to this public infrastructure and purpose. The static site generators popular in DH, for example, proved to be a powerful and lightweight means of rapidly setting up a website for PCW in its early days. But the learning curve required to contribute to the PCW website or edit its content, involving a working knowledge of tools like the command line, GitHub, Hypertext Markup Language (HTML), Cascading Style Sheets (CSS), and markdown, created a barrier for volunteers unfamiliar with creating and maintaining static sites. More broadly, PCW members discussed making *minimal computing* the ethos of the project's approach to technology. In DH, "minimal computing" refers to "computing done under some set of significant constraints of hardware, software, education, network capacity, power, or other factors" (Minimal Computing). But minimal computing's emphasis on constraint seemed to reproduce the very deficit at the heart of the digital divide that PCW was trying to address, especially for users who did not want less than optimal, minimal resources (Wythoff). As Nabeel Siddiqui writes in a *Digital Humanities Quarterly* issue devoted to the topic, minimal computing champions tools that often create more problems than they solve (as in the case of markdown, which actually "requires a large infrastructure to support it and is far from minimal"). Siddiqui goes on to argue that minimal computing "leads to few of the supposed benefits advocates profess, and in many cases, worsens inequalities" (Siddiqui). Minimal computing might work as a model for some forms of software development and DH pedagogy, but when it comes to telecommunications infrastructure, a more complicated paradigm is needed.

An approach that was more familiar to the larger group of activists and organizers (and not just DH participants) among PCW's volunteers, and that came to the fore as a closer fit for the project's goals, was *appropriate technology*. A movement that coincided with "the end of the Vietnam War, a major energy crisis, and the first years of the environmental movement," appropriate technology was a loosely defined philosophy of the 1970s that emphasized the situatedness of any given technology within a community and its environment (Pursell, 629). As Mario Pansera and Mariano Fressoli observe, "In essence, proponents of appropriate technology sought a more situated, environmentally concerned and socially just set of design and operational principles for diverse technology choices by involving local communities. Appropriate technology was a reaction against wholly blueprint developments involving imported Western technologies, whose industrial contexts were ill-suited to the poor, and ended up lying idle for lack of supportive supplies, infrastructure, and relevant skills (388)." With its emphasis on community autonomy and the unique needs of specific communities, appropriate technology resonated far more for the members of PCW than did minimal computing, with its emphasis on constraint and the use of specific tools regardless of context.

These are just a few examples of how working on a community technology project has encouraged the digital humanists within PCW to rethink some of the canonical concepts in their field. Going forward, we argue that more digital humanists should join community technology projects and organizations, especially those neighboring their places of employment and residence, not just to see how their theories hold up to the rigor of praxis, but also to ensure that their research and academic institutions help empower the local communities upon which they depend. For instance, ongoing efforts by DH practitioners at the universities neighboring PCW's service area to map its expanding network coverage involve a feedback loop between on-the-ground work to build the network and evolving DH methods used to visualize the city's infrastructure and the network's growth. DH projects focused on problems in technology, culture, and society rarely involve such practical immersion in the development—not just the study—of the infrastructures underlying those problems; nor do they often focus on the local communities that their academic institutions both depend on and affect through such processes as gentrification. Projects like PCW provide fertile soil for the development, refinement, and teaching of DH and cultural analytics methods that can map hidden, pervasive digital inequities and visualize how the internet works, so that we can change it.

Human Rights and the Internet

Given the proliferation of technology across all aspects of contemporary society, it is often assumed that because the infrastructure for high-speed broadband exists in cities and rural communities alike, its use must be ubiquitous in a "wealthy" nation like the United States. In reality, as with any resource vital to human development (e.g., water, food, housing, and medical care), access to the internet is tenuous in communities and areas of the United States where its infrastructure is deemed unprofitable (Vogels; Sandvig; Muller and Aguilar).

The internet should be a human right. That is because not just businesses but nonprofits, community organizations, government agencies, churches, and schools increasingly require the use of online platforms to access services and resources. And it is also because inequity in internet access and its underlying infrastructure not only correlates with, but helps lock in, broader social inequity by age, income, neighborhood, race, and other differences.

Capitalist society in the United States today ignores this fundamental right to internet access. Telecommunication conglomerates and ISPs, such as Comcast, Verizon, and AT&T, enjoy cozy relations with U.S. legislators and politicians at all levels of government, resulting in policies and legislation favorable for their profits. Through capital creation and extraction, private industry ultimately siphons resources out of poor and nonwhite communities and into the hands of a few wealthy families. This trend was presaged by Martin Luther King, Jr. in his 1968 "To Minister the Valley" speech, in which he said that the United States provides "socialism for the rich and rugged free enterprise capitalism for the poor" (King). Rugged, free-enterprise capitalism leads to the poorest communities being unable to afford expensive internet plans, while socialism for the rich leads to the cozy relationships that ISPs enjoy with elected officials and policymakers, whose gifts of tax breaks and other profit-driven measures serve to prop up a technological ecosystem that locks in the digital divide.

Community-controlled internet networks offer a unique alternative to deploy against the widening digital divide at a time when broadband technology, left to itself and its business proprietors, would otherwise automate and exacerbate the injustices that communities have always faced. By empowering communities to grow and maintain their own alternative infrastructures, projects and organizations like PCW shift communities' relationships to technology, and ultimately to themselves, opening the potential for new relations and distributions of power across the city.

NOTES

1. See https://phillycommunitywireless.org/. PCW is a fiscally sponsored project of the Movement Alliance Project, a 501c3 that provides crucial critical infrastructure, including legal, administrative, financial, and strategic support for local organizations working toward social change. Panjwani et al. define collective impact initiatives as "a group of community partners and/or organizations who work together to achieve a common goal. This collaborative action often aims to address disparities that emerge from existing social and economic differences within a community" (406–7).

2. This figure is according to the latest data from the US Census Bureau measuring poverty rates in the nation's ten most populous cities. In 2023, 20.3 percent of Philadelphia's residents were below the poverty threshold, which the Bureau defines by size and age of family.

3. The Federal Broadband Equity, Access, and Deployment (BEAD) program "prioritizes unserved locations that have no internet access or that only have access under 25/3 Mbps and underserved locations only have access under 100/20 Mbps" (BroadbandUSA). As a result of this short-sighted categorization, most urban areas of the United States are being excluded from receiving federal funding to expand broadband access.

4. See https://www.sucho.org/; https://nimbletents.github.io/.

5. See https://www.nycmesh.net/; https://detroitcommunitytech.org/eii; https://www.projectwaves.net/.

6. See https://www.communitytechny.org/ and https://huntspoint.nyc/ NYC Mesh is the most famous and largest mesh network in North America; it also presents a model for decentralized governance and technical infrastructure that serves as an inspiration to many new mesh networks cropping up all over North America, from Tucson, Arizona, to Vancouver, Canada. Community Wi-Fi projects in other parts of the world are largely beyond the scope of this chapter, but much is to be learned about how these projects can be adapted to a U.S. context. Prime examples of large mesh networks outside the United States are Freifunk in Germany, Guifi.net in Spain, AlterMundi in Argentina, and Mpumalanga Mesh in South Africa.

7. The United States presents unique networking challenges; a close study of the mesh networks in the Northeast requires an in-depth discussion that we can only skim in this chapter.

8. NYC Mesh's website contains a plethora of information about their process and strategy. Freifunk's Github also contains open-source code used by many other mesh projects. The textbook *Wireless Networking in the Developing World: A Practical Guide to Planning and Building Low-Cost Telecommunications Infrastructure* (edited by Jane Butler) is one of the preeminent publications in this field.

9. Most community Wi-Fi projects use hardware manufactured and sold by Ubiquiti, a private company based in New York City (https://ui.com), which provides an industry-standard product line for wireless mesh networks. But NYC Mesh increasingly has turned to using a hardware line made by MikroTik (https://mikrotik.com/), a Latvian company that offers more freedom to use open-source software.

10. Arriaga continues that "it is important to keep in mind that many of these Afro-descendant communities and networks in the Americas have existed for more than five hundred years. In that sense, the expanded vision of infrastructure constructed by Proyecto and C.N.O.A. taps into preexisting processes and relations, and so it is, itself, most usefully understood as a process that uses and reuses existing sociocultural and technological structures in order to construct more open and diverse conceptions of humanity."

11. For a list of PCW's local partner organizations, see https://phillycommunitywireless.org/about/people/.

12. Breitbart, Lakshmipathy, and Meinrath, "The Philadelphia Story, 2007."

13. See https://phillywisper.net/. We should note here that the City of Philadelphia's Digital Literacy Alliance has been a powerful supporter of PCW's efforts and countless other digital inclusion partners throughout the city, especially since the Covid-19 pandemic. See https://www.phila.gov/programs/digital-literacy-alliance/.

14. See https://lenfestcenter.temple.edu/Digital-Equality-Center; https://holobiontlab.org. The air quality monitoring sensors were installed for a community-based research project with Temple University associate professor of geography and urban studies Christina Rosan, entitled Acting on Air, in close collaboration with the Clean Air Council.

BIBLIOGRAPHY

Ahmed, Syed Ishtiaque, Nusrat Jahan Mim, and Steven J. Jackson. "Residual Mobilities: Infrastructural Displacement and Post-colonial Computing in Bangladesh." In *Proceedings of the 33rd Annual ACM Conference on Human Factors in Computing Systems, CHI '15*, 437–46. Association for Computing Machinery, 2015. https://doi.org/10.1145/2702123.2702573.

AlterMundi. Home page, 2024. https://altermundi.net/.

Arriaga, Eduard. "Epistemological Inclusion in the Digital Humanities: Expanded Infrastructure in Service-Oriented Universities and Community Organizations." In *People, Practice, Power: Digital Humanities Outside the Center*, edited by Anne McGrail, Angel David Nieves, and Siobhan Senier, University of Minnesota Press, 2022. https://dhdebates.gc.cuny.edu/read/people-practice-power/section/a763c995-2178-4113-a633-fb587b1ad5aa##ch12.

Breitbart, Joshua (author), Naveen Lakshmipathy (appendices), and Sascha D. Meinrath (editor). *The Philadelphia Story: Learning from a Municipal Wireless Pioneer*. New America Foundation, 2007. https://technical.ly/wp-content/uploads/2017/03/wireless-philadelphia-report-breitbart-et-al.pdf.

BroadbandUSA. "Broadband Equity, Access, and Deployment Program Overview," May 2022. https://broadbandusa.ntia.doc.gov/funding-programs/broadband-equity-access-and-deployment-bead-program#initialproposal.

Butler, Jane, ed. *Wireless Networking in the Developing World: A Practical Guide to Planning and Building Low-Cost Telecommunications Infrastructure*. 3rd ed. Wireless Network in the Developing World, 2013. https://wndw.net/download/WNDW_Standard.pdf.

Coen, Rachel. "The FTAA Is None of Your Business." Fairness and Accuracy in Reporting (FAIR), January 1, 2003. https://fair.org/extra/the-ftaa-is-none-of-your-business/.

Community Internet Solutions (Pittsburgh). Home page, n.d. https://web.archive.org/web/20240430042039/https://www.bringtheweb.org/.

Community Technology Collective (CTC). "CTC Principles," 2020. https://www.ctcollective.org/principles/.

Conde, Ximena. "Philadelphia Buys $1 Million in Liens to Protect Community Gardens from Sheriff's Sale." *Philadelphia Inquirer*, June 20, 2023. https://www.inquirer.com/news/philadelphia/philadelphia-community-gardens-sheriff-sale-liens-us-bank-20230620.html.

Cooper, Tyler. "Municipal Broadband 2023: 16 States Still Restrict Community Broadband." *BroadbandNow*, April 11, 2023. https://broadbandnow.com/report/municipal-broadband-roadblocks/.

Denham, Diana, and the C.A.S.A. Collective (eds.). *Teaching Rebellion: Stories from the Grassroots Mobilization in Oaxaca*. P.M. Press, 2008.

Detroit Community Technology Project. n.d. "Equitable Internet Initiative." https://detroitcommunitytech.org/eii.

Freifunk. n.d. "What Is Freifunk About?" https://freifunk.net/en/what-is-it-about/.

Freifunk. 2024. GitHub repository. https://github.com/freifunk.
Galloway, Alexander. "Introduction." In *Protocol: How Control Exists after Decentralization,* 2–28. MIT Press, 2004.
Guefi. "What Is Guefi.net?" July 6, 2009. https://guifi.net/en/what_is_guifinet/.
iNethi Technologies. 2024. "iNethi Background." https://www.inethi.org.za/about/.
Jones, Layla A., and John Duchneskie, "Philly Poverty Rate Sees Largest Drop in 10 Years, But We're Still the Poorest Big City." *Philadelphia Inquirer,* September 12, 2024. https://www.inquirer.com/politics/philadelphia/philadelphia-poverty-rate-decline-household-income-20240912.html.
King, Martin Luther, Jr. "To Minister to the Valley." Ministers Leadership Training Conference, March 31, 1968. Emory Special Collections and Archives. https://findingaids.library.emory.edu/documents/sclc1083/series11/subseries11.2/.
Kravets, David. "U.N. Report Declares Internet Access a Human Right." *Wired,* June 6, 2011. https://www.wired.com/2011/06/Internet-a-human-right/.
Minimal Computing. Home page, n.d. GO::DH, https://go-dh.github.io/mincomp/.
Muller, Charlie, and João Paulo de Vasconcelos Aguilar. "What Is the Digital Divide?" Internet Society (blog), March 3, 2022, https://www.internetsociety.org/blog/2022/03/what-is-the-digital-divide/.
Nield, David. "How Amazon Sidewalk Works—and Why You May Want to Turn It Off," *Wired,* May 11, 2021. https://www.wired.com/story/how-amazon-sidewalk-works/.
Nimble Tents Toolkit. Edited by Alex Gil, Francesca Giannetti, Vika Safrin, and Jason Jones. Home page, n.d. https://nimbletents.github.io/people/.
Nucera, Diana J. "Teaching Community Technology Handbook." Detroit Community Technology Project, 2016. https://detroitcommunitytech.org/teachcommtech.
NYC Mesh. Home page, n.d. https://www.nycmesh.net/.
Panjwani, Sonya, Taylor Graves-Boswell, Whitney R. Garney, et al. "Evaluating Collective Impact Initiatives: A Systematic Scoping Review." *American Journal of Evaluation* 44, no. 3 (September 2023): 406–23. https://doi.org/10.1177/10982140221130266.
Pansera, Mario, and Mariano Fressoli. "Innovation Without Growth: Frameworks for Understanding Technological Change in a Post-growth Era." *Organization* 28, no. 3 (May 1, 2021): 380–404. https://doi.org/10.1177/1350508420973631.
Parks, Lisa. "Stuff You Can Kick: Toward a Theory of Media Infrastructures," In *Between Humanities and the Digital,* edited by Patrik Svensson and David Theo Goldberg, 355–73. MIT Press, 2015. https://doi.org/10.7551/mitpress/9465.003.0031.
Philly Community Wireless (PCW). 2024. "About." https://phillycommunitywireless.org/about/.
Pursell, Carroll. "The Rise and Fall of the Appropriate Technology Movement in the United States, 1965–1985." *Technology and Culture* 34, no. 3 (1993): 629–37. https://doi.org/10.2307/3106707.
Roose, Kevin. "Maybe There's a Use for Cryptocurrency After All." *The New York Times,* February 6, 2022. https://www.nytimes.com/2022/02/06/technology/helium-cryptocurrency-uses.html.

Sanchez, Alvaro. "Toward Digital Inclusion: Broadband Access in the Third Federal Reserve District." *Cascade Focus* (Federal Reserve Bank of Philadelphia), March 2020. https://fraser.stlouisfed.org/title/cascade-focus-6890/toward-digital-inclusion-628554.

Sandvig, Christian. "Connection at Ewiiaapaayp Mountain: Indigenous Internet Infrastructure." In *Race After the Internet,* edited by Lisa Nakamura and Peter Chow-White, 168–200, Routledge, 2013.

Saving Ukrainian Cultural Heritage Online (SUCHO). Home page, n.d. https://www.sucho.org/.

Siddiqui, Nabeel. "Hidden in Plain-TeX: Investigating Minimal Computing Workflows." *Digital Humanities Quarterly* 16, no. 2 (2022). https://www.digitalhumanities.org/dhq/vol/16/2/000588/000588.html.

Vogels, Emily A. "Digital Divide Persists Even as Americans with Lower Incomes Make Gains in Tech Adoption," Pew Research Center (blog), June 22, 2021, https://www.pewresearch.org/short-reads/2021/06/22/digital-divide-persists-even-as-americans-with-lower-incomes-make-gains-in-tech-adoption/.

Wythoff, Grant. "Ensuring Minimal Computing Serves Maximal Connection." *Digital Humanities Quarterly* 16, no. 2 (2022). https://www.digitalhumanities.org/dhq/vol/16/2/000596/000596.html.

Zepeda, Manuel Garza. "The Popular Movement of Oaxaca, Ten Years Later." *Open Democracy.* December 8, 2016. https://opendemocracy.net/manuel-garza-zepeda/popular-movement-of-oaxaca-ten-years-later.

PART II][Chapter 9

Understanding Multilingualism in Digital Humanities Infrastructures

PAUL SPENCE

Multilingualism has long been recognized as one of the major social and cultural challenges facing the internet, where even today, only a small proportion of the world's 7,000 living languages has a meaningful presence despite considerable efforts on behalf of language communities, policy organizations, digital activists, global media companies, and language researchers alike. Reports such as "META-NET White Paper Series," "Digital Language Survival Kit" (Ceberio Berger et al.), *State of the Internet's Languages: Summary Report*, and "Disrupting Digital Monolingualism" (Spence)[1] have mapped out the status of digital language diversity from numerous perspectives, highlighting future challenges and proposing numerous agendas and road maps to achieve greater linguistic diversity in digital spaces. Projects driven by both language activists and language technology researchers in fields such as language documentation, endangered languages, machine translation, and speech technologies have made significant progress in some areas, such as keyboard support for languages (Esch et al.), grassroots natural language processing (NLP) support for research in African languages (Masakhane), community support for minority or endangered languages (Internet Languages),[2] and the ambitious European Language Equality (ELE) work toward a "roadmap for achieving full digital language equality in Europe."[3] Meanwhile, landmark publications such as Brenda Danet and Susan Herring's edited volume *Multilingualism on the Internet* (Danet and Herring), Laurent Vannini and Hervé Le Crosnier's edited book *Net.Lang: Towards the Multilingual Cyberspace* (Vannini and Le Crosnier), and Carmen Lee's book *Multilingualism Online* (Lee) have challenged anglophone bias in the study of language interactions online, proposing new conceptual frameworks for studying multilingualism across different media modes, digital writing systems, and sites for knowledge production.

While the broader topic of "cultural diversity" has emerged as a major area for debate and action for the digital humanities (DH) in recent years, DH has only lately started to engage in earnest with the theoretical and practical challenges that language diversity and multilingualism bring to the field. It is now over ten years since Domenico Fiormonte posed the question, "Is there a non-Anglo-American digital humanities (DH), and if so, what are its characteristics?" (Fiormonte, 59) and since then, we have seen an extensive and ongoing dialogue around the global identities of the field.[4] DH communities operating in languages other than English have become increasingly visible and vocal (whether as associations such as the francophone Humanistica or as research networks such as the hispanophone TTHub), while initiatives such as Global Outlook::Digital Humanities (GO::DH)[5] have shaped discussion around how we represent and cultivate geolinguistic diversity in DH through the "Around DH" community global mapping exercises,[6] "The Translation Toolkit,"[7] and minimal computing[8] endeavors (Risam and Gil).

Part of a wider reappraisal of DH's attention to inclusion and diversity, these activities have attempted to redress geolinguistic imbalance in the field from numerous perspectives such as decolonial/postcolonial studies, modern languages, biocultural diversity, and Indigenous pedagogies, but the multilingual nature of DH global interactions is still relatively marginalized in such discussions. In "Towards Language Sensitivity and Diversity in the Digital Humanities," Renata Brandão and I argued that DH currently lacks a coherent global/community strategy for how to engage with language diversity, and it is yet to achieve a mature understanding of how its research might contribute to key global multilingual challenges (Spence and Brandão). While we have seen a late proliferation of resources fostering language diversity in DH, including projects such as Programming Historian (PH) and OpenMethods (OM),[9] which we will examine later in the chapter, the anglophone core of the field still fails to engage meaningfully with the realities of carrying out DH research beyond English (plus, arguably, a handful of other globally dominant languages), whether that be in technical areas such as non-Latin script (NLS) interface support or social/cultural areas such as the geolinguistic dynamics of scholarly communications.

This chapter assesses how multilingualism is understood and practiced in global-leaning DH, in particular through DH sociotechnical infrastructure, and how this in turn shapes the way that the DH field contributes to global knowledge production. Drawing on DH literature and project case studies, I examine how those building or advocating for DH infrastructures understand multilingualism, as well as how this is framed within wider efforts to improve geocultural and linguistic diversity within the field.

Multilingualism Design in DH Infrastructures

A critical infrastructural perspective on multilingualism in the DH is fundamental, given the rapid spread of language technologies and research infrastructures dedicated to language resources in and around the DH community in the last few years. While many of these technical developments are welcome (and in some cases essential) for filling huge gaps in multilingual support, they have sometimes overshadowed the need for a more balanced (critical-social) evaluation of the relationship between DH, multilingualism, and infrastructure.

The field of critical infrastructure studies has expanded in recent years as a site for exploring global culture from the perspective of sociotechnical infrastructures, and this has increasingly been applied to study research infrastructures in the humanities and social sciences (and culture-based knowledge infrastructures more broadly) (Critical Infrastructure Studies.org).[10] Alan Liu appraises the potential role of DH in treating infrastructure as an object of study. Noting the "convergence between infrastructure and culture for humanistic critique," he argues that DH is "uniquely placed to interpret and critique culture at the level of infrastructure" (Liu, 2–4), and here I propose that this infrastructural critique urgently needs to be extended to the multilingual plane.

Whereas critical studies have devoted considerable attention to diversity in digital infrastructures in recent years (McPherson; Noble; Ricaurte), critical engagement with geocultural/multilingual diversity in infrastructure studies has been relatively scarce up to now. Work by authors such as P. P. Sneha has provided a welcome corrective to Global North–slanted studies on infrastructural diversity (Sneha), while David Wrisley (in exploring scholarly reactions in the Arab world to open scholarship models) has called for us to move beyond a narrow focus on European/North American models of knowledge creation and to engage more fully with "global infrastructural difference" (Wrisley).

Part of the overall challenge here is that in countries where digital infrastructure is designed and developed, multilingualism is roundly marginalized in public and academic discourse, and scholarly or professional fields addressing digital multilingualism are spread across radically diverse research networks with weak mutual engagement. The recent and considerable increase in attention to multilingualism and language diversity has led DH to contest the assumed normative nature of monolingualism, especially in countries with influential major digital industries such as the United States, where monolingual ideology is pervasive. Nevertheless, multilingual DH research in an anglophone context has attended to monolingual (and especially anglophone) bias in DH research infrastructures on largely practical grounds.

What different models exist for approaching multilingual challenges in DH research infrastructures, and what do they tell us about how the field views multilingualism? The next sections explore multilingual DH from four perspectives: (1) as access to language resources, (2) as literacy and ideation, (3) as translation, and (4) as tactical response.

Multilingualism and Language Resources

Common Language Resources and Technology Infrastructure (CLARIN) describes itself as the "research infrastructure for language as social and cultural data"[11] and is one of the best-known infrastructural responses to language diversity in digital scholarship. It primarily consists of datasets, tools, and services developed by corpus and computational linguists to support research in the humanities and social sciences and is based on the model of a "single sign-on online environment" to facilitate the use of "digital language resources and tools from all over Europe and beyond" (CLARIN, "About CLARIN").[12]

Rather than creating new resources centrally, CLARIN integrates access to resources generated either by national constituent organizations or elsewhere. Like other European research infrastructures, CLARIN operates in a highly multilingual space (the European research community) with a strong community identity, bolstered by regional policies and strategies that firmly promote multilingualism. While the general operating language (what Andreas Witt calls the "meta language") is English, each national participating consortium makes its own choice regarding operating language (the German and Croatian consortia use both English and their respective national languages, for example; CLARIN, "CLARIN Participating Consortia").[13] In creating a networked federation of language data, service, or knowledge expertise centers, its choices with regard to language support are to some extent influenced by a combination of national funding priorities and national/regional community research interests and priorities.

Despite having a clear European bias—in the top ten languages listed at its Virtual Language Observatory,[14] the only non-European ones are Japanese (eighth), Chinese (ninth), and, depending on our classification, Turkish (tenth)—CLARIN is open to all languages, and the CLARIN model has been an important reference point for research infrastructures beyond Europe, such as the South African Centre for Digital Language Resources (SADiLAR).[15] Although there is no formal strategic directive regarding language choice, CLARIN does have knowledge centers for individual languages, language families, or other language groupings, as well as Knowledge Centres for linguistic diversity and language documentation.[16] In addition, its Spanish K-Centre aims to offer language resources in four co-official languages of Spain: Castilian (Spanish), Catalan, Basque, and Galician, which points to growing awareness of the specific needs of lower-resourced languages.

The scale and variety of resources that CLARIN makes available clearly benefits language diversity in digital research practices in the humanities and social sciences, and it provides a valuable model for multilingual resource exchange and collaboration between language communities in DH.

Multilingual Literacy and Ideation

"Language is . . . always plural, always a place of difference," argues Polezzi in her introduction to the concept of "language indifference" (Polezzi). Normalized monolingualism tends to render multilingual work invisible or to treat it as something narrowly technical or culture-free, and the challenge for multilingual DH has been to expand debate and action beyond the very valuable work being carried out in linguistics and language technologies. Brandão and I, in the previously mentioned publication, explore frameworks that DH might use to foster "a languages-centric agenda for DH" (Spence and Brandão), while Aliz Horváth emphasizes "sensitivity to multilingualism as an overarching concept," which moreover needs to move DH infrastructure ideation beyond a Western perspective (Horváth, 1). Here, I wish to expand on how various DH infrastructural initiatives have fostered multilingual literacy and ideation.

Projects have attended to a broad range of multilingual challenges in DH in recent years, including historical and multilingual optical character recognition (OCR; Smith and Cordell), right-to-left language scripts,[17] and multilingual NLP. The Multilingual DH network launched by Quinn Dombrowski in 2019 has served to aggregate non-English language resources (NLP and OCR resources in more than twenty-five historic and modern languages or language families), and it is also one of many efforts to disseminate methods-based DH resources *across* languages.[18] Similarly, the New Languages for NLP workshops in 2021/2022 both trained researchers to create data and language models for low-resource (Quechua or Kannada) and historic languages (Classical Arabic or Old Chinese) and drew attention to the pitfalls of generic approaches (such as browser support limitations or the dangers of making assumptions about character/word/sentence boundaries when processing text in different languages). In publishing annotated linguistic datasets in an open repository, the project will also be one of a number of multilingual DH projects making a broader contribution to greater linguistic diversity in NLP.[19]

One major hurdle facing digital research is how to support NLS languages. At present, there are a number of barriers to carrying out DH research on NLS language content, including OCR quality, discovery in general information systems, and interface usability (Lee and Wagner). DH researchers working with East Asian and Middle Eastern languages and scripts have been among the most active in this area, and a series of workshops and projects have contributed to critical reviews of DH infrastructure through the NLS workshops organized by Cosima Wagner and Martin Lee in 2019 (Asef and Wagner), the German DH association's Special Interest Group Arbeitsgruppe Multilingual DH,[20] and the Disrupting Digital Knowledge Infrastructures (DDKI) collective. The DDKI's plan "to formulate a set of guidelines for universities, libraries, relevant organizations, and developers to consider towards a more language-inclusive digital environment, sensitive to the needs of scholars working with non-Latin scripts" will make an important contribution here (Horváth, 7). More recently, this collective published "Six User Personas for the Multilingual DH Community," which has helped lay the groundwork for further collective work (Horváth et al.).

Much attention in DH is currently focused on other globally dominant languages, such as Spanish, Arabic, and Chinese. There is far less consideration at present of endangered, minority, or heritage languages, and there is also a need for greater focus in DH on methods and resources operating in spoken forms of language, which is so crucial to languages with no (or little) textual tradition.

Whereas these projects draw attention to different *language-mediated mechanics,* projects such as Programming Historian (PH) and OpenMethods (OM) spotlight the challenges of modeling *multilingual workflows.* PH currently offers open-access, peer-reviewed tutorials in digital methods in four languages (English, Spanish, French, and Portuguese) within a geoculturally diverse editorial structure. In the PH multilingual workflow model, none of the operating languages are privileged (original creation can occur in any language), and this also attends to cultural/intralingual variation in the terms used (e.g., avoiding geographic dominance of any Spanish-speaking country/region in both the editorial board composition and policies). Translated content (and, where relevant, examples/data) are adapted to different linguistic and cultural contexts—and each language version has its own International Standard Serial Number (ISSN) and each PH lesson its own digital object identifier (DOI), which foregrounds linguistic or translation labor.

While PH offers a model for multilingual editing/resource *creation,* the OM platform presents a framework for multilingual research *dissemination.* OM functions both as a metablog for republishing micropublications, which present DH tools and methods, and as a forum for critical reflection on their use. One of the core aims of OM is to provide greater visibility to the application of DH methods and tools in different linguistic contexts, and its editorial team currently curates content in eleven languages. Each post is still available in its original language, but it is preceded by a short introduction in English, as part of an attempt to improve discovery of non-English content in DH research. The platform has recognized the barriers that this kind of multilingual approach has to overcome—such as lesser availability of non-English content, the slower workflow for non-English content due to fewer available editors who speak those languages, the preference for some language communities to speak English in a global setting, and the continued centralization of English as a lingua franca—but it does at least provide a provisional counterbalance to monolingual distortions within DH scholarly communications. While it was initially developed with European languages in mind, the platform has expanded to cover non-European languages.[21]

While, taken together, these initiatives still do not represent a holistic vision or strategy for multilingualism, they do characterize an attempt to build and strengthen multilingualism within the fabric of DH at various levels. In a study that I carried out as part of wider research into attitudes toward multilingual DH, one interviewee connected to the Digital Research Infrastructure for the Arts and Humanities (DARIAH)[22] highlighted the importance of attending to multilingualism in research infrastructures as a "vital condition for rich and healthy humanities research."[23] This interviewee traced the implications for multilingual DH at various stages of the research ecosystem through various DARIAH collaborations, from its work with Europeana and other cultural heritage organizations (to facilitate access to multilingual source materials), to the OPERAS-P project (which examines and facilitates the conditions for open-access multilingual publishing), to its involvement in the GoTriple discovery service (which connects research data through multilingual vocabularies). DARIAH represents a loose federation model, and one can see how this approach could easily be threaded into a coherent strategy to address multilingual issues throughout the digital scholarly research life cycle.

Multilingual DH as Translation

"Cultural and Linguistic Variety—Transnational RIs," a section in the 2011 European Science Foundation report "Research Infrastructures in the Digital Humanities," provides an early (and rare) treatment of multilingualism and intercultural exchange within literature on digital research infrastructures. Recognizing that "theoretically informed, comparative and transnational research gains from access to large datasets" of diverse linguistic and cultural origin, the report highlights the need for "sensitivity to the need for many-level translations (translation taken in the broad sense of the word)" in infrastructure design (*Research Infrastructures in the Digital Humanities*, 32–34). What precisely that "broad sense" entails in practice is left to our imagination, but the report does signal the need not only to translate in the linguistic sense, but also to translate between different ontologies, taxonomies, and the different meaning-making cultures that they embody. As Wrisley remarks, "Crossing borders into different knowledge cultures can be confusing business, indeed" (Wrisley), and here translation skills and intercultural literacy play a key role.

How has the concept of translation been conceived within DH? In her examination of the "plurality of DH approaches: what I call a DH *ecology of knowledges*" (Ortega, 180), Élika Ortega builds on Mary Louise Pratt's term "zones of contact" to argue for the importance of "translation work" in defining the connections among different geolinguistic communities. Recognizing DH as "a horizontal exchange of referents and nonfixed positions in regard to one another" rather than a "unitary" model, she describes projects such as RedHD in Translation and DH Whisperers, each being a tactical intervention to disrupt anglophone monolingualism in scholarly events and publication venues (Ortega, 180–83). She understands these initiatives to be translational not only in a linguistic sense, but also in the sense of "movement," which renders DH accessible and receptive to nonhegemonic cultural and linguistic communities.

Earlier, we examined how PH provides a possible countermodel for multilingual workflows within DH. Here, I would like to turn to how it models the concept of "translation" within digital research infrastructure. Jennifer Isasi and Antonio Rojas Castro ground their analysis of the Spanish-language edition of the PH platform in translation theory (specifically contrasting the notion of "equivalence" and Skopos theory, which privileges a translation's purpose) in their discussion of a translation strategy for DH pedagogical materials that facilitates multilingual understanding while also acknowledging multicultural diversity. They present three translation strategies that have been used on the project, which are respectively labeled *linguistic* (simple translation of equivalent terms, without adapting tutorial design or sample data), *expressive* (tutorial is partially adapted to a hispanophone audience, but there may still be English-language dependencies in some aspects, such as the tools used), and *substantial* (major changes in both the tutorial design and the materials used).

While we might wish to explore further the labels and divisions used, at the substantial end of the spectrum, according to their analysis, the material may have been reworked with references and examples designed specifically for the Spanish-language context and may involve reworking entire sections of the original text. This approach embodies the "broader sense" of translation mentioned earlier and helps the reader to better understand/digest the content (in this case, a tutorial) in their own personal linguistic and cultural context. As the authors point out, some tutorials present specific translation challenges due to linguistic differences (which lead to different results in different languages), as well as a lack of appropriate content or tool support for a given language. Theirs is a strategy that recognizes the differences in how cultures operate in digital spaces and what digital traces they leave behind, which in turn determines the effectiveness of a digital method in a particular language (Isasi and Rojas Castro).

Multilingual DH as Tactical Response

In their analysis of "the many languages of digital infrastructures," Sneha and Sengupta signal the urgency in engaging with "knowledge and infrastructural gaps in order to make the web more multilingual, accessible and safe, particularly for marginalized and non-dominant communities." Partly based on analysis of the *State of the Internet's Languages* report, which they contributed to, they highlight the "content and participation gap," based on numerous social and technical barriers that inhibit members of non-dominant language community from active participation in knowledge spaces (Sneha and Sengupta). Moving to a DH perspective, Eduard Arriaga cites the work of the historian-anthropologist Michel-Rolph Trouillot as Trouillot assesses the "silences" in DH "objects, tools and infrastructures" produced by design decisions in "architecture and engineering,"[24] which we are only starting to engage with critically (Arriaga, 544).

Launched in 2012, GO::DH, a special interest group within the Alliance of Digital Humanities Organizations (ADHO), has played a key role in addressing language diversity in DH as part of its wider mission to promote collaboration among DH researchers across the world. Strongly focused around agile/pop-up projects, advocacy, and community-building, GO::DH has embodied a multilingual and Global South–facing ethos, creating both the Translation Commons and the Translation Toolkit initiatives to foster multilingual collaboration, as well as the DH Whisperers drive to encourage informal multilingual interpretation and interaction at DH conferences. Influential well beyond its limited means, GO::DH has enjoyed success partly due to the expansive, open, low-regulation, and community-driven approach that it has adopted. GO::DH has often embodied an "activist" ethic, a "tactical" response to mainstream DH infrastructures (with their historic anglophone/monolingual and Global North bias). This kind of approach speaks to David Berry's concept of "tactical infrastructures"—adapted from the concept of tactical media—as "counter-infrastructures," which, in contrast to the often "instrumental" perspective that DH has taken on infrastructure, enables "new modes of knowing and thinking, assembling and acting" (Berry). On a broader level, we might situate this within a wider concept of "multilingual DH as community," where a heterogenous block of stakeholders including language activists, modern languages/area studies, and most important of all, DH research communities (operating in their own languages) has become significantly more active and visible within DH in recent years.

The numerous multilingual DH initiatives that have sprouted are starting to address multilingual research challenges in advanced humanities and social sciences research, but they do not generally engage extensively with stakeholder language communities (as opposed to research communities operating within a language). In his study of projects addressing the needs of Afro-Latinx and Afro–Latin American digital culture, Eduard Arriaga assesses the role of infrastructure in consolidating the hegemony of text and explores the representational erasure of marginal communities along racial and ethnic lines, which has often occurred in digital archives and publications. In describing their capacity for "representation and self-representation," reparation and reclassification, which emphasizes infrastructure as social process, he introduces the concept of "expanded digital infrastructure" as a counterinfrastructure consisting of minimal computing, hybrid technologies, and the "use of digital dynamics based on the humanisation of the historically dehumanized person" (Arriaga, 548, 546). Crucially, the concept of expanded digital infrastructure necessarily integrates the agency and perspectives of both academic and marginal communities, and while not specifically multilingual in nature, the principles that Arriaga proposes are highly relevant for marginal linguistic communities.

How should multilingual action in DH be resourced? Some infrastructural initiatives promoting multilingualism described here enjoy medium- to long-term funding (e.g., European initiatives like CLARIN), but in most cases, they have so far relied on the vision and dedication of a small but committed group of people working on a largely voluntary basis. The social aspect of this multilingual infrastructure is driven by an increasingly coherent policy and research agenda, but it faces complex credit and workload issues, not least because multilingualism typically operates in liminal spaces in the academy. The recent growth of formalized multilingual DH collectives such as the ADHO and DARIAH community groups perhaps also points to a growing realization that (notably set against the long autumn of academic Twitter and convulsions in the wider social mediasphere) some aspects of the tactical approach have limitations, and that multilingual strategy needs to be embedded more deeply into DH infrastructures through policy action, such that the benefits (and costs) of acting multilingually are structurally integrated into their underlying fabric.

DH multilingual initiatives may have a combination of motivations that can influence their success: acting on specific digital research needs (diversity in NLP), a research interest *on* particular languages, the desire to promote research *in* particular languages, or a wider commitment to geolinguistic diversity at an activist or policy level (as made by GO::DH). Some projects may be shorter term in principle (to provoke discussions around multilingual exchange at research conferences, as in the DH Whisperers project), whereas other initiatives may have longer-term aspirations (such as multilingual editions of PH).

It is difficult to generalize about such a wide spectrum of multilingual moves in DH, but it is likely that infrastructural initiatives with a well-defined focus (whether that consists of fixed outcomes, such as a set of PH tutorials, or an infrastructure to foster DH research in a particular language) are more likely to endure over time than initiatives with looser commitment to global diversity in general.

There are clearly limits to what a volunteer community can support in an already overcommitted research space, and the challenge going forward will be how DH can operationalize multilingual transformation on a greater scale. In particular, there is a pressing need for research contrasting the incentives and barriers for individual language (DH) communities, including low-resourced ones, because they are the natural drivers of multilingual transformation in digital scholarship. What we know already, though, is that the linguistic hurdles in non-Anglophone DH labor systems are complex. Beyond the common pressure to publish in English (or other dominant language) for so-called global impact, a given DH language community needs to expend extra energy toward articulating its own research identity and culture within wider digital platforms and standards with a strong anglophone imprint. In my interview study, more than one respondent highlighted the obstacles in getting DH research in their language recognized because (in a situation analogous to what happens in linguistics) English-language research is often treated as the "gold standard" for advanced computational studies, due to its access to larger datasets, more developed tools, and more extensive literature.

Tactical approaches to digital multilingualism in DH mirror broader attempts to disrupt digital monolingualism in wider digital studies and practice. DH has much to offer, but also to learn, from wider community-driven and participatory approaches to the challenge. The changing terms of participation by language communities in digital knowledge production are enabling us to question how infrastructures shape and are shaped by language power dynamics, and in particular anglophone digital cultural dominance. It makes us attend to the dangers of infrastructural discourse in DH, which can easily facilitate, even if unconsciously, a kind of monolingual and "center-dominant" platformization where the (here-linguistic) margins passively adopt anglophone/major language epistemological designs.

A tactical multilingual response would see DH become a disruptor, rather than a magnifier, of anglophone cultural hegemony and digital monolingualism in an expanded multilingual view of infrastructure.

In his analysis of global knowledge dynamics through social technology, Thomas Petzold notes that these dynamics are "not a matter of scale only" and "we need to examine the underlying cultural dynamics more closely" (64). If "the digital universe is still to be turned, from a global inequality amplifier with pockets of knowledge, into a truly global network of knowledge," rather than simply adding new features in multiple languages to an anglophone root infrastructure, it should instead be understood as "a problem-solving network that identifies very precise actions to tackle or resolve any specific issue" (Petzold, 136). This chapter has examined how such "problem-solving networks" within DH have understood and addressed multilingual challenges, whether addressing access to language resources, ideating DH's own multilingual strategies, enabling translation between languages and cultures, or using multilingualism as a tactical response to infrastructural needs.

David Gramling has charted how the "race to multilingualize trade logistics infrastructure has been profoundly rearranging both public discourses *about* multilingualism and the pathways of global content diffusion *by way of* multilingual technologies" (5); a charge that DH (and digital studies more broadly) have just begun to contemplate. This obfuscation, or appropriation for commercial interests, of a languages-focused agenda in digital research can easily lead to marginalization of multilingual expertise—in fields such as modern languages or area studies—privileging instead an agenda driven largely (if not solely) by machine translation and technolinguistic solutionism. Linguistic diversity in digital knowledge production is a useful proxy for knowledge creation, broadly speaking, and DH research infrastructures are thus representative of the ongoing conflict in knowledge infrastructures more generally between geolinguistic diversity and the dynamics of dominant language consolidation. Critical *multilingual* infrastructural studies play an important role in reorienting digital practice toward greater epistemic diversity and cultivating richer intralingual and geocultural dynamics in digital scholarship.

NOTES

1. See http://www.meta-net.eu/whitepapers/overview, http://www.dldp.eu/sites/default/files/documents/DLDP_Digital-Language-Survival-Kit.pdf, https://internetlanguages.org/media/pdf-summary/EN-STIL-SummaryReport.pdf, https://zenodo.org/record/5743283.

2. See resources at https://internetlanguages.org/en/resources/.

3. See https://european-language-equality.eu/.

4. See Pawlicka-Deger for discussion on how this relates to relational infrastructure.

5. See https://www.humanisti.ca, https://tthub.io/, http://www.globaloutlookdh.org/executive-board/.

6. See https://arounddh.org/.
7. See https://go-dh.github.io/translation-toolkit/.
8. See https://go-dh.github.io/mincomp/.
9. See https://programminghistorian.org, https://openmethods.dariah.eu.
10. See https://cistudies.org.
11. See https://www.clarin.eu/.
12. See https://www.clarin.eu/content/about-clarin.
13. See https://www.clarin.eu/content/participating-consortia.
14. See https://www.clarin.eu/content/virtual-language-observatory-vlo.
15. See https://www.sadilar.org/.
16. See https://www.clarin.eu/content/knowledge-centres.
17. See https://dhsi.org/course-archive/.
18. See http://multilingualdh.org/.
19. See https://newnlp.princeton.edu/.
20. See https://dig-hum.de/ag-multilingual-dh.

21. See also Horváth. On the epistemological implications of technical standards and the dangers of normative "global" DH practices "eliminating local variants of research," see Priani Saisó.

22. See https://www.dariah.eu/.

23. The interviews form part of ongoing research into multilingual attitudes within DH that started in 2022. The interviewees (twenty in the first stage) were selected on the basis of their involvement in DH infrastructure design and/or involvement in multilingual DH research.

24. Throughout the chapter, translations from Arriaga's article have been done by the author.

BIBLIOGRAPHY

AG Multilingual DH. Home page, 2023. https://dig-hum.de/ag-multilingual-dh.
Arbeitsgruppe Multilingual DH, the German DH association's special interest group. Home page, 2023. https://dig-hum.de/ag-multilingual-dh.
Around DH. Home page, 2023. https://arounddh.org.
Arriaga, Eduard. "Culturas Digitales Afrolatinxs y Afrolatinoamericanas: De La Recuperación Histórica a La Humanización Digital." *Hispania* 104, no. 4 (2021): 543–55. https://doi.org/10.1353/hpn.2021.0125.
Asef, Esther, and Cosima Wagner. "Workshop Report 'Non-Latin Scripts in Multilingual Environments: Research Data and Digital Humanities in Area Studies.'" *Das Blog der Universitätsbibliothek der Freien Universität Berlin* (blog), January 18, 2019. https://blogs.fu-berlin.de/bibliotheken/2019/01/18/workshop-nls2018/.
Berry, David M. "Tactical Infrastructures." *stunlaw* (blog), September 15, 2016, http://stunlaw.blogspot.com/2016/09/tactical-infrastructures_94.html.

Ceberio Berger, Klara, Antton Gurrutxaga Hernaiz, Paola Baroni, et al. "Digital Language Survival Kit: The DLDP Recommendations to Improve Digital Vitality." Digital Language Diversity Project. 2018. http://www.dldp.eu/sites/default/files/documents/DLDP_Digital-Language-Survival-Kit.pdf.

CLARIN Participating Consortia. 2023. https://www.clarin.eu/content/participating-consortia.

Common Language Resources and Technology Infrastructure (CLARIN). About CLARIN. 2023. https://www.clarin.eu/content/about-clarin.

Common Language Resources and Technology Infrastructure (CLARIN). Home page, 2023. https://www.clarin.eu.

Critical Infrastructure Studies.org. Home page, 2023. https://cistudies.org.

Danet, Brenda, and Susan C. Herring, eds. *The Multilingual Internet: Language, Culture, and Communication Online.* Oxford University Press, 2007.

Digital Research Infrastructure for the Arts and Humanities (DARIAH). Home page, 2023. https://www.dariah.eu.

Esch, Daan van, Elnaz Sarbar, Tamar Lucassen, et al. "Writing Across the World's Languages: Deep Internationalization for Gboard, the Google Keyboard." arXiv, 2019. https://arxiv.org/abs/1912.01218.

European Language Equality (ELE). Developing an Agenda and a Roadmap for Achieving Full Digital Language Equality in Europe by 2030. Home page, 2023. https://european-language-equality.eu.

Fiormonte, Domenico. "Towards a Cultural Critique of the Digital Humanities." *Historical Social Research* 37, no. 3 (2012): 59–76.

Global Outlook::Digital Humanities (GO::DH). Home page, 2023. http://www.globaloutlookdh.org/executive-board/.

Gramling, David. *The Invention of Multilingualism.* Cambridge University Press, 2021.

Horváth, Aliz. 2021. "Enhancing Language Inclusivity in Digital Humanities: Towards Sensitivity and Multilingualism: Includes Interviews with Erzsébet Tóth-Czifra and Cosima Wagner." *Modern Languages Open* 1, no. 26 (2021): 1–21. https://doi.org/10.3828/mlo.v0i0.382.

Horváth, Alíz, Cornelis van Lit, Cosima Wagner, and David Joseph Wrisley, et al. "Six User Personas for the Multilingual DH Community." Zenodo, 2023. https://doi.org/10.5281/zenodo.7811800.

Humanistica. Home page, 2023. https://www.humanisti.ca.

Internet Languages. 2023. "Resources and Inspiration." https://internetlanguages.org/en/resources/.

Isasi, Jennifer, and Antonio Rojas Castro. "¿Sin Equivalencia? Una Reflexión Sobre La Traducción al Español de Recursos Educativos Abiertos." *Hispania* 104, no. 4 (2021): 613–24. https://doi.org/10.1353/hpn.2021.0130.

Knowledge Centres. CLARIN. 2023. https://www.clarin.eu/content/knowledge-centres.

Lee, Carmen. *Multilingualism Online.* Routledge, 2017.

Lee, Martin, and Cosima Wagner. "Towards Multilingualism in Digital Humanities: Achievements, Failures and Good Practices in DH Projects with Non-Latin Scripts (Workshop)." Multilingual DH, July 2019. https://multilingualdh.org/en/dh2019/.

Liu, Alan. "Toward Critical Infrastructure Studies." Paper presented at University of Connecticut, Storrs. February 23, 2017. http://cistudies.org/wp-content/uploads/Toward-Critical-Infrastructure-Studies.pdf

Masakhane. A Grassroots NLP Community for Africa, by Africans. Home page, 2023. https://www.masakhane.io.

McPherson, Tara. "Designing for Difference." *Differences* 25, no. 1 (2014): 177–88. https://doi.org/10.1215/10407391-2420039.

"META-NET White Paper Series." META Multilingual Europe Technology Alliance. Home page, 2021. http://www.meta-net.eu/whitepapers/overview.

Minimal Computing. Home page, 2023. https://go-dh.github.io/mincomp/.

Multilingual DH. Home page, 2023. http://multilingualdh.org.

New Languages for NLP. Building Linguistic Diversity in the Digital Humanities, workshops. Center for Digital Humanities, Princeton University. 2021–2022. https://newnlp.princeton.edu.

Noble, Safiya Umoja. *Algorithms of Oppression: How Search Engines Reinforce Racism*. Illustrated ed. NYU Press, 2018.

OpenMethods. Highlighting Digital Humanities Methods and Tools. DARIAH-EU. Home page, 2023. https://openmethods.dariah.eu.

Ortega, Élika. "Zonas de Contacto: A Digital Humanities Ecology of Knowledges." In *Debates in the Digital Humanities 2019*, edited by Matthew K. Gold and Lauren F. Klein, 179–87. University of Minnesota Press, 2019.

Pawlicka-Deger, Urszula. "Infrastructuring Digital Humanities: On Relational Infrastructure and Global Reconfiguration of the Field." *Digital Scholarship in the Humanities* 37, no. 2 (2022): 534–50. https://doi.org/10.1093/llc/fqab086.

Petzold, Thomas. *Global Knowledge Dynamics and Social Technology*. Palgrave Macmillan, 2017.

Polezzi, Loredana. "Language Indifference." In *Translating Cultures: A Glossary*, edited by Charles Forsdick. Forthcoming.

Priani Saisó, Ernesto. "Codificación y buenas prácticas. Crítica a la delimitación de las humanidades digitales en América Latina." *Relaciones Estudios de Historia y Sociedad* 40, no. 158 (2019): 129–44. https://www.scielo.org.mx/scielo.php?pid=S0185-39292019000200129&script=sci_abstract.

Programming Historian. Home page, 2023. https://programminghistorian.org.

"Research Infrastructures in the Digital Humanities." European Science Foundation, 2011. http://archives.esf.org/index.php?eID=tx_nawsecuredl&u=0&g=0&t=1694513415&hash=acbd9e3d67f8d72ddbcedae325990c612fb9c248&file=/fileadmin/be_user/research_areas/HUM/Strategic_activities/RIs_in_the_Humanities/SPB42_44p-5oct_FINAL.pdf.

Ricaurte, Paola. "Data Epistemologies, The Coloniality of Power, and Resistance." *Television & New Media* 20, no. 4 (2019): 350–65. https://doi.org/10.1177/1527476419831640.

Right-to-Left Language Scripts. DHSI Archive, 2021. https://dhsi.org/course-archive/.

Risam, Roopika, and Alex Gil, eds. "Minimal Computing," special issue. *Digital Humanities Quarterly* 16, no. 2 (2022). http://www.digitalhumanities.org/dhq/vol/16/2/index.html.

Smith, David A., and Ryan Cordell. "A Research Agenda for Historical and Multilingual Optical Character Recognition." NULab for Texts, Maps, and Networks. Northeastern University. 2018. https://ocr.northeastern.edu/report/.

Sneha, P. P. "Mapping Digital Humanities in India." Centre for Internet and Society, 2016. https://cis-india.org/papers/mapping-digital-humanities-in-india.

Sneha, P. P., and Anasuya Sengupta. "The Many Languages of Digital Infrastructures." India-Seminar, 2021. https://www.india-seminar.com/2021/742/742_puthiya_and_anasuya.htm.

South African Centre for Digital Language Resources (SADiLAR). Home page, 2023. https://www.sadilar.org.

Spence, Paul. "Disrupting Digital Monolingualism: A Report on Multilingualism in Digital Theory and Practice." Language Acts and Worldmaking Project. Zenodo, 2021. https://doi.org/10.5281/zenodo.5743283.

Spence, Paul, and Renata Brandão. "Towards Language Sensitivity and Diversity in the Digital Humanities." *Digital Studies/Le Champ Numérique* 11, no. 9 (2021): 1–29. https://doi.org/10.16995/dscn.8098.

State of the Internet's Languages: Summary Report. Internet Languages. 2022. https://internetlanguages.org/media/pdf-summary/EN-STIL-SummaryReport.pdf.

Translation Toolkit. Home page, 2023. https://go-dh.github.io/translation-toolkit/.

TTHub. Home page, 2023. https://tthub.io.

Vannini, Laurent, and Hervé Le Crosnier eds. *Net.Lang: Towards the Multilingual Cyberspace.* Coordinated by Maaya Network. C & F Editions, 2012. https://unesdoc.unesco.org/ark:/48223/pf0000216692.

Virtual Language Observatory (VLO). CLARIN. 2023. https://www.clarin.eu/content/virtual-language-observatory-vlo.

Witt, Andreas. "CLARIN as a Multilingual Infrastructure." CLARIN Café presentation, May 31, 2021, video. https://www.youtube.com/watch?v=ZsIhFHFqHZ8.

Wrisley, David. "Enacting Open Scholarship in Transnational Contexts." *Pop! Public. Open. Participatory* 1 (2019). https://doi.org/10.21810/pop.2019.002.

PART II][Chapter 10

What's Missing
Studying Digital Humanities and Critical Infrastructure in India

MAYA DODD AND SHARIKA PARMAR

Imagining Infrastructures for Whom

In imagining the infrastructural futures for digital humanities (DH) in India, we must first understand whom we are building for. Most importantly, conceptualizing infrastructural futures for DH in India requires us to move beyond traditional academic boundaries. As with many countries in the Global South, DH in India is not simply a product of academic innovations or robustly funded ecosystems but a response to public need and demand. The profusion of actors engaged in the interplay of digital and human activities goes beyond institutionally sanctioned spaces, producing a vibrant and conceptually demanding space (Dodd and Kalra 2021). This practical and epistemic diversity is, of course, characteristic of many countries in the Global South (Gobbo and Russo 2020). If we are to imagine a future for DH infrastructures in India, we must first assume a broader conception of DH, which implies a parallel assumption that infrastructures for DH in India need to be infrastructures for *public humanities* in India. Hence the infrastructures required presuppose a different trajectory than countries in the Global North, charting a course that embraces multimodal ways of designing, thinking, and building.

The first wave of DH in India (and in other South Asian states) over the past two decades has been characterized by the emergence of individual and crowdsourced initiatives without the benefit of official mandates in terms of funding or policy. This activity has, nevertheless, generated an important first generation of digital resources designed by several actors for the public good.[1] Such practices, aligned to but distinguishable from the practice of minimal computing, have produced far more dispersed national DH infrastructure than would have resulted from initiatives by individual institutions (Roy and Dodd 2024, 248). Diasporic and national publics, representing broad aspects of society, have used digital methods to establish infrastructure in support of public memory and community history. This activity has, for instance, produced religious textual corpora for digital publics,[2] with patronage frequently found outside of academic structures.

As Roy and Dodd remind us, practitioners and infrastructures of DH in India lie within and outside institutional systems (Roy and Dodd 2024, 243). The story of Indian DH is an extension of public humanities in India; infrastructural futures need to encompass this broad landscape. It is useful to view the emergence of DH in India as rhizomatic, where "the realisation of DH in research, pedagogy, and practice is [gained] primarily through acts of self-identification and not through previous affiliations to the existing big tents" (Roy and Dodd 2024, 240). That perspective should also animate imagined futures for DH labs and courses in India, that need to be informed by postcolonial histories and the diverse contexts of Indian classrooms.[3]

As with most countries around the world, in the Global North as well as South, design and delivery of DH curricula in India relies to a great degree on free and open-source software and extra-institutional resources and support (Shanmugapriya and Menon 2020, 7). Both inside and outside the university system, this has allowed for the scaffolding of DH work, despite continued challenges to connectivity and funding. The imaginative deployment of digital infrastructure, tools, and methods can help overcome these constraints. It is only by apprehending the uneven terrain of DH in India, and understanding its functional drivers, that we can delineate the challenges it faces. Infrastructures then need to address practitioners both inside and outside the university lab and are further informed by a broader ecosystem of internet freedom based on an emerging landscape of legal rights.

Digital India and Disconnected India

The futures of DH in India rely to a remarkable degree on an ability to contend with a rapidly developing and sociotechnically complex national digital infrastructure. In this regard, it is salient to note that for the past five years, India has topped the charts for internet shutdowns globally.[4] This is not—it is important to stress—a result of technical failures but rather administrative direction. Denial of service has become a flash point for conversations about national internet governance, with providers reportedly shutting down services in the name of public safety. The effect is to deny users the right to connectivity by citing security concerns. According to Access Now and the #KeepItOn coalition's new report, India imposed eighty-four internet shutdowns in 2022—the highest number globally for the fifth year in a row (Skok et al; Rajvanshi). The world's largest democracy has been responsible for approximately 58 percent of the total number of shutdowns documented since 2016.

Reliable service and open access would seem to be obvious requirements for a developing national internet infrastructure, but this is not always viewed as the case by a state apparatus with complex motivations. As Nishant Shah observes, the patchiness of internet connectivity influenced the first wave of DH activity in India, focusing on questions of access and the provision of technological solutions for the preservation and circulation of data (169). This first wave sought to bridge the "connected" from the "to-be-connected," which is most evident in the case of Facebook Basics (see Sosale). This developed into a politics of the disconnected subject. The notion of infrastructure-as-development has continually shifted, molded by parallel debates about citizenship rights. Uncertain legal privileges undermine connectivity rights; this basic reality needs to be factored into any discussion of DH infrastructural futures.

Impending legislation suggests that individual privacy, enshrined in frameworks like that of the European Union's General Data Protection Regulation (GDPR), are not a priority for Digital India. The definition of high-quality infrastructural development is defined by the imagined user, and in India, this implies a late capitalist maximization of a vast emerging market. The rollout of 5G is a case in point, with national aspirations far exceeding reality, as the price point of the hardware was unattainable to the vast majority of consumers (ET Telecom). The desire for the creation of ever-larger markets is creating an untenable equation where affordances and affordability are not in sync. The problem crosses sectors and service categories: cheap data, smartphone growth, and the rollout of artificial intelligence (AI) technologies also assumes that digital inclusivity is around the corner. Conversations about the risks of emergent technologies in India are stuck in limbo, between concern about growing digital divides that impede human flourishing (Thirumal and Tartakov) and national anxiety about increasing access for digital belonging. And although there are legitimate anxieties around basic issues of access and privacy for the common citizen or heightened threats related to AI, more profound issues exist in draft legislation poised between surveillance (such as a blanket deployment of facial recognition software in the name of national security) and inadequate attempts to bridge often-substantial gaps between intention and implementation. On the eve of the proposed Digital India Act, there is no doubt that we are at an inflection point that divides an internet that we have known, as a space of potential empowerment albeit with uneven access, from a brave new digital world that we fear.

Digital India, inaugurated as state policy in 2015,[5] represents a comprehensive vision for using technology to drive social and economic development, improve governance, and empower citizens across the country. India Stack, initiated as a set of digital infrastructure components designed to enable a secure and efficient ecosystem for online transactions and service delivery across the country, is central to this policy; and since November 2023, it has been termed "Digital Public Infrastructure."[6] Technology stakeholders often boast about the success of the initiative, noting that "what began with the Aadhaar digital identity way back in 2009—joined later by services like the Unified Payments Interface (UPI), JAM (Jan Dhan Yojana, Aadhaar and Mobile number) trinity, and Co-WIN (for managing the Covid-19 vaccination programme), among others—helped India achieve 80 per cent financial inclusion in just six years, which one paper from the Bank for International Settlements estimated would have otherwise taken 47 years to achieve" (Singal). While ordinary citizens seek personal data protection, a lot more is expected to be built on India Stack.

As India grapples with the challenges of delivering high-quality digital services to the last mile, it also increases engagement with what will simultaneously be the world's most populous and youngest nation. This means that any analysis of digital infrastructure in India must contend with the reality of *increasing* state power, which seeks to deliver development through digital platforms and services, and the *use* of state power to rein in existing freedoms through the same technology. India Stack is touted as the state's success story because of its contribution to digital inclusion, and it is marketed as a solution for developing countries needing to provide solutions at scale for identity, payments, health care, and education. Eight countries have signed memoranda of understanding (MoUs) with India: Armenia, Sierra Leone, Suriname, Antigua, Barbados, Trinidad and Tobago, Papua New Guinea, and Mauritius, which allow them to access Indian digital infrastructure with no-cost open-source access (Raghavan, Jain, and Varma 2019).

As Adrian Athique and Akshaya Kumar have noted, "The digital development race in Asia has impelled infrastructure investments that entail close partnerships between national governments and local firms capable of operating universal digital infrastructure at scale" (1421). This reflects the "conceptually unruly" nature of infrastructures that Brian Larkin identifies: Whether viewed "as a system of substrates, networks or interconnected technological systems . . . [infrastructures] need to be seen as an amalgam of technical, administrative, and financial techniques" (328). In India, as elsewhere, digital platforms are market systems that connect domains of commerce, technology, sociality, and logistics (Athique and Parthasarathi). Adrian Athique and Akshaya Kumar demonstrate how this is playing out in India, noting that the "ambitions of Digital India are gargantuan, with multiple projects grouped under three core headings: 'digital infrastructures', 'governance service' and 'digital empowerment'" (1421). While the story on state schemes continues to unfold, the legal landscape around markets, individuals, and rights remains nebulous.

Because DH (in India as much as elsewhere) do not represent a singular cultural phenomenon but rather a range of practices, its conditions of future possibility need to be connected to discourses related to the state and the market, and to private and personal limits to freedom. And this larger canvas needs to be positioned alongside the seemingly unstoppable processes of global digitalization that proliferate around us. While we were writing this chapter, many facts on the ground changed—people's access to India's digital infrastructure, reformulating state laws, and machinic possibilities are all fluidly recasting a future faster than can be apprehended. Laws are being redrafted (as with the much-revised Digital Personal Data Protection Bill 2022, the impending Digital India Act, and the National Data Governance Policy). These are among a plethora of initiatives, ranging from definitions of digital liability to new taxation, and symbolize a creeping centralization with its own dangers. Further, when all these initiatives occur in a fast-evolving landscape of AI technology, that itself seems to elude real-time regulatory thinking, and these rapid changes obscure who the real subject of these efforts indeed is.

Imagined and Excluded Subjects

The dizzying speed of regulatory changes can obfuscate the subject at the heart of all these infrastructural affordances, requiring us to recast the question again: Who are we building for? It is useful to return to the imagined subject that this mega-scaffolding is intended for. The challenges are more complex than those that can be met with promises of providing material infrastructures for bringing people online or providing an internet device. Digital divides are not simply remedied by providing connectivity and access. In a time in which we are all "data citizens" (Bowker, xiii) questions of trade off are inevitable. Just as Facebook Basics[7] proved to be an unworthy deal (Mukerjee), we need to question whether Faustian bargains are being entered into in exchange for mass connectivity. If DH in India has a single message to offer the wider DH world, it is that digital public infrastructures are profoundly political, and arguments about what constitutes the public good can rarely be resolved in purely academic contexts (Larkin, 189; Appel, Anand, and Gupta, 2). The potential of critical infrastructure studies (CIS), in this sense, lies in its ability to problematize and potentially surmount academic, public, and private boundaries.

The relationship between CIS and DH in an Indian context is fraught with difficulty too, however. Although infrastructures are central to the practice of development, limits to digital technology render many DH practices fragile in India. In the Sisyphean task of building capacity, elements necessary for the development of DH as an inclusive field remain missing. Are connectivity and development alone adequate to bridge the digital divide, without an imaginative regulation that is capacious enough to handle a variety of needs on scale? Critical theory places such an emphasis on defining the problem (as if it were static) that we are often left holding fossils of meaning that were once accurate but inadequately reflect the reality of the present moment.

To explain this, we turn to the ideas of Prathama Banerjee. In her recent work, *Elementary Aspects of the Political: Histories from the Global South*, she challenges the givenness of certain assumptions as universal, such as the definition of the political. Just as the definition of what constitutes "political" can itself be unraveled, so can assumptions about the connections between infrastructure and development in the Indian context. Banerjee makes visible and unravels our "contemporary assumptions about the self-evidence, universality and primacy of the political" to show how there is "no essence to the political" and that "what we call political emerges only in terms of its differentiation from the non-political . . . and in terms of its delimitation by the extra-political that always returns to haunt it" (215). Her gesture informs our unpacking of the basic elements of infrastructure and serves as a heuristic device to guide our understanding.[8]

Viewed in its full sociotechnical complexity, the discourse of infrastructure-as-development comes perilously close to info-solutionism (Dodd, "Querying Info-Solutionism"), suggesting lessons for the global DH community. It is easy to position DH infrastructures as merely in need of more development—as suffering from a resource deficit problem—rather than being imbricated in broader patterns of sociopolitical and technical control. This has flow-on effects that obscures clear thinking on other topics. We assume that we need more formal funding structures, for example, but ignore other less resource-rich models that are capable of meeting DH needs (Roy and Dodd, 248). These sorts of problems probably stem from the fact that development narratives are embroiled in the modernist teleology that always demands a singular temporality, instead of being open to multiple ones (Gupta 2018, 70). Modernity impels us to look at infrastructural solutions in terms of possessing a phone, having Wi-Fi or 5G. And yet a phone and an internet connection do not guarantee connectivity or imply access, representation, or equality, any more than a DH lab does. Viewed in the broader context of digital colonization (Dhapola), we need to continually ask how many people are automatically excluded not only from connectivity but also representation (which, certainly in the context of a DH lab, could occur *without* individual connectivity). Beyond the legal landscape and the connectivity question lies the human reality of India's current internet architecture. Are we building DH *for* them? Whom are we building these imagined futures for?

Exclusions by/of Language

For a country with the largest population in the world, it is striking that not a single Indian language appears on the list of the top ten most used languages on the internet (Brandom). The structural exclusion of non-English-speaking peoples is a defining impediment to curricular DH in higher education in India, as it is in many countries around the world, framing institutional possibilities (Roy and Dodd). Consider how many common Indian languages are present online. The first-ever platform survey to collect and analyze global interface languages, the pioneering "State of the Internet's Languages Report,"[9] throws some light on this. The report found that "in South Asia, almost half of the platforms surveyed do not offer interface support for any regional language, and major South Asian languages such as Hindi and Bengali, spoken by hundreds of millions of people, are not as widely supported as we might expect. Support for South-East Asian languages is similarly mixed: while Indonesian, Vietnamese, and Thai tend to be very well supported by the platforms we surveyed, most other South-East Asian languages are not" (Whose Knowledge?, Oxford Internet Institute, and the Centre for Internet and Society [India]).

In their survey of eleven websites, twelve Android apps, and sixteen iOS apps (selected across widely used platforms), one finding stands out. Even though Wikipedia ("the largest collaborative effort in human history") began with a single English-language edition, it is now available in over 300 language editions. Wikipedia's user interface has been translated into more languages than any of the commercial platforms, including Google (available in 150 languages) and Facebook (106 languages).

It is important to recognize that this situation also extends to the code that underwrites global digital infrastructure as well. In an article for *Wired* magazine in 2019, Gretchen McCulloch bluntly noted that "even huge languages that have extensive literary traditions and are used as regional trade languages, like Mandarin, Spanish, Hindi, and Arabic, still aren't widespread as languages of code. I've found four programming languages that are widely available in multilingual versions."

Comparing the journey from Latin to English also serves as a reminder that the first website was not written in just Hypertext Markup Language (HTML)—it was written in *English* HTML. The lack of bridges between languages and code has resulted in a warped representation of lived realities online and offline. As Mark Graham of the Oxford Internet Institute noted in *The Guardian*'s feature on the digital language divide, "rich countries largely get to define themselves and poor countries largely get defined by others" (Young). Even if it is heartening that Wikipedia's 300 languages and several translation tools enabled the transfer of content across these languages, the fact is that what has been left behind is staggering. The point to remember here is that we are not only referring to the absence of mother tongues and semantic worlds, but entire worldviews that cannot be accurately portrayed by mere translation. So, where is the imagined nation of Indian-language speakers online?

The Absent Training Set

As Anasuya Sengupta has noted, only 7 percent of the world's languages are found in published materials, and even fewer are represented on the internet (Vrana, Sengupta, Pozo, and Bourterse; Spence). English and Chinese dominate the languages available online, which include just 500 of the 7,000-plus languages of the world. This is significant to note, especially since 75 percent of internet users are from the Global South, indicative of an expansive gulf between the users and producers of digital content and infrastructures (Vrana et al.). This gains further urgency when one considers that the training sets of large language models (LLMs) used by AI models are primarily trained on the English-dominated internet for their capacities. Kalika Bali is a principal researcher at Microsoft India, whose work is featured in a report noting that India is "a profoundly multilingual country with 16 languages enjoying primary official language status at state level and 29 languages with more than a million native speakers" (Spence, 10). Given this, and noting the reliance of many AI, natural language processing (NLP), and speech technology systems on a very small number of languages, Bali highlights the challenges of building technology for low-resourced languages. "In global terms, no indigenous Indian language qualifies in her category of the fifteen 'highly-resourced' global languages in terms of digital support, and most sit in the 'under resourced' or 'no source' categories" (Spence, 10).

To create in one's own image is to tell one's own story (and also, in a time of generative AI, to multiply it), rendering cultural and linguistic absence deeply impactful. Acknowledging the profound biases of AI training sets draws attention to the specter of the stillborn user. With the rapid development of AI, quickly increasing its capacity from one generation to the next, these absences are dire. It is telling that India provides ChatGPT (the AI chatbot service that famously earned 100 million downloads within two months of its launch), with its second-largest user base (at 7.1 percent, compared with 15.1 percent from the United States) (Upadhyay), but without data from Indian sources, LLMs will merely extend the biases of the internet into the future.

Missing Infrastructures and AI

Indigenous innovation will be needed to truly harness the power of AI for India, and the Indian state is aware of this. To accelerate AI, in May 2022 the Ministry of Electronics and Information Technology (MEITY) launched Bhashini, an Indian languages database that can be used by research institutes (Das). The government is also setting up a draft National Deep Tech Startup Policy to explore the development of LLMs (Koshy). As two office-bearing government representatives opined in a recent newspaper piece,

> The current landscape of LLM development is dominated by US-centric models. . . . akin to the global impact of social media platforms like Facebook, Google . . . the risk of perpetuating this US-centric paradigm is immense . . . For reference, GPT-3 was trained on 45 terabytes of text data, roughly equivalent to 500 billion tokens . . . and OpenAI has invested in the neighbourhood of $10 billion to 'train' GPT-4 . . . The Nilekani Centre at IIT Madras' AI4Bharat has collected over 21 billion open text tokens [and] also collected 100,000 hours of YouTube videos in Indian languages that have been published under Creative Commons (CC) 4.0 licence. However, even this massive data collection is about two orders of magnitude too small to build GPT-4-style models (D'Monte and Kolla).

The assumption is that by unlocking existing Indian-language data, developing an Indigenous LLM, and forging public-private partnerships, India can create an indigenous LLM that counterbalances the influence of dominant players and ensures a more equitable distribution of AI benefits. The spirit of nation-building is invoked for this, as they point out that "last year, the Bhasha Daan initiative of the Bhashini project was launched in which citizens could 'donate' their existing voices to help build datasets for Indian-language AI. This programme may be reimagined and revamped to crowdsource data for Indian languages that may be relevant to collect data to 'train' an LLM" (Vempati and Raghavan).

In line with these policies, Indian academia is initiating an increasing number of generative AI research projects, many of which are seeking to understand how technology can help create tools similar to ChatGPT using Indian languages. Researchers indicate a host of challenges for such projects, however, the biggest of which lies in sourcing ample amounts of Indian languages data. The cost of such projects and the scale of computing power needed are equally challenging problems to resolve (Das). The scale of the training sets needed for developing AI across India will require crowdsourcing to resolve. While the state has already turned to AI to take on questions of scale, such as filtering tax returns and generating court transcripts (*Economic Times*), the need for greater diversity in input also requires citizen activity and other forms of social datasets (Haini).

The Future of Infrastructure

DH curricula could support the development of crowdsourced and inclusive language datasets if designed with this goal in mind; the process of decolonizing a curriculum and developing content for the curriculum are one and the same. Reflecting on what is missing in our thinking about infrastructures for DH in India requires us to think along three interconnected planes. First, we look at the *subject* and *content* of DH: Who are the audiences for DH infrastructures? Whom are we building for? Is there enough recognition of DH in India? Second, what are some of the *challenges* in building for this audience, and what factors might influence or shape the ecosystem? Because building infrastructures for DH in India intersects with public humanities, infrastructures for DH are imbricated in the existing and shifting legal and rights infrastructures (not to mention the existing language inequalities) of the internet. Third, *imagining futures* for DH infrastructures in India requires a critical understanding of infrastructures outside the teleological logics of modernity. Is it possible to draw out the contours affording DH possibilities in India as a function of both imagination and infrastructure?

The simple fact is that the use of digital tools rests on a global access system predicated on the English (and to a significant extent, Chinese) language, something that both DH and CIS need to contend with. The need to develop infrastructures goes beyond commercial and instrumentalist ends. By returning this challenge to the lived practices of DH in India, we might yet accord digital citizenship to those who do not live in the English-speaking world. Until then, the view that we have of the future is but an elementary aspect of temporary infrastructures that shall always fall short.

NOTES

1. See Project Madurai, founded in 1988 (https://www.projectmadurai.org/); the Punjab Digital Library, established in 2003 (http://www.panjabdigilib.org/webuser/searches/mainpage.jsp); the Indian Memory Project, established in 2010 (https://www.indianmemoryproject.com/); Sahapedia, established in 2016 (https://www.sahapedia.org/); pad.ma (https://pad.ma/); the Nepal Public Library (https://www.nepalpicturelibrary.org/); and the South Asian American Digital Archive (https://www.saada.org/).

2. See the Tibetan and Himalayan Digital Library (https://www.thlib.org/) and Pali Tipitaka (https://tipitaka.org/).

3. See Roy and Dodd; Shanmugapriya and Menon.

4. India has ranked first in the world for shutting off the internet over the past five years, according to SFLC.in and other digital rights watchdogs. For the case of Manipur, see Gupta; and for the larger context, see Gupta and Shih and Ali.

5. This program was launched on July 1, 2015. See https://csc.gov.in/digitalIndia.

6. India Stack (https://indiastack.org/) was conceptualized to support the government's Digital India initiative and promote financial inclusion. The key components of India Stack are a unique identity number known as Aadhar, a number that is issued to Indian residents; an electronic identification process known as e-KYC, the paperless governance of a Digital Locker that serves as an online vault for verification documents, backed by electronic signature or e-sign; and a United Payments Interface (UPI), a digital payment methods that use Aadhaar-enabled Payments System (AePS). For more on India Stack as a digital public infrastructure, see Singal.

7. The saga that began ten years ago in India in 2013 is well delineated (Mukherjee). Almost a decade later, instead of the state, Reliance Jio and Meta entered a deal for chat-based shopping (enabling grocery delivery on WhatsApp) and are now said to be in talks to set up physical infrastructures for large AI applications (Kar).

8. There is a longer history of the use of the term *elementary* in Indian scholarship. We build upon the form of *elementary* in the same vein as Prathama Banerjee (examining elementary aspects of the political) and Ranajit Guha (examining elementary aspects of peasant insurgency in colonial India).

9. Begun in 2019 to advocate against language inequality, the State of the Internet's Languages Report is a collaboration among three organizations: Whose Knowledge?, Oxford Internet Institute, and the Centre for Internet and Society (India).

BIBLIOGRAPHY

Ali, Arbab. "No Internet Means No Work, No Pay, No Food." *Human Rights Watch*, June 14, 2023. https://www.hrw.org/report/2023/06/14/no-internet-means-no-work-no-pay-no-food/internet-shutdowns-deny-access-basic.

Appel, Hannah, Nikhil Anand, and Akhil Gupta. "Introduction: Temporality, Politics and the Promise of Infrastructure." In *The Promise of Infrastructure*, edited by Nikhil Anand, Akhil Gupta, and Hannah Appel, 1–38. Duke University Press, 2018.

Athique, Adrian, and Akshaya Kumar. "Platform Ecosystems, Market Hierarchies and the Megacorp: The Case of Reliance Jio." *Media, Culture and Society* 44 no. 8 (2022): 1420–36.

Athique, Adrian, and Vibodh Parthasarathi. "Platform Economy and Platformization." In *Platform Capitalism in India*, edited by A. Athique and V. Parthasarathi, 1–19. Palgrave Macmillan, 2020.

Banerjee, Prathama. *Elementary Aspects of the Political: Histories from the Global South*. Duke University Press, 2020.

Bowker, Geoffrey C. "Introduction: The Infrastructural Imagination." In *Information Infrastructure(s): Boundaries, Ecologies, Multiplicity,* edited by Alessandro Mongili and Giuseppina Pellegrino, xii–xiii. Cambridge Scholars Publishing, 2014.

Brandom, Russell. "What Languages Dominate the Internet?" Rest of World, June 7, 2023. https://restofworld.org/2023/internet-most-used-languages/.

Das, Shouvik. "Indian Colleges Accelerate Work on Indic Languages Gen AI." *Mint,* April 16, 2023, https://www.livemint.com/companies/start-ups/indian-engineering-colleges-lead-generative-ai-research-projects-in-indic-languages-facing-challenges-in-data-sourcing-and-computing-power-11681663553393.html.

Dhapola, Shruti. "Digital Colonisation: How Indian Languages Lost out to English on the Internet." *Indian Express* (blog), November 19, 2020. https://indianexpress.com/article/technology/tech-news-technology/how-digital-colonisation-is-still-keeping-internet-away-from-indian-languages-7057030/.

D'Monte, Leslie, and Jayanth N. Kolla, *AI Rising: India's Artificial Intelligence Growth Story.* Jaico Publishing House, 2023.

Dodd, Maya. "Digital Cultures in India." In *Literary Cultures and Digital Humanities in India,* edited by Nishat Zaidi and A. Sean Pue, 19–37. Routledge, 2023.

Dodd, Maya. "Querying Info-Solutionism." *Biblio India,* January 6, 2023. https://www.flame.edu.in/in-the-media/querying-info-solutionism.

Economic Times. "Supreme Court Launches AI for Live Transcript of Court Hearings." *Economic Times,* February 22, 2023. https://economictimes.indiatimes.com/news/india/supreme-court-launches-ai-for-live-transcript-of-court-hearings/articleshow/98131111.cms.

ET Telecom. "India's 5G Smartphone Sales Cross 100-Million Mark in May, Driven by Entry-Level Devices: Counterpoint." ET Telecom, July 8, 2023. https://telecom.economictimes.indiatimes.com/news/devices/5g-smartphone-shipments-cross-10-cr-in-india-for-1st-time/101573866.

Exploring Digital Humanities in India: Pedagogies, Practices, and Institutional Possibilities, edited by Maya Dodd and Nidhi Kalra. Routledge, 2021.

Gobbo, Federico, and Federica Russo. "Epistemic Diversity and the Question of Lingua Franca in Science and Philosophy." *Foundations of Science* 25 no.1 (2020): 185–207.

Guha, Ranajit. *Elementary Aspects of Peasant Insurgency in Colonial India.* Oxford University Press, 1983.

Gupta, Akhil. "The Future in Ruins: Thoughts on the Temporality of Infrastructure." In *The Promise of Infrastructure,* edited by Nikhil Anand, Akhil Gupta, and Hannah Appel, 62–79. Duke University Press, 2018.

Gupta, Anant, and Gerry Shih. "India Uses Widespread Internet Blackouts to Mask Domestic Turmoil." *Washington Post,* October 19, 2023. https://www.washingtonpost.com/world/2023/10/18/india-internet-blackouts/.

Gupta, Apar. "Apar Gupta Writes: On Manipur, Internet Shutdown Is No Cure." *Indian Express,* August 4, 2023. https://indianexpress.com/article/opinion/columns/apar-gupta-writes-on-manipur-internet-shutdown-is-no-cure-8873758/.

Haini, Siti Isnaine, Nor Zairah Ab. Rahim, and Norziha Megat Mohd. Zainuddin. "Adoption of Open Government Data in Local Government Context: Conceptual Model Development." In *Proceedings of the 2019 5th International Conference on Computer and Technology Applications,* 193–98. ICCTA '19. Association for Computing Machinery, 2019.

Jaiswal, Anuja. "Over 3.5 Crore Pages from History Now Available at Click of a Mouse." *The Times of India,* November 15, 2023. https://timesofindia.indiatimes.com/city/delhi/over-3-5-crore-pages-from-history-now-available-at-click-of-a-mouse/articleshow/105221096.cms.

Jha, Girish Nath, Sobha L, Kalika Bali, and Atul Kr. Ojha, eds. "Proceedings of the LREC 2022 Workshop Language Resources and Evaluation Conference." In *6th Workshop on Indian Language Data: Resources and Evaluation (WILDRE-6).* Marseille, 2022. https://aclanthology.org/2022.wildre-1.pdf.

Jindal, Siddharth. "The Dire Need for LLM Benchmarks in India." *Analytics India Magazine,* September 29, 2023. https://analyticsindiamag.com/the-need-for-llm-benchmarks-in-india/.

Kar, Ayushi. "Reliance May Extend Partnership with Meta for Homegrown Large Language Model." BusinessLine, September 12, 2023. https://www.thehindubusinessline.com/info-tech/reliance-may-extend-partnership-with-meta-for-homegrown-large-language-model/article67299532.ece.

Koshy, Jacob. "India Will Explore Building Large Language Models: Principal Scientific Advisor Ajay Sood." *The Hindu,* October 18, 2023. https://www.thehindu.com/news/national/india-to-set-up-a-committee-to-explore-deep-tech-scientific-advisor/article67435070.ece.

Larkin, Brian. "The Politics and Poetics of Infrastructure." *Annual Review of Anthropology,* 42 no. 1 (October 21, 2013): 327–43.

Majumder, Shayak. "AI for India, by India: What It Takes to Build an Open-Source Indic-Language Model." *ABP Live.* September 20, 2023. https://news.abplive.com/technology/project-indus-ai-llm-tech-mahindra-what-it-takes-to-build-an-open-source-indic-language-model-1630757.

McCulloch, Gretchen. "Coding Is for Everyone—as Long as You Speak English." *Wired,* April 8, 2019. https://www.wired.com/story/coding-is-for-everyoneas-long-as-you-speak-english/.

Mukerjee, Subhayan. "Net Neutrality, Facebook, and India's Battle to #SaveTheInternet." *Communication and the Public* 1, no. 3 (1 September 2016): 356–61.

Pawlicka-Deger, Urszula. "Infrastructuring Digital Humanities: On Relational Infrastructure and Global Reconfiguration of the Field." *Digital Scholarship in the Humanities,* 37, no. 2 (2022).

Raghavan, Vivek, Sanjay Jain, and Pramod Varma. "India Stack—Digital Infrastructure as Public Good." *Communications of the ACM* 62, no. 11 (November 2019): 76–81.

Rajvanshi, Astha. "How Internet Shutdowns Wreak Havoc in India." *TIME,* August 15, 2023. https://time.com/6304719/india-internet-shutdowns-manipur/.

Roy, Dibyadyuti, and Maya Dodd. "Digital Humanities Laboratories and their Discontents: Experiments and Perspectives from India." In *Digital Humanities and Libraries: Perspectives on Knowledge, Infrastructure and Culture*, edited by Urszula Pawlicka-Deger and Christopher Thompson, 239–52. Routledge, 2024.

Shah, Nishant. "Digital Humanities on the Ground: Post-access Politics and the Second Wave of Digital Humanities." *South Asian Review* 40, no. 3 (July 3, 2019): 155–73.

Shanmugapriya, T., and Nirmala Menon. "Infrastructure and Social Interaction: Situated Research Practices in Digital Humanities in India." *Digital Humanities Quarterly,* 14, no. 3 (2020).

Singal, Nidhi. "Here's How India's Digital Public Infrastructure Is Going Global." *Business Today,* November 9, 2023. https://www.businesstoday.in/magazine/deep-dive/story/heres-how-indias-digital-public-infrastructure-is-going-global-405177-2023-11-09.

Skok, Zach Rosson, et al. "Internet Shutdowns in 2022: The #KeepItOn Report." *Access Now* (blog), February 28, 2023. https://www.accessnow.org/internet-shutdowns-2022/.

Sosale, Sujata. "National Authority and Public Interest in Facebook's Free Basics Indian Encounter: A Case of 'Slowbalisation.'" *Journal of International Communication,* 28, no. 2: 169–87.

Spence, Paul. "Disrupting Digital Monolingualism: A Report on Multilingualism in Digital Theory and Practice." *Zenodo,* December 9, 2021. https://doi.org/10.5281/zenodo.5743283.

Thirumal, P., and Gary Michael Tartakov. "India's Dalits Search for a Democratic Opening in the Digital Divide." In *International Exploration of Technology Equity and the Digital Divide: Critical, Historical and Social Perspectives,* edited by Patricia Randolph Leigh, 20–39. IGI Global, 2011.

Upadhyay, Shyam Nandan. "9 Groundbreaking GPT-Based Models Developed in India." *Analytics India Magazine,* April 14, 2023. https://analyticsindiamag.com/9-groundbreaking-gpt-based-models-developed-in-india/.

Vempati, Shashi Shekhar, and Vivek Raghavan. "India Needs to Develop a Large Language Model That Reflects Its Linguistic, Cultural Tapestry." *Economic Times,* April 7, 2023. https://economictimes.indiatimes.com/opinion/et-commentary/india-needs-to-develop-a-large-language-model-that-reflects-its-linguistic-cultural-tapestry/articleshow/99325316.cms.

Vrana, Adele, Anasuya Sengupta, Claudia Pozo, and Siko Bouterse. "Decolonizing the Internet's Languages: Summary Report," 2020. https://whoseknowledge.org/wp-content/uploads/2020/02/DTIL-Report.pdf.

Whose Knowledge?, Oxford Internet Institute, and the Centre for Internet and Society (India). "State of the Internet's Languages: Summary Report," 2022. https://internetlanguages.org/media/pdf-summary/EN-STIL-SummaryReport.pdf.

Young, Holly. "The Digital Language Divide." *The Guardian Labs,* 2015. http://labs.theguardian.com/digital-language-divide/.

PART II][Chapter 11

Connecting Digital Systems by Whom and for Whom?
Taking Stock of the Digital Humanities Infrastructures in China

LIK HANG TSUI AND JING CHEN

Background and Historical Context

The term *cyberinfrastructure* is derived from the natural sciences. The U.S. National Science Foundation put the term into play in a 2003 report (Atkins) and followed in later reports by describing cyberinfrastructure as "an idea that has emerged . . . from some basic technological realities" that "have led researchers to envision a tightly integrated, planet-wide grid of computing, information, networking and sensor resources" (Hart, para. 1). As such, a cyberinfrastructure is more akin to a new knowledge environment or ecology of virtually connected organizational resources than only an amassing of research materials. Later, the American Council of Learned Societies (ACLS) specified the role of cyberinfrastructure in humanities research (Courant, Fraser, Goodchild, et al.). Currently, more than fifteen years have passed, and it seems that the concept of research infrastructure in the digital humanities (DH) still fits such criteria—at least in the planning strategy of the American scholarly community. Other strategies have been devised in Europe, the United Kingdom, Australia, and elsewhere (ESFRI; Australian Department of Education; UKRI). However, is the idea of cyberinfrastructure or a DH infrastructure readily applicable to research communities in non-Western contexts? How should we evaluate this issue, and who should serve as the target users and participants of such infrastructure?

Participating in globalizing and diversifying DH, this chapter offers a critical analysis of efforts to create a DH infrastructure in mainland China, where since the end of the decade of the 2000s, the need for such infrastructure has increasingly been recognized and more efforts have been made to build it.[1] The focus has been on creating research infrastructures in particular—especially ones that can link various resources and systems not just for digital scholarship generally, but for the study of China specifically. We feel that these two concerns of DH infrastructure in China are substantially intertwined, and hence we integrate both into our discussion.

To set the issues in context, we start by noting that since the 1990s, the humanities in China have faced two main challenges. First, there has been a mismatch between the demand for talent required by society and the mode of education provided by the higher education system. Second, and in great part accounting for this mismatch, scholarly research and education have fragmented between the so-called two cultures—or, better, the "three cultures" of the humanities, the natural sciences, and the social sciences. Various reforms in Chinese higher education have responded to these conditions, including policies such as "quality-oriented education" (*suzhi jiaoyu*, 素质教育) instead of examination-oriented education; "general education" (*tongshi jiaoyu*, 通识教育) or liberal arts education, instead of professional education; "audiovisual education or 'Electrified' Education" (*dianhua jiaoyu*, 电化教育) instead of educational activities in traditional formats; and, most recently, the New Liberal Arts (*xin wenke*, 新文科) instead of the "traditional" humanities.[2] Of these, it is the New Liberal Arts, proposed by the Ministry of Education at roughly the same time as the policies for the New Engineering and New Medicine, that can be seen as a response by the Chinese higher education system to the new information revolution impacting China since the second half of the 2010s.

The development of DH in China in the late 2010s and early 2020s coincided with the introduction of the New Liberal Arts (Chen, "*Xin yilun*, 新一轮"). Concepts about DH and its practices were thus incorporated into this national-level educational policy, which provides the context for our analysis of China's recent infrastructural efforts and specifically its development of DH. Keeping in mind the particular context of China in this regard is important because, although in one sense DH was a Western import, Chinese scholars also established their own foundations for the field as early as the 1980s (Tsui, "*Hua wen* 华文"). Their efforts, which included initiatives for building an infrastructure for digital research, raised issues unique to the Chinese academy because the development of DH and its infrastructure had to relate to China's institutions for humanities research in the higher education system as a whole. Any infrastructure involves a substantial investment of financial, human, and educational resources, and in China, these resources almost always rely on their larger institutional system, which is often state-led. This inevitably led to a need for coordination and adaptation, resulting in development paths for digital research infrastructure that are quite different from earlier paths elsewhere. This brought Chinese scholars today to a point where, even though some standard procedures and workflows for such an infrastructure are already under discussion, there are still many challenges for the basic development of DH (Zhu, Benjun, and Jiuzhen Zhang). For example, there are still some barriers for smooth and streamlined text mining of Chinese data, even for issues such as word segmentation for data in the Chinese language. Conditioned by the need to coordinate and adapt to larger institutional configurations and government-led initiatives, China's DH infrastructure is emerging as a complex and diverse environment in which DH practitioners must survive, adapt, and grow.

Defining "Digital Humanities Infrastructure" in China

The concept of a DH infrastructure is a product of negotiation and adaptation in China. The Chinese academic community, including scholars in China and in digital Chinese studies, have only relatively recently (since the mid-2010s) begun to pay attention to such an infrastructure, adopting various understandings of its definition and main function and emphasizing different components and resources.

For example, Wei Liu, Rong Xie, and Lei Zhang suggest that DH infrastructure supports humanities research activities in general. This infrastructure then refers to all the documents, data, relevant software tools, public utilities, and services for scholarly communication and global publishing. Liu et al. thus emphasize the basic and instrumental role of DH infrastructure as an integrative scholarly function for users. They propose that such an infrastructure should be the foundation for humanities research activities that include working with research literature, data, software tools, utilities for communication and publication, and related services. In their view, the ongoing transition toward a national data infrastructure for the DH field is prompted especially, though not solely, by significant advances in data science, which allows researchers to work with larger, more complex, and more fully digital (i.e., digitally born) datasets than before. Data science–driven infrastructures can also facilitate collaboration across institutions and disciplines, making it easier for scholars in China to share resources and know-how and to work together on research projects. In addition, it can provide tools and techniques for analyzing, visualizing, and even crowdsourcing data, thus helping researchers identify patterns and connections that they might have missed using other methods. A national data infrastructure, therefore, has the potential to transform the way that scholars conduct research. Because Liu and his coauthors come from a library background (Liu was then the deputy director of the Shanghai Library), they are also very aware of the role of digital infrastructures in preserving knowledge, especially in meeting the challenge of permanently storing and providing access to materials.

In contrast to Liu et al.'s more general understanding of DH infrastructure as including but not necessarily focused on data, Cuijuan Xia (also based at the Shanghai Library then) narrows her definition to specifically emphasize data infrastructure. She focuses on the creation and organization of data, content, and knowledge as the crucial "part of the digital humanities research infrastructure" (Xia). In particular, she explains that there is a need to build a data layer that is independent of specific application development and domain research initiatives. Such a layer should also follow technical specifications for the long-term preservation and sharing of and public access to data. Xia thus centers on the knowledge-organization function of DH infrastructure in data science.

In a rather similar way, Zixin Rao, Luxiang Deng, and Xin Xu also present an alternative, data-oriented perspective on DH infrastructure, proposing not a DH infrastructure as such, but rather a "data infrastructure for humanities research," to be precise. They define this infrastructure as the collaboratively shared tools, systems, platforms, software, and services that facilitate the preservation, acquisition, reuse, and publication of data in humanities work. They therefore emphasize the importance of data reusability, correlation, and aggregation, which they consider to be integral to any humanities research procedures. Their approach thus highlights the role of digital data as the primary driver of DH infrastructure in China.

These slightly varying definitions of DH infrastructure in China reflect the ongoing negotiation and adaptation of the concept within the academic community in China, demonstrating the need for further discussion and collaboration. Other, more specific recent discussions on similar topics have different foci and deserve some discussion here too. To mention some examples, Tao Chen, Rina Su, and Xun Sun, as well as Jia Yan, Min Yang, and Mei Peng, focus on image data infrastructure; Jiaqin Jin and Cuijuan Xia on the construction of an ontology for DH institutions; Dan Lu, Xin Li, and Jinchuan Chen on DH infrastructure based on application programming interface (API) technology; Lihua Wang and Yike Zhang on the significance of data competitions for DH infrastructures; and Chenfei Xu and Ping Bao propose a DH infrastructure specifically for advancing the field of agricultural history. With the exception of Liu, Xie, and Zhang's paper in 2016, all these studies were published after 2019, and their authors are primarily based in libraries or information management, mainly in institutions such as the Shanghai Library, East China Normal University, and Nanjing Agricultural University. These institutions have dedicated DH teams, centers, or relevant DH projects, which explains why their conceptions of DH infrastructure converge and why they devote extra attention to the role of the gallery, library, archive, and museum (GLAM) sector in their writings. It is worth noting also that their definitions of DH infrastructure are not specific to any one context or nation; instead, they seek to introduce a universal definition in their works.

Certain pertinent discussions may not be confined to the Chinese national context but instead focus predominantly on Chinese studies, which as a field is unsurprisingly deeply rooted in mainland Chinese and Taiwan academia. Based on their experiences building the China Historical Geographic Information System (CHGIS) and the China Biographical Database (CBDB), Hongsu Wang, Lik Hang Tsui, and Peter Bol point out in a comprehensive overview of challenges faced by researchers in historical China studies that DH infrastructures serving as cyberinfrastructure can link two basic components: underlying technologies for computing, storage, and communication on the one hand, and various platforms, tools, software, and services running on those technologies on the other. They explore the need for a cyberinfrastructure integrating these levels to support the study of Chinese history and Chinese studies, arguing for a more coordinated approach to research infrastructure development to address the proliferation of databases and the increasing number of researchers involved in their development. Addressing challenges that include lack of standardization in data formats and metadata and the reluctance to share resources across projects. Wang and his coauthors propose a cyberinfrastructure that provides a common platform for data sharing and collaboration and discuss several tools that could be used to facilitate resource sharing, including object-oriented databases and software for processing Chinese-language data that can now be operated on networked personal computers. They underscore the importance of designing and implementing metadata standards specifically tailored to historical China studies that would enable researchers to easily discover and access relevant resources and, where possible, share their data with other stakeholders. They then turn their attention to the CBDB, which they describe as an example of a database project that has helped develop metadata standards that contribute to cyberinfrastructural development—for instance, data standards (which within this project are stored in "code tables") for Chinese names, calendrical dates, geographical places, imperial offices, and other data for historical research. Their study also discusses the utilization of APIs in historical China studies, specifically mentioning the MARKUS tagging system developed by Hou Ieong Brent Ho and Hilde De Weerdt as an example of how APIs can facilitate the markup of Chinese texts by drawing data from other utilities and online databases to analyze, map, and compare texts (MARKUS). Beyond data curation and management, Wang, Tsui, and Bol assert that a cyberinfrastructure should offer platforms for scholarly communication, both online and offline, in both digital and analog formats, emphasizing the necessity of continued collaboration and communication among researchers, libraries, and database companies. In conclusion, they present a compelling argument for resource sharing in the Chinese history field, supported by concrete examples of tools and standards.

These discussions can be seen as an effort by Chinese DH academics to define a DH infrastructure. However, the current discussions are usually either pitched at the macro level of conceptualization or at the micro level of focusing on specific project cases. There is no systematic construction program like the Digital Research Infrastructure for the Arts and Humanities (DARIAH) or the European Research Infrastructure Consortium (ERIC). This could mean that DH in China is still lacking in the development of an infrastructure, and more systematic approaches will first require addressing fundamental issues such as who builds the DH infrastructure, which we will turn to next.

Who Builds the Digital Humanities Infrastructure?

It is only in the past two decades that the concept of infrastructure has turned from the original context of physical facilities providing public services for everyday social life to being associated with terms such as *science and technology*, *data*, and *research* in the Chinese context. *Infrastructure* has become one of the new Chinese words frequently mentioned in state-of-the-field articles, acquiring additional prefixes to create new Chinese terms, such as *shuzi jijian* (数字基建, digital infrastructure), *xin jijian* (新基建, new infrastructure), and *lüse jijian* (绿色基建, green infrastructure). The development of science and technology infrastructure, in particular, is deemed an important national strategic policy in China. The Chinese central government has issued several strategic plans on such infrastructure, among which the most important official documents include the "Outline of National Medium- and Long-Term Science and Technology Development Plan (2006–2020)" (《国家中长期科学和技术发展规划纲要 (2006–2020 年)》), the "Medium- and Long-Term Plan for the Construction of National Major Science and Technology Infrastructure (2012–2030)" (《国家重大科技基础设施建设中长期规划 (2012–2030 年)》), the "Thirteenth Five-Year Plan for the Construction of National Major Science and Technology Infrastructure" (《国家重大科技基础设施建设"十三五规划"》), and other items addressing long-term planning and investment in science and technology infrastructure, especially major scientific and technological installations. Regarding the latter, for example, official statistics indicate that the fifteen science and technology infrastructure projects that have actually been carried out during "The Twelfth Five-Year Plan for National Economic and Social Development of the People's Republic of China" and "The Thirteenth Five-Year Plan for National Economic and Social Development of the People's Republic of China." (2011–2020) involved a total investment of over 30 billion yuan. In plans for such infrastructure projects, *data infrastructure* began to be frequently mentioned around 1998 in relation to geographical mapping and spatial science, especially as these approaches were associated with concepts such as "digital earth," "digital city," and related research topics that caught the attention of policymakers. The phrase *research infrastructure* appeared much later and did not become a standard phrase for Chinese stakeholders until after academic articles introduced the term, mainly in reference to research infrastructures in Europe, the United States, Japan, Australia, and elsewhere.

Meanwhile, throughout this period of growing awareness of the importance of research infrastructure, the humanities appeared quite distant from anything to do with infrastructure. Some scholars have discussed the reasons behind this longstanding disconnect. Xiao Long of Macau University of Science and Technology, formerly of the Peking University Library and Shanxi University and also the former deputy director of the China Center for Humanities and Social Sciences in Higher Education, points out that due to the uniqueness of humanities and social science research, a research infrastructure that aids the humanities is distinct from that for the natural sciences (Xiao). For example, requirements for hardware facilities in the humanities are usually smaller, and most humanities scholars do not need laboratories for their work. Instead, demanding academic requirements for scholarly documentation and information resources make the humanities and social sciences "literature-dependent" disciplines, not infrastructure-dependent ones. This argument seems bolstered by the fact that a large amount of money has been invested in China on the construction of text databases that archive and disseminate scholarly literature in the humanities and social sciences. Those familiar with the context of academic papers in China will immediately think of the China National Knowledge Infrastructure (CNKI), an online publishing platform that owns the largest database of academic papers originating in Chinese academia and was developed by a Chinese state-owned software company linked to Tsinghua University. Even though Xiao has acknowledged the need for a humanities research infrastructure, she has referred to infrastructure only metaphorically in discussing the function of libraries.

However, the emergence of DH in China now creates a higher awareness of the need for a specifically humanities-oriented Chinese research infrastructure, even if it is still nascent by comparison with scientific and technological infrastructure. To elevate cultural digitization as a national strategy, the Chinese government in May 2020 issued its "Notice on the Proper Establishment of National Culture Big Data System" (*guanyu zuohao guojia wenhua dashuju tixi jianshe gongzuo tongzhi* 关于做好国家文化大数据体系建设工作通知)[3] and in 2022, its "Opinions on Promoting the Implementation of National Culture Digitization Strategy" (*guanyu tuijin shishi guojia wenhua shuzihua zhanlüe de yijian* 关于推进实施国家文化数字化战略的意见) came out.[4] These policies aim to construct a fundamental cultural digital infrastructure and service platform by the end of the "Fourteenth Five-Year Plan for National Economic and Social Development of the People's Republic of China" in 2025, and a national cultural data system by 2035. The national policy objective is thus a "cultural big data" infrastructure at all levels of government, with a standardized management system of data storage, production, processing, utilization, and exchange providing "physical distribution, logical correlation, rapid linking, efficient searching, comprehensive sharing, and key integration" for a panoramic digital view of Chinese culture (State Council, "关于推进实施," para. 2). In essence, this will mean a mega-infrastructure built on top of various types of digital subinfrastructures.

Exploratory discussions and scholarly work on building DH infrastructure in China must occur against this backdrop of government policy. Yet a comparison with research infrastructures (particularly in the humanities) in Europe, the United Kingdom, Australia, and the United States reveals that a more bottom-up, adaptive humanities research infrastructure—one that does not rely solely on top-down, policy-level interventions—is relatively underemphasized. The need for a more bottom-up, adaptive humanities research infrastructure is a lesson learned from past problems in constructing DH infrastructures in China. The national campaign to digitize Chinese ancient books is one example. Such digitization began in the 1970–1990s, with strong national policy support and aligned with the recognition by scholars that adequate DH infrastructure was necessary for research on Chinese documentology and the digitization of ancient books (Tsui, "*Hua wen* 华文"). In early 2007, the Ministry of Culture and Tourism officially launched the Chinese Ancient Books Protection Project, which received a 25 million yuan grant from the central government to protect Chinese ancient books and sustain traditional

Chinese culture.⁵ Participating in the project (and responsible for hosting the project's administrative office), the National Library of China collaborated in 2016 with other provincial and municipal libraries to create a digital platform (*Zhonghua guji ziyuanku pingtai* 中华古籍资源库平台) that released more than 102,000 digital images of ancient book items. Another case is the National Platform of Digital Resource of Ancient Books (*Guojia guji shuzihua ziyuan pingtai* 国家古籍数字化资源总平台), hosted by Gulian (Beijing) Media Tech. Co., Ltd., a subsidiary of the Zhonghua Book Company, a major publisher in the humanities. The platform is "guided and managed by the Office of the Leading Group of National Planning for the Collection and Publication of Ancient Books (*quanguo guji zhengli chuban guihua lingdao xiaozu bangongshi* 全国古籍整理出版规划领导小组办公室)."⁶ It aims "to solve the problems of dispersed ancient books digitization resources, low utilization rate, and duplicated construction . . . to build a national authoritative, integrated and public welfare ancient books digitization resource platform, to realize the convergence, concentration and effective utilization of the ancient books digitization resources, and to integrate the results of digitization of ancient books with related databases and digitization platforms."⁷ The ambitious project echoes the series of national policies issued in 2022 in two main documents: the "Opinions on Promoting the Work of Ancient Books in the New Era" (*guanyu tuijin xinshidai guji gongzuo de yijian* 关于推进新时代古籍工作的意见), issued by the General Office of the CPC Central Committee and the General Office of the State Council, and the "National Work Plan for Ancient Books 2021–2035" (2021–2025 *guojia guji gongzuo guihua* 2021–2035 年国家古籍工作规划). These documents, along with other instances of state directives, currently represent a top-down approach to the development of DH infrastructure, especially for premodern Chinese books. However, as the National Platform of Digital Resource of Ancient Books mentions on its website, problems such as lack of integrated planning, the absence of uniform standards, and various technical limitations cannot be solved solely through top-down policy.

How best, then, to approach these challenges? As previously mentioned, discussions on DH infrastructure in Chinese academia often involve adapting concepts from Western contexts, while also attempting to establish a definition and approach for China. However, building infrastructure for DH work is not merely a technical matter akin to creating a data service system. Such infrastructure must be founded on a combination of scientific and technical concepts, hardware equipment, software development, workflows, team building, and a broader social-cultural ecology encompassing all stakeholders and participants. Equally important is the human "wetware"—the interactions between individuals and their engagement with systems—an aspect that extends beyond the scope of hardware and software infrastructure. Equally important, therefore, are human and social factors that contribute to the success and sustainability of infrastructure. This tenet is the core of the discussion that follows, about who is responsible for building China's DH infrastructure and for whom is it built.

A Digital Humanities Infrastructure for Whom?

For whom is DH infrastructure in China being built? Answers to this question in China often align with state-led initiatives and their predetermined objectives for particular institutional and individual users. Three kinds of initiatives and their intended beneficiaries have been most important.

The first is the national, government-led initiative in the late 2010s and into the 2020s to build digital infrastructure, especially "new digital infrastructure" (*xinxing shuzi jichu sheshi* 新型数字基础设施). Part of China's efforts to build "new infrastructures" (*xin jijian* 新基建), including infrastructures that are "digital, smart, and innovative" (State Council, "关于推进实施,") this initiative was a crucial development priority for the Chinese government after the outbreak of the Covid-19 pandemic because of the importance of digital technologies in controlling and mitigating the pandemic's effects and for supporting economic growth. New infrastructures created for initiatives such as this provide resources and support for government bodies, businesses, and individuals to thrive in a rapidly changing digital landscape. They are public-facing in creating 5G networks, data centers, artificial intelligence (AI), the Internet of Things, and so on.

The second kind of initiative that we identify here are infrastructural projects spearheaded by actors in the GLAM sector in China. These serve the needs of the GLAM institutions themselves and their patrons and often focus on specific types of collections or data. An example at the national level is the National Science and Technology Library (NSTL) (*Guojia keji tushu wenxian zhongxin* 国家科技图书文献中心), an information network infrastructure and virtual institution set up in 2000. It links national-level libraries and major institutions while coordinating standards for scholarly information resources and administering the sharing of resources among national research institutes. NSTL also focuses on collecting scholarly papers, databases, and documents in science and technology published in China to provide research support services. Approved as part of this initiative in 2001 was the National Science and Technology Digital Library (*Guojia keji shuzi tushuguan* 国家科技数字图书馆), which led to the establishment of a network of centers offering digital library services focused on scientific literature.

The third kind of initiative, and the one most immediately relevant to DH, concerns developing academic-led DH infrastructures. These projects are mainly developed by academic institutions, researchers, and research groups, and they address the scholarly community through outputs in such traditional scholarly formats as articles in DH journals. Notably, there are now at least four DH periodicals in Chinese published in the China region. These include one from Taiwan, *Shuwei diancang yu shuwei renwen* 数位典藏与数位人文 (*Journal of Digital Archives and Digital Humanities*), and three from mainland China: *Shuzi renwen* 数字人文 (*Digital Humanities*), *Shuzi renwen yanjiu* 数字人文研究 (*Digital Humanities Research*), and the homonymous (in Chinese) *Shuzi renwen yanjiu* 数字人文研究 (*Digital Humanities Studies*).

Beyond producing traditional scholarship, however, academic-led DH infrastructures have the key benefit of enabling diverse outputs adapted to digital scholarship. As digital technologies continue to advance, researchers in China are increasingly exploring new, creative ways to share research findings—including through such born-digital materials as historical GIS data generated and curated by scholars. For this purpose, DH platforms have been developed to enable uploading, analyzing, visualizing, and sharing Chinese data (and data in Chinese). Some of these platforms are geared specifically toward premodern Chinese humanities. Examples include DocuSky, Jihe Net (籍合网), MARKUS, Shanghai Library's Open Data Platform, and Zhejiang University's Chinese Academic Map Publishing Platform, among others currently under development (Tsui, "Charting the Emergence of the Digital Humanities in China").

To better understand the significance of academic-led digital research infrastructures, we will briefly survey four projects with infrastructural functions. The first two are projects dealing with the intersection of textual and geospatial data that emerged from efforts to map literary writers and texts in premodern China. One, Sou-yun (搜韵)—the largest website for Chinese poetry—incorporates numerous functions for accessing, analyzing, and mapping traditional Chinese poems. With more than 5.5 million users by 2020, it testifies to the high demand for accessing such poetry for literary appreciation and education.[8] While academics curate Sou-yun's core data (as in the case of its "Chronological Map of Tang-Song Literature," led by literary scholar Wang Zhaopeng), the project is a community-driven resource—one more successful than most academic databases in promoting a specific field of humanities study (Wang and Qiao). The second project that we discuss along with Sou-yun is the Chinese Academic Map Publishing Platform (AMAP) from Zhejiang University, which also relies on crowdsourced data and allows users to map their own data on the platform. AMAP enables the accumulation of spatiotemporal datasets and provides integrated visualization tools for these datasets.

The second pair of cases briefly considered here are MARKUS and DocuSky, which fulfill multiple infrastructural functions for DH research. Led by digital humanists working with primarily Chinese-language data, both platforms focus on the online storage, processing, visualization, and sharing of texts. MARKUS, mentioned earlier, is a "reading and text analysis platform with a wide range of functionality," which helps researchers tag and annotate Chinese and Korean texts (MARKUS). DocuSky is an "online platform for humanities scholars to organize, use, and analyze personalized materials to meet their research needs."[9] Designed especially with the needs of humanities researchers in mind, these projects facilitate tracking and intervening in data analysis. Enabling conversion between text-markup formats and locally processed downloaded texts, they do not "lock in" users, in the manner of other projects that make it easy to ingest data but then difficult to export that data to other platforms in other formats. In addition, both MARKUS and DocuSky offer user-friendly environments with interface designs and how-to manuals notably attentive to the needs of humanities scholars. Both platforms also work with APIs for interoperability with other databases, facilitating data sharing and integration.

The development of the four projects mentioned here adopts a model of collaboration between humanities scholars and technical experts. For example, Sou-yun's developer is an engineer and poetry enthusiast who collaborates with humanities scholars led by the literary scholar Wang Zhaopeng; AMAP is built by scholars of premodern literature inspired by a collaboration with Harvard University's Center for Geographic Analysis team; MARKUS is supported by a specialized team that includes both humanities scholars and software engineers; and DocuSky is constructed by the first batch of Taiwan's DH scholars, who combine computational expertise and scholarly literacy in the humanities.

Broadly, these projects show that multidisciplinary collaboration is feasible for humanities research. One could even say that they represent to some extent a possible model for DH infrastructure initiatives. But sadly, not all digital infrastructure projects for humanities scholars in China learned from such a collaborative model. For a long time, building a database or developing a digital tool was taken to be the work of software engineers and could be delegated to them. This is one reason that digital infrastructure for the humanities has often been outsourced to companies with technical experts or libraries with a dedicated staff in technical services. Lacking technical skills, humanists were not traditionally involved in creating or evaluating infrastructure in these scenarios. This practice is rooted not only in the division of the "two cultures" or "three cultures," but also in the underlying disciplinary structures of higher education in China. One view is that the work of making digital infrastructures is exclusively that of a different branch of knowledge work done by IT engineers or librarians, not the domain of humanists. However, there are promising signs of change in the DH landscape in China, particularly in how interdisciplinary research groups operate. For example, a research group led by James Lee and Cameron Campbell based mainly at the Hong Kong University of Science and Technology has shown that humanists are definitely able to conduct humanities and social science research involving the building of databases and digital infrastructure (Liang, Dong, and Li). More humanists are thus now recognizing the importance of participating in developing these infrastructures to support their scholarly work. To build a broad-purposed digital infrastructure that meets the needs of a diverse range of academic users, a multidisciplinary and collaborative approach is needed—one that involves humanists, artists, and designers in ensuring that user experience and design are optimized for humanities research. In addition, the participation of community members is essential to ensure that infrastructure addresses the needs of diverse communities and enables the sharing of data.

Future Challenges for the Chinese DH Community

One main future challenge for DH and its infrastructure in China concerns the resources and attention that will be needed for growth within the broader institutional infrastructures of higher education. Here, it is imperative to examine DH specifically in light of undergraduate education. According to the Ministry of Education, there were 47.63 million enrolled students (including undergraduate and graduate students) in the country in 2023, making China's higher-education system the largest in the world.[10] If the objective for DH infrastructure in China is to improve what we earlier called the overall "wetware" and not only the hardware and software of DH, then creating suitable curricula, teaching materials, and training courses for undergraduates should be a central focus. But DH education in China has yet to receive the attention and support necessary to be established on a solid foundation.

Leading Chinese universities—including Peking University, Nanjing University, Renmin University of China, Wuhan University, and Zhejiang University—all offer some DH courses to undergraduates (Tsui, Zhu, and Chen). For instance, Nanjing University, which one of us works at, has extended its offerings to a massive open online course (MOOC) in DH and open to the public,[11] while Renmin University of China has developed a more comprehensive honors program in digital humanities consisting of eighteen courses (School of Information Resource Management, Renmin University of China).[12] With the exception of the MOOC, these courses are usually electives. Such courses cover essential DH theories and such areas as digital history, digital art history, historical GIS, data visualization, and the usage of relevant software. Universities in China also provide short-term DH training programs in the form of one- or two-day workshops, one- or two-week summer schools, or one-off lectures offered by academic departments (Tsui et al.).

Yet on the whole, the institutional infrastructure for DH teaching in Chinese higher education remains inadequate, and DH courses have yet to be truly incorporated in the general undergraduate education curriculum. At the beginning of this chapter, we discussed general education and New Liberal Arts as the current state-sponsored models for the humanities within the Chinese higher education system. But DH has not been formally recognized as part of either paradigm. Instead, it is viewed as training in digital skills or as a driver for innovation.[13] Relatively narrow views of the scope and importance of DH teaching are compounded by the general shifting of resources and attention in universities away from the humanities, toward increasingly prominent professional schools such as those in engineering and business and law. These are perhaps not specific to China, but such factors certainly constrain the advancement of humanities education in Chinese universities generally and DH education specifically.

Another infrastructural challenge for the DH community in China concerns the lack of coherent organization of DH researchers in relation to available infrastructures. Currently, DH scholars are very decentralized, with some clustered in the library and information science fields and others gathered around interdisciplinary initiatives or settings (such as DH centers or publication projects, including academic journals). In this framework, researchers and technological infrastructures cannot bond together into the kind of cohesive community envisioned in early, optimistic views of DH such as that expressed in the "Digital Humanities Manifesto 2.0," which argues that DH "dreams of models of knowledge production and reproduction that leverage the increasingly distributed nature of expertise and knowledge and transform this reality into occasions for scholarly innovation, disciplinary cross-fertilization, and the democratization of knowledge" (Schnapp and Presner, 5). Such a vision of DH is inspiring, but transforming institutional and technological infrastructures particularly in a higher education framework toward this goal is still hugely challenging, especially when the framework distributes resources strictly by disciplines. Creating interdisciplinary communities and cross-fertilizing the knowledge of their members is a process that requires constant, persistent work.

One effort in China to address such challenges is the Chinese Alliance of Digital Humanities Institutions (CADHI), an institutional consortium and special committee on DH that (because it is difficult to set up entirely new professional scholarly associations in China with official standing) launched under the umbrella of the already established Chinese Society of Indexing. The CADHI currently comprises DH institutions from more than ten universities, publishes annual reports, and holds an annual conference (most recently, CDH2023 in Wuhan, CDH2024 in Shanghai, and CDH2025 in Guangzhou).[14] However, its infrastructural potential has yet to be fully realized, as not many regular activities have been implemented. Nonetheless, the alliance represents an important step toward establishing a more cohesive and organized DH community in China.

A third significant infrastructure-related challenge for the Chinese DH community is the lack of sufficient assistance by professional staff—such as engineers, librarians, artists, designers, and others—acting as supporting teams or teaching auxiliaries. This problem can be attributed in part to the structural makeup of Chinese universities, which are heavily staffed by faculty and administrators while offering inadequate opportunities for engineers or researchers in project-funded positions, especially in the humanities fields. In addition, financial-management policies in universities make it difficult to hire project-based staff; and, of course, even if available, project funding is not guaranteed in the long term. For DH, long-term planning is thus very challenging and project sustainability is imperiled. As a result, many DH projects in China have historically taken the form of partnerships with companies. Researchers often delegate the technical implementation of projects to companies or collaborate with them on spin-off projects. This practice of outsourcing technology infrastructure has led to technical inertia among academic researchers and hindered their ability fully to control or lead the technical development required for scholarly advances in the humanities. Moreover, this practice inhibits scholarly community building since the interests and priorities of corporate partners do not always align with those of scholars. An example of recent efforts to address this issue is the partnering of Peking University's Research Center for Digital Humanities with ByteDance (a large Chinese tech company) to support research infrastructure. In 2022, the center received from ByteDance Public Welfare a donation to conduct research on the intelligent development and utilization of ancient books. The PKU Digital Humanities Lab operating under the Peking University Institute of Artificial Intelligence was then renamed the Peking University–ByteDance Digital Humanities Open Lab. Because the contribution by ByteDance was in the form of a donation, and thus less strictly defined in its uses than government or university research funds, the lab has more flexibility compared to other university and company initiatives.[15] But such examples remain uncommon.

How public communities can engage in the work of DH community is also a challenge that warrants exploration at the level of infrastructure building. The GLAM sector recognizes the importance of public engagement in DH. For example, the Shanghai Library has been actively working for many years to establish a DH infrastructure for public use. The Chinese Genealogy Knowledge Service Platform is a notable example. Similarly, the Shanghai Museum has introduced an online interactive platform to accompany its exhibit of works by Dong Qichang (1555–1636), a famous Chinese artist of the Ming Dynasty. The platform employs digital methods to showcase the artist's life, works, and social networks. Public cultural institutions with resources like these are experienced in creating public-facing DH projects. But by contrast, many university-based DH projects address only the research needs of specific scholars, making it difficult due to intellectual property and other concerns to provide access to their outcomes even to other scholars, let alone the public. Largely absent in university-based DH projects is any opportunity for the involvement of public communities, even when platforms that might allow for this are in place (Chen, "*Dangxia* 当下"). The goal of public access to DH work is increasingly emphasized within the international DH community. China's university-based DH initiatives also need to make more of an effort in this direction.

This chapter cannot by itself cover all aspects of DH infrastructure in China, of course. We have tried to provide a brief overview of the development of DH infrastructure in the nation by surveying how Chinese DH infrastructure is defined and who its builders and beneficiaries are. However, we can do no more than give a glimpse of the spots, as when "a leopard is seen through a tube" (as in the Chinese saying: *guanzhong kuibao* 管中窥豹). Even such a glimpse, however, allows us as Chinese DH researchers to conclude by sharing our hopes and concerns for the future development of DH infrastructure in China.

On a positive note, we hope that the Chinese government's determination and ability to build infrastructure from the top down may benefit DH. The association of DH with digital technology, AI, and other sci-tech areas may make it an object of attention for the government. This should be a good development for DH, and possibly the humanities more broadly, given the rather difficult situation of the humanities globally at the moment. Second, we are encouraged by rising attention within the Chinese DH community itself to the need for infrastructural development. As we are revising this chapter, a number of new centers and platforms have appeared and will have an impact in the near future.

But, of course, it is impossible to ignore what may be negative impacts of the expectations mentioned in this discussion. Government support comes with policy guidance, and the result may be that the development of DH infrastructure will for some time be centralized and top-down. This would impose inflexibility on the whole system, constraining academic users of DH infrastructure from addressing their needs and working on their ideas in responsive and effective ways. Also, the rise of interest in DH infrastructure (in association with an enthusiasm for all things scientific and technological in our day and age) may in the worst case result in a bubble that is destined to burst for lack of a stable, integrated, and sustainable infrastructural foundation that can truly support humanities research and teaching. It will require the strength of the whole Chinese DH community to act on its hopes by addressing its challenges.

NOTES

Acknowledgment: Jing Chen's work as a coauthor of this chapter is sponsored by the National Social Science Foundation of China (No. 21BA026) and Lik Hang Tsui's work is sponsored by Hong Kong's Research Grants Council General Research Fund project "An Analytical History of the Early Years of Digitising Chinese Historical Sources (1980–2009): The View from Five Foundational Database Projects" (No. 9043453 [CityU 11604422]).

1. See also our recent chapter in *Global Debates in the Digital Humanities*, which in some parts touches on issues of cyberinfrastructure in relation to Chinese DH (Chen and Tsui, esp. 82).

2. "Audiovisual education" or "Electrified education" (*dianhua jiaoyu*, 电化教育) was promoted by the central government of the Republic of China in the 1930s as a national policy, as in such cases as "The Principal Law for Electrified Education" (*dianhua jiaoyu zhongyao faling*, 电化教育重要法令) from 1932–1936. Following the development of audiovisual education in Western countries, the campaign aimed to take advantage of film and broadcasting as mass media to transform the education system from a traditional model into a modernized one. (For more information, see Guangsheng Du and Jing Zhu.) This was also taken up under the People's Republic of China. The campaign for "quality-oriented" education (*sushi jiaoyu*, 素质教育) was then launched around 1995. The Ministry of Education announced a series of polices to guide higher-education institutions in enhancing and promoting humanities education that is integrated with science education, which for a long time prior to the 1990s was considered much more important

than the humanities. General education (*tongshi jiaoyu*, 通识教育), inspired by the U.S. model of higher education grounded in the liberal arts, was then introduced as a reform of humanities education in several top universities in China beginning in the 2000s. (The universities affected include Sun Yat-sen University, Tsinghua University, and Nanjing University, among others.) Subsequently, general education became a national campaign in higher education recognized by the Ministry of Education. When the "Implementation Plan for the Audit and Evaluation of Undergraduate Education Teaching in General Colleges and Universities (2021–2025)" was issued in 2021, general education was included in the undergraduate curriculum in colleges and universities. (More detail can be found in the papers of Shuzi Yang and Yang Gan.) The New Liberal Arts (*xin wenke*, 新文科) is the latest policy in the humanities. Announced in 2020, its purpose is to encourage cross-disciplinary research in the humanities, science, and technology. The policy especially emphasizes using new technologies such as IT and AI for research and training in the humanities. For more information, see "The Manifesto of Construction of the New Liberal Arts" (*xin wenke jianshe xuanyan*, 新文科建设宣言, November 2020), https://jwc.cuc.edu.cn/2022/0114/c6974a190755/page.htm.

3. See http://www.gsass.net.cn/zdxm/whypt/zlhj01/content_3972.

4. See http://www.gov.cn/xinwen/2022-05/22/content_5691759.htm.

5. This project is guided principally by the official document of the "Opinions on Further Strengthening the Protection of Ancient Books" (*guanyu jinyibu jiaqiang guji baohu gongzuo de yijian*, 关于进一步加强古籍保护工作的意见) issued by the General Office of the State Council. The National Center for the Protection of Ancient Books of China was under the National Library and assumes the responsibilities of professional guidance center, training center, and research center for the protection of ancient books of the country. For more information, see the official report on the Chinese Ancient Books Protection Project (zhonghua guji baohu jihua, 中华古籍保护计划) in 2008: https://zwgk.mct.gov.cn/zfxxgkml/ggfw/202012/t20201205_916516.html.

6. The Leading Group of National Planning for the Collection and Publication of Ancient Books (quanguo guji zhengli chuban guihua lingdao xiaozu, 全国古籍整理出版规划领导小组) was set up in 1958, halted during the Cultural Revolution in the 1960s and 1970s, and resumed in 1981. The main function of the group is to be responsible for the collation and republication of material from ancient Chinese books. This Leading Group office is led by the National Press and Publication Administration (Guojia xinwen chubanshu, 国家新闻出版署) and is considered to be the highest administrative office for ancient book preservation and digitization in China.

7. See http://read.nlc.cn/.

8. User statistics are taken from the Sou-yun site: https://sou-yun.cn/about.asp.

9. See https://docusky.org.tw/DocuSky/.

10. See http://www.moe.gov.cn/jyb_sjzl/sjzl_fztjgb/202410/t20241024_1159002.html.

11. For example, see https://www.icourse163.org/course/NJU-1465603161.

12. See https://irm.ruc.edu.cn/rcpy/bss/b_zyjs/szrwzy/index.htm.

13. When we applied to design a new DH course for undergraduates at Nanjing University, we were told that such a course should be listed under the category of "innovation and entrepreneurship" (*chuangxin chuangye*, 创新创业) because it imparts new digital skills. Instead of calling it "Introduction to Digital Humanities," therefore, we should call it something that includes the term *innovation* in order to fit the course catalogue and its categories. In the end, we called our course "Innovative Thinking and Methods in the Digital Humanities" (*shuzirenwen chuangxin siwei yu fangfa*, 数字人文创新思维与方法).

14. See https://simjwz.whu.edu.cn/info/1073/13481.htm; https://schim.shu.edu.cn/info/1375/8863.htm; https://ischool.sysu.edu.cn/zh-hans/event/718.

15. See https://pkudh.org/intro.html.

BIBLIOGRAPHY

Atkins, Daniel E., et al. "Revolutionizing Science and Engineering Through Cyberinfrastructure: Report of the National Science Foundation Blue-Ribbon Advisory Panel on Cyberinfrastructure." National Science Foundation, 2003. https://web.archive.org/web/20121017100852/http://www.nsf.gov/od/oci/reports/atkins.pdf.

Australian Government Department of Education. "2021 National Research Infrastructure Roadmap Exposure Draft." 2022. https://www.education.gov.au/national-research-infrastructure/resources/2021-national-research-infrastructure-roadmap-exposure-draft.

Bol, Peter, Cuijuan Xia, and Hongsu Wang (包弼德、夏翠娟、王宏甦). "数字人文与中国研究的网路基础设施建设." 图书馆杂志 37, no.11 (2018): 18–25.

Chen, Jing (陈静). "当下中国'数字人文'研究状况及意义."山东社会科学, no. 7 (2018):59–63.

Chen, Jing (陈静). "新一轮科技革命与新文科发展." 中国社会科学报. August 31, 2020, http://www.nopss.gov.cn/n1/2020/0831/c219544-31842562.html.

Chen, Jing (陈静). "数字人文创新思维与方法." China University MOOC, February 16–June 30, 2025. https://www.icourse163.org/course/NJU-1465603161.

Chen, Jing, and Lik Hang Tsui. "Debating and Developing Digital Humanities in China: New or Old?" In *Global Debates in the Digital Humanities*, edited by Domenico Fiormonte, Sukanta Chaudhuri, and Paola Ricaurte, 71–86. University of Minnesota Press, 2022. https://dhdebates.gc.cuny.edu/read/global-debates-in-the-digital-humanities/section/2662518e-42ff-4026-9605-a1c27b4aed27#ch06.

Chen, Tao, Rina Su, and Xun Sun (陈涛、苏日娜、孙逊). "数字人文基础设施中图像中台设计与探讨." 图书馆杂志 40, no.10 (2021): 124–32.

China Biographical Database (CBDB). Home page, n.d. https://projects.iq.harvard.edu/cbdb/home.

Chinese Genealogy Knowledge Service Platform. Home page, 2016. https://jiapu.library.sh.cn/.

China Historical GIS (CHGIS). Home page, n.d. https://chgis.fas.harvard.edu/.

China National Knowledge Infrastructure (CNKI). Home page, 2024. https://oversea.cnki.net/index/.

Chinese Academic Mapping Platform (AMAP). Home page, n.d. http://amap.zju.edu.cn/.

Courant, Paul N., Sarah E. Fraser, Michael E. Goodchild, et al. *Our Cultural Commonwealth: The Report of the American Council of Learned Societies Commission on Cyberinfrastructure for the Humanities and Social Sciences.* American Council of Learned Societies, 2006. https://www.acls.org/wp-content/uploads/2021/11/Our-Cultural-Commonwealth.pdf.

DocuSky Collaboration Platform. Home page, 2024. https://docusky.org.tw/DocuSky/.

Du, Guangsheng (杜光胜). "民国时期江苏省电化教育发展研究." PhD diss., (Inner Mongolia Normal University, 2013).

European Strategy Forum on Research Infrastructures (ESFRI). Home page. 2024. https://www.esfri.eu/.

Gansu Academy of Social Sciences (甘肃省社会科学院). "关于做好国家文化大数据体系建设工作的通知." May 11, 2020. http://www.gsass.net.cn/zdxm/whypt/zlhj01/content_3972.

Gan, Yang (甘阳). "大学人文教育的理念、目标与模式." 北京大学教育评论4. no. 3(2006): 38–65.

General Office of the CPC Central Committee and the General Office of the State Council (中共中央办公厅、国务院办公厅). "关于推进实施国家文化数字化战略的意见." 中华人民共和国中央人民政府, May 22, 2022. https://www.gov.cn/xinwen/2022-05/22/content_5691759.htm.

Hart, David. "Cyberinfrastructure: A Special Report." National Science Foundation. 2006. https://www.nsf.gov/news/special_reports/cyber/Cyberinfrastructure%20_NSF.pdf.

Horváth, Alíz, and Hilde De Weerdt. "Special Issue on Digital Humanities and East Asian Studies." International Journal of Digital Humanities 4, no. 1(2023): 1–4.

Jin, Jiaqin, and Cuijuan Xia (金家琴、夏翠娟). "数字人文数据基础设施建设中机构本体的构建:研究和应用." 图书馆论坛 40, no. 4 (2020): 30–39.

Liang, Chen, Hao Dong, and Zhongqing Li (梁晨、董浩、李中清). "从看一幅画到做一幕戏:互联网时代历史教研新动向探微." 文史哲, no. 6 (2018):121–134.

Liu, Wei, Rong Xie, and Lei Zhang (刘炜、谢蓉、张磊). "面向人文研究的国家数据基础设施建设." 中国图书馆学报 42, no. 5 (2016): 29–39.

Lu, Dan, Xin Li, and Jinchuan Chen (鲁丹、李欣、陈金传). "基于API技术的数字人文基础设施的建构." 图书馆学研究, no.13 (2019): 42–46.

MARKUS. "MARKUS," n.d. https://dh.chinese-empires.eu/markus/.

Ministry of Education of the People's Republic of China. "2020年全国教育事业统计主要结果." March 1, 2021. http://www.moe.gov.cn/jyb_xwfb/gzdt_gzdt/s5987/202103/t20210301_516062.html.

Ministry of Education of the People's Republic of China (中华人民共和国教育部). "2023年全国教育事业发展统计公报." October 24, 2024. http://www.moe.gov.cn/jyb_sjzl/sjzl_fztjgb/202410/t20241024_1159002.html.

Nanjing University. n.d. "数字人文创新思维与方法." https://www.icourse163.org/course/NJU-1465603161.

National Digital Library of China (读者云门户). Home page, n.d. http://read.nlc.cn/user/index.

National Science and Technology Digital Library (国家科技数字图书馆). Home page, 2024. https://sthj.nstl.gov.cn/.

National Science and Technology Library (NSTL) (国家科技图书文献中心). Home page, 2024. https://www.nstl.gov.cn/index.html.

Rao, Zixin, Luxiang Deng, and Xin Xu (饶梓欣、邓璐芗、许鑫). "国际视野下面向人文研究的数据基础设施分析与探讨." 图书情报知识 39. no. 5 (2022): 31–41+11.

Research Center for Digital Humanities, National Taiwan University (國立臺灣大學數位人文研究中心). "DocuSky 數位人文學術研究平台." Home page, n.d. https://docusky.org.tw/DocuSky/.

Research Center for Digital Humanities of PKU (北京大学数字人文研究中心). "北京大学-字节跳动数字人文开放实验室." n.d. https://pkudh.org/intro.html.

Schnapp, Jeffrey, and Todd Presner, with Peter Lunenfeld, Johanna Drucker, et al. "Digital Humanities Manifesto 2.0." Humanities Blast (blog), 2011. https://www.humanitiesblast.com/manifesto/Manifesto_V2.pdf.

School of Cultural Heritage and Information Management, Shanghai University (上海大学文化遗产与信息管理学院). " 'Integration of Arts and Sciences: Digital Humanities in the Age of AGI' CDH2024 November 8th-10th, 2024, Shanghai University Notice of Conference (No.2)." June 5, 2024. https://schim.shu.edu.cn/info/1375/8863.htm.

School of Information Resource Management, Renmin University of China (中国人民大学信息资源管理学院). "中国人民大学信息资源管理学院2024年'数字人文荣誉辅修项目'招生公告." April 16, 2024. https://irm.ruc.edu.cn/xydt/tzgg/7f1a296a6eb2480eb6d271e082d427b8.htm.

School of Information Management, Sun Yat-sen University (中山大学信息管理学院). "会议通知 | 人文嬗变：数字人文的智慧奇点"学术研讨会暨2025年中国数字人文年会（CDH2025）." April 11, 2025. https://ischool.sysu.edu.cn/zh-hans/event/718.

School of Information Management, Wuhan University (武汉大学信息管理学院). "数实共生：预见数字人文未来图景——第五届中国数字人文年会（CDH2023）通知." June 12, 2023. https://simjwz.whu.edu.cn/info/1073/13481.htm.

School of Undergraduate Studies, Communication University of China (中国传媒大学本科生院). "新文科建设宣言." January 14, 2022. https://jwc.cuc.edu.cn/2022/0114/c6974a190755/page.htm.

Shanghai Museum. n.d. "董其昌书画艺术大展——上海博物馆." https://www.shanghaimuseum.net/museum/dongqichang/index.html.

Sou-yun. Home page, n.d. https://sou-yun.cn/.

State Council, People's Republic of China. "中共中央办公厅国务院办公厅印发《关于推进实施国家文化数字化战略的意见》." May 22, 2022. https://www.gov.cn/xinwen/2022-05/22/content_5691759.htm.

State Council, People's Republic of China. "China's Central SOEs Up Investment in New Infrastructure." June 22, 2022. https://english.www.gov.cn/statecouncil/ministries/202206/22/content_WS62b2c660c6d02e533532c976.html.

Tsui, Lik Hang. "Charting the Emergence of the Digital Humanities in China." In Chinese Culture in the 21st Century and its Global Dimensions: Comparative and Interdisciplinary

Perspectives, edited by Kelly Chan Kar Yue and Garfield Lau Chi Sum, 203–16. Springer, 2020.

Tsui, Lik Hang (徐力恒). "華文學界的數位人文探索：一種「史前史」的觀察角度." 中國文哲研究通訊 30, no. 2 (2020): 107–27.

Tsui, Lik Hang, Benjun Zhu, and Jing Chen. "Finding Flexibility to Teach the 'Next Big Thing': Digital Humanities Pedagogy in China." In What We Teach When We Teach DH: Digital Humanities in the Classroom, edited by Brian Croxall and Diane Jakacki, 274–91. University of Minnesota Press, 2023.

UK Research and Innovation (UKRI). "The UK's Research and Innovation Infrastructure: Opportunities to Grow Our Capability." 2020. https://www.ukri.org/wp-content/uploads/2020/10/UKRI-201020-UKinfrastructure-opportunities-to-grow-our-capacity-FINAL.pdf.

Wang, Hongsu, Lik Hang Tsui, and Peter Bol (王宏甦、徐力恒、包弼德). "用於中國歷史研究的網路基礎設施：對相關探索的建議和展望." 數位典藏與數位人文 6, (2020): 1–35.

Wang, Lihua, and Yike Zhang (王丽华、章亦可). "面向数字人文的开放数据竞赛研究—基础设施的角度." 高校图书馆工作 5, (2022): 1–7.

Wang, Zhaopeng, and Qiao Junjun. "Geographic Distribution and Change in Tang Poetry: Data Analysis from the 'Chronological Map of Tang-Song Literature'." Translated by Thomas J. Mazanec. Journal of Chinese Literature and Culture 5 (2) (2018): 360–74.

Xia, Cuijuan (夏翠娟). "面向人文研究的'数据基础设施'建设—试论图书馆学对数字人文的方法论贡献." 中国图书馆学报 46, no.5 (2020): 24–37.

Xiao, Long (肖珑). "人文社会科学繁荣发展的软性基础设施建设." 图书情报工作 5, no. 11 (2011): 5–9.

Xu, Chenfei, and Ping Bao (徐晨飞、包平). "面向农史领域的数字人文研究基础设施建设研究—以方志物产知识库构建为引." 中国农史 38, no. 6 (2019): 40–51.

Yan, Jia, Min Yang, and Mei Peng (颜佳、杨敏、彭梅). "面向数字人文的图像数据基础设施建设研究—以我国图博档领域为视角." 图书馆, no. 5 (2021): 51–58.

Yang, Shuzi (杨叔子). "文化素质教育与通识教育之比较." 高等教育研究 28. no. 6 (2007): 1–7.

Yu, Li, and Jiawa Guan (余力、管家娃). "我国古籍数字化建设现状分析及发展研究." 数字图书馆论坛 162, no. 11 (2017): 41–47.

Zhu, Benjun, and Jiuzhen Zhang. "Digital Humanities Cyberinfrastructure for Ancient China Studies: Past, Present, and Future." Library Trends 69, no. 1 (2020): 319–33.

Zhu, Jing (朱敬). "早期电化教育中国特色探源." 电化教育研究 29, no. 2 (2008): 92–96.

PART II][Chapter 12

Reproducibility and Contestation in Humanities Digital Infrastructure

DEB VERHOEVEN, MIKE JONES, TOBY BURROWS, AND ANN BORDA

Contestation is a social activity.

—Antje Wiener

Albert Einstein purportedly quipped (but probably didn't) that "the definition of insanity is doing the same thing over and over again and expecting different results" (Calaprice, 474). To which a humanities scholar might reply, the definition of insanity is doing the same thing over and over again and expecting the same result. Alexander Nehamas expresses a distinct but related sentiment: "What to me is truly frightful is not the quality of what everyone agrees on, but the very fact of universal agreement" (211).

Contestability and contestation are among the underlying precepts of humanities research that help distinguish it from other disciplines that find evidence in processes of consensus. Methods of contestation might suggest the antagonistic excesses of "cancel culture" or the underhand insinuations of popularly deployed strawman arguments. We have in mind something that is less about binary conflicts or one-sided interrogations and more about a disposition to inquiry that is sometimes playful, sometimes pensive, or sometimes introspective. We envisage contestation as a generative practice of creating alternative suggestions for the world as it has been understood and described. Contestation, then, is wrought in the way that ideas sidle up to one another, moving with and against and beyond possibilities to produce new and imagined affinities. Done well, contestation is additive and annotative rather than purely contradictory. Contestation, in this sense, is a form of collaborative knowledge production that can serve to multiply without accumulating, a way of appreciating the substantial work of revision, resistance, and resilience. There is also a humility in practices of contestation that implies that we alone are not the end of the story, or the only ones sanctioned to complete the picture.

Adopting a disposition to contestation and its associated values in digital infrastructure means implementing a distributed or collective approach to who is authorized to speak and contribute, and appreciating the necessity of different, diverse, and competing knowledge systems and views of the world without silencing other points of view by speaking for them or over them or accumulating them to one's own argument. Conversation rather than conversion. We believe that contestation can be more than an extension of liberal theories of recognition that are defined by a pluralistic reverence for cultural differences (as seen in versions of "multiculturalism"). We also want to move beyond the idea that contestability is an operative form of correction and that, as scholars, we are working toward something true or perfect or "objective" (Suchman). In a contemporary era marked by polarized, "post-truth" politics and the promulgation of "alternative facts," it is easy to retreat into outmoded positivist traditions and singular or universalizing (or even Enlightenment) notions of what is true.

Instead, this chapter explores how contestability and contestation bear on digital research and information infrastructure, much of which is explicitly built to support what Christian Bueger called "confirmatory research," by examining in detail some exceptions to convention (Bueger, 131). In this endeavor, we want to stress that we are not proposing a structural antagonism between the humanities and the sciences, but there are disciplinary distinctions that can be meaningfully drawn between their approaches to contestation. Scientific practices of replication and refutation, whether they are applied at the level of procedure (questioning the generality of a method) or validity (disputing empirical accuracy or predictive success, and so on), constitute a scholarly regime of "optimistic negation" in which knowledge practices reach perpetually forward toward a more truthful future (Edelman). For us, contestation most certainly does not have at its core a utopic futurity based on reproductive generation. Paraphrasing Lauren Berlant, we are committed to a political project of imagining how to resile from the limitations of digital infrastructures, by thinking through and with existing alternatives that signal toward a flourishing, "not later, but in the ongoing now" (Berlant and Edelman, 5). In the spirit of our own definition of contestation, we want to recognize and articulate the social conditions of power in which both distinctive knowledges and their suppression result from hegemonic information systems. For us, contestation is an encounter with the social nature of knowledge itself.

In the first part of this chapter, we address the way in which contestability has been minimized in favor of infrastructure focused on reproducibility and consensus. Principal among the key issues identified by scholarly infrastructure organizations (De Weerd-Wilson and Gunn; UKRN; Munafò et al.) is the need to support transparency, efficiency, reproducibility and replication (often synthesized in the term "Open" or "FAIR" practice) in scientific research. And yet these fundaments are not themselves without complexity, especially for humanities disciplines. We argue that rather than focusing on efficiency and replication, humanities digital infrastructure should include support for additive and annotative contestation as a foundational concept. The term *infrastructure* can refer to physical, organizational, institutional, methodological, or technical systems that enable the possibility of acts of interconnection (Verhoeven, "As Luck Would Have It").

The remainder of this chapter looks at two examples of humanities driven digital infrastructure that foster this approach to contestation in different ways. The Humanities Networked Infrastructure (HuNI)[1] platform emphasizes the potential of digital infrastructure to encapsulate collaborative, relational meaning-making; and the digital heritage access platform Mukurtu[2] provides a framework supporting different, sometimes contradictory knowledge systems (especially Indigenous community knowledge). Mukurtu in particular draws on gallery, library, archive, and museum (GLAM) infrastructures. Despite this focus—and a tendency within some communities to conflate humanities infrastructure and GLAM infrastructure more broadly—we see the question of how information archives can better encapsulate contestation as relevant to humanities digital infrastructure across all domains and disciplines. However, as with contestation itself, infrastructural possibilities are contingent, multiple, and composite rather than a single set of specifications or functions. Therefore, while championing the high-level idea that contestation should be a foundational consideration for humanities digital infrastructures, how this might be achieved is necessarily context specific and, in our view, cannot be simply replicated.

Contesting Reproducibility

> *Instead of conducting confirmatory research, we should play with the concepts and be open to surprise and the actual messiness of practice.*
>
> —Christian Bueger (131)

A reproducibility crisis (also known as a *replication crisis*) has been ongoing for over a decade in the sciences, in which large-scale experimental studies across several scientific communities were reexamined in the context of the failure to replicate (Pashler and Wagenmakers; Baker; Hicks). Reproducibility here is broadly considered a means of obtaining the same results from the conduct of an independent study whose procedures are as closely matched to the original experiment as possible (Goodman, Fanelli, and Ioannidis).

The source of this crisis was located particularly in the fields of social psychology and biomedical research. For example, the Reproducibility Project (Open Science Collaboration) could confirm only thirty-nine of one hundred published social psychology studies were reproducible, and similarly preclinical cancer studies were identified as irreplicable (Begley and Ellis). A poll undertaken by the journal *Nature* reported that 52 percent of the scientists surveyed were convinced that science was facing a replication crisis (Baker).

Several types of poor practice (some of which continue) variously identified in the scientific literature suggest that a crisis is at hand—namely, the failure to reproduce the results of published studies in large-scale systematic replication projects (e.g., Open Science Collaboration); evidence of publication bias (Fanelli, "Negative Results Are Disappearing from Most Disciplines and Countries"); a high prevalence of questionable research practices (Fraser et al.); and a general lack of transparency in the reporting of methods, data, and analysis (Nuijten et al.).

Researchers in other fields have referred to the reproducibility crisis in science and a resulting lack of confidence (Pashler and Wagenmakers). Such conditions have led to a growing desire to accommodate support for reproducibility in research infrastructure (including scholarly publishing), as well as to consider the role of reproducibility in other scientific and scholarly fields, such as in the humanities (Peels and Bouter; Peels; Sikk), archaeology (Marwick), and linguistics (Berez-Kroeker et al.). An international cohort of linguists put forward a position statement on reproducibility in 2018 and expressed an overall need to establish protocols for verification and accountability, such as through data citation and attribution (Berez-Kroeker et al., 12). For other academics, however, this crisis has simply been mislabeled (Fanelli, "Opinion: Is Science Really Facing a Reproducibility Crisis, and Do We Need It To?," 2630; Lash, Collin, and Van Dyke). They point to the fact that criticism and comments about reproducibility (real and perceived) largely focus on statistical and methodological approaches (Larregue).

The replication "crisis" has been instrumental in the move toward the adoption of open science, in which open data and open-source software and hardware are critical to enabling reproducible results by making the original data analysis and related processes more transparent (Terras; Munafò et al.; Schofield, Whitelaw, and Kirk). The open science movement is also underpinned by well-developed standards for open data, such as the FAIR (findable, accessible, interoperable, and reusable) principles,[3] which have become increasingly adopted across scientific disciplines (Wilkinson et al.) and are now commonly invoked in library and archive settings (Koster and Woutersen-Windhouwer) and increasingly in the humanities. The European-funded All European Academies (ALLEA)[4] initiative (Harrower et al.) is a recent sectoral example of key recommendations toward aligning digital humanities (DH) with the FAIR principles. Although data-sharing approaches have been ubiquitous in humanities computing (Terras; Sikk), there also have been criticisms of FAIR's applicability in the humanities (Verhoeven, "Scholarship in a Clopen World").

There is a smaller debate emerging over the definition, process, and value of reproducibility itself. The notion of reproducibility has been largely based on empirical, quantitative sciences where data are reified and viewed as objective. However, the terms *reproducibility* and *replicability* are not interchangeable in this sense; actually, they have variable meanings and requirements across different science domains (Plesser). Sabina Leonelli focuses on replicability and reproducibility as different processes that exhibit different characteristics. More broadly, the climate crisis and the complex disciplinary intersections in climatic data collection and reporting flag the limitations of a singular approach to reproducibility and replication (Bush et al.).

In the *Reproducibility and Replicability* report undertaken by the National Academies of Sciences, Engineering, and Medicine (National Academies), there is a concerted move to clarify the definitional boundaries of the two terms. Reproducibility involves the original data and code, and replicability involves new data collection to test for consistency with previous results of a similar study (46). Both processes differ in the type of results that may be expected. Of particular relevance to this chapter is that the contributors to this report consider reproducibility as being synonymous with "computational reproducibility" (obtaining consistent results using the same input data, computational workflow, methods, documentation and code, and conditions of analysis). This definition aligns with the acknowledgment that computational methods are becoming integral to research investigations across digitally transforming and "big data" disciplines, such as DH (Eijnatten, Pieters, and Verheul; Kitchin; Schofield, Whitelaw, and Kirk; McGillivray et al.). Although some scholars argue that computer replicability is possible and necessary in the humanities (Eijnatten et al.; Peels and Bouter; Peels; Sikk), the rationale supporting the practice remains open to challenge. The question around the value of reproducibility in the humanities is further exacerbated by evidence of unreliable, even unusable research in the sciences arising from the use of machine learning and artificial intelligence (AI). This has given rise to a full-circle debate around a renewed crisis in reproducibility (Ball), not least because many variants of AI remain "black boxes"—a system theory debated since the 1960s (Bunge, 346).

Scholars further point out, for example, that reproducibility can be a limited (and often forced) epistemic criterion for research quality and validity, articulating a specific technology of accountability and diverting humanities researchers from assuming the agency to explore innovative approaches or to add intrinsic forms of value to their data (O'Sullivan; Penders, de Rijcke, and Holbrook, "Rinse and Repeat: Understanding the Value of Replication Across Different Ways of Knowing"; Sui and Kedron). Furthermore, humanities researchers typically work within highly contextual and situated findings (Suchman; Cecire; Posner). Bart Penders, Sarah de Rijcke, and J. Britt Holbrook ("Rinse and Repeat"; "Science's Moral Economy of Repair: Replication and the Circulation of Reference") suggest that the wholesale adoption of epistemic standards from the sciences disregards humanities scholars as legitimate knowing subjects or subjects of agency in their own field, thus incurring a form of epistemic injustice (Suchman; Fricker).

In the digital information systems that matter to many humanities scholars, the collections of data describing the work of galleries, libraries, archives, and museums, an emphasis on positivist principles is especially pernicious. Typically, GLAM institutions organize their collections into structured hierarchies of knowledge that serve to reassert the patriarchal and colonial logics that underpinned their formation

(Burrows, Jones, and Verhoeven). Seemingly in contradiction to this is the way that museums and libraries are positioning themselves as increasingly "participatory" knowledge organizations (Shilton and Srinivasan; Simon), engaging in high-profile advocacy addressing alternative truths, racism, homelessness, and migration, among other issues (Borda and Bowen; Lynch; Message; Sieg). In the next sections, we explore how contestability and contestation, as social and epistemological architectures, have been incorporated into alternative designs for digital infrastructure.

Contestability and Complexity in Galleries, Libraries, Archives, and Museums

For GLAM institutions, and museums in particular, the concept of "participation" is now so familiar as to be considered a defining characteristic of modern institutions. The Extraordinary General Assembly of the International Council of Museums in Prague on August 24, 2022, approved a new definition for museums that included the following: "Open to the public, accessible and inclusive, museums foster diversity and sustainability. They operate and communicate ethically, professionally and with the participation of communities" (ICOM).

Digital infrastructure is seen as an important space for inviting such participation, for example through crowdsourcing initiatives. When putting collections online, institutions can provide tools that allow users to tag, annotate, transcribe, describe, or otherwise contribute. As Trevor Owens writes: "[W]e might actually think about crowdsourcing as one of the most precious experiences we can offer our users. Instead of simply giving them the ability to browse or poke around in digital collections, we can invite them to participate. We are in a position to let the users of these collections leave a mark on the collections. Instead of browsing through a collection they literally become authors of our historical record" (128).

Such language emphasizes the agency of contributors. But as a portmanteau of *crowd* and *outsourcing, crowdsourcing* retains much of its original meaning: "The act of taking work once performed within an organization and outsourcing it to the general public through an open call for participants" (Ridge, 1). While institutions might benefit from external expertise, there is often little recognition that there are perspectives outside the institution that are not found within, let alone a move toward spaces outside the institution where people can challenge or contest the authority of that institution. Instead, contributing to "our historical record" becomes a precious gift from those select institutions that "let the users . . . leave a mark."

What is more, these marks and the infrastructures used to capture them sometimes have little visible longevity. In 2014, Shelley Bernstein wrote about the Brooklyn Museum's online collection, which at that time included opportunities for users to tag, comment, and "favorite" items, and contribute to the augmentation of collection records. She cites a simple example where a user's comment allowed them to correct the orientation of an image displayed in collections online, providing a link to their comment in the footnotes (Bernstein, 18). Eight years later, the link returns a 404 error, and no comments are visible on this or any other collection record. Although tagging options remain, it appears that other user contributions have either been assimilated into the institutional record or rendered invisible (assuming that they have not been discarded altogether).

For many First Nations communities, control over records about artefacts and archives is yet another example of the ways in which large heritage institutions deny them agency, knowledge, and voice (Sentance). Therefore, the development of digital infrastructures that support more complex, polyvocal readings of Indigenous artifacts and archival material mostly have been developed separate from GLAM collection management systems. One of the most prominent examples of alternative infrastructures for representing collections is the digital access platform Mukurtu.

Mukurtu started as a grassroots collaboration between the Warumungu Aboriginal community in Australia's Northern Territory and cultural anthropologist, ethnographer, and "accidental archivist" Kimberly Christen (Christen, "Opening Archives: Respectful Repatriation," 185). The primary impetus behind the development of the tool was an identified need to manage access at a granular, contextual level based on "dynamic social and cultural systems, relationships, and cultural protocols" (Christen, Merrill, and Wynne). For example, communities might manage access based on gender, familial relationships, or social status. The name Mukurtu means "dilly bag" in the Warumungu language: "The dilly bags held sacred materials and elders kept and protected them as part of their obligations to care for their communities, relatives, places and ancestors, but the elders could not be 'stingy' and had to open them up when younger generations asked respectfully [. . .] like the dilly bag—Mukurtu—the digital archive we built should be 'a safe keeping place'" (Christen, Merrill, and Wynne). The first version of the platform was deployed for the Warumungu community in 2007, followed by beta versions with Plateau tribes in the United States. This was followed in 2010 by the launch of Mukurtu CMS, an open-source platform "aiming to empower communities to manage, share, narrate, and exchange their digital heritage in culturally relevant and ethically-minded ways" (Mukurtu CMS).

In addition to access controls, Mukurtu includes options for capturing multiple perspectives. According to Christen ("Does Information Really Want to Be Free?," 2887), it allows users "to infuse their voices, their cultural concerns, and their notions of sociality and historicity into the system." Unlike typical crowdsourcing initiatives, the platform is designed to prioritize the unique knowledge of First Nations peoples regarding their own culture and collections. Where institutional collection records already exist, this is achieved by appending a parallel record that can include alternative narratives and classification structures, as well as additions or corrections to the existing record without altering or expunging those records. As Christen ("Opening Archives: Respectful Repatriation," 201) writes: "We all understood the multiple benefits to tribes, scholars, and public users of the site, of seeing that history is indeed made, unmade, and negotiated over time; whereas 'records' often seem official and irrefutable, they are malleable and susceptible to change."

Plateau Peoples' Web Portal[5] is an example of a collaboratively curated archive built on the Mukurtu CMS. It includes records created by the community and augmented records for digitized artifacts found in GLAM institutions such as the National Museum of the American Indian (NMAI). Whereas the catalogue record for a woman's basket hat at NMAI includes little more than an image of the item and a set of fielded metadata (Spino), the Mukurtu record includes some of the NMAI metadata, as well as a separate Confederated Tribes of Warm Springs record that includes an extended cultural narrative section with a video and transcript where three Warm Springs women discuss the basket hat and its use as part of traditional food gathering practices (Plateau Peoples' Web Portal; Spino).

Mukurtu takes important steps beyond annotation, tagging, and comments, providing opportunities for communities to capture and preserve their significant knowledge as part of a well-structured digital archive. However, some of the conceptual limitations of GLAM description remain evident. Visitors to the Plateau Peoples' Web Portal looking for items like the woman's basket hat are directed to the museum record first, and then they must select the community record to move beyond the institutional perspective. This gives a sense that the community record is appended to the museum record; community knowledge here is additional rather than foundational. Although users can compare the two records and see what (if anything) is different, this only allows indirect contestation. And there is no clear link to the source NMAI record, or to other relevant items or records in the Portal or elsewhere.

Perhaps most significantly, community records in Mukurtu remain aligned with the object-based segmentation of the world imposed by existing collections documentation, rather than providing space for alternative, more relational information structures and ways of knowing. As Michael Christie argues: "There is much stripping and splicing to be done to fit Aboriginal representations of ecological knowledge into an official archive" (62). Discrete object-based description does little to capture the complexities of stories, histories, and knowledge that weave together artifacts, places, people and other beings, events, and diverse temporalities. Writing about Indigenous thinking, Tyson Yunkaporta from the Apalech Clan says: "How might we identify and utilize the various sets of Indigenous Knowledge scattered throughout this kaleidoscope of identities? Not by simplistic categorization, that's for sure. Through the lens of simplicity, historical contexts of interrelatedness and upheaval are sidelined, and the authenticity of Indigenous Knowledge and identity is determined by an illusion of parochial isolation, another fragment of primitive exotica to examine, tag and display" (13). More expansive, interconnected, relational approaches to collection documentation are required (Jones, 118–37), not just as a means for annotating and contesting the content of existing GLAM records at the level of the field and record, but as a first step toward opening up

GLAM data, metadata structures, and infrastructures to new modes of thinking and ways of working.

Still, Mukurtu takes valuable steps in this direction. There is no requirement for consensus, nor is it considered necessary to try and resolve differences in perspective or disagreements about interpretation. Multiple voices, knowledge domains, and historicities sit alongside each other, sometimes with little duplication (let alone replication). The complexity of the layers and relationships grows with each new addition. Access control is a key part of the system, placing limits on who can contribute; but though access controls embody notions of collective rather than individual ownership, Mukurtu and the communities who use it show no desire to try and develop or replicate a singular collective voice.

Contestation and Collaboration: Humanities Network Infrastructure Project

One example of research infrastructure designed for the relational trail-making approach to humanities knowledge realization is the Humanities Networked Infrastructure (HuNI; pronounced "honey"). HuNI is a knowledge graph with eighteen million nodes, drawn from more than thirty data sources that reflect the perspectives of various disciplines across the humanities and creative arts (Verhoeven and Burrows). The data comes from libraries and museums, including Museums Victoria, the Australian National Maritime Museum, the Australian Film Institute (AFI) research collection, the library of the Australian Institute for Aboriginal and Torres Strait Islander Studies (AIATSIS), and the National Library of Australia (specifically the Trove digitized newspaper collections), as well as from research groups focused on subjects like literature (AustLit), performance (AusStage), art and design (Design and Art Australia Online, or DAAO), and endangered languages (PARADISEC). While the primary focus has been on Australian culture, a growing body of Canadian data has recently been added and annotated.

HuNI was developed by a consortium of humanities scholars in part to redress the way that collecting institutions minimized the possibilities for contestation through infrastructures with highly restricted access and an authoritative approach to information organization. It provides an environment for data collection that enables researchers to creatively reconfigure data relationships to refute hierarchies and enable alternative interpretations. In their original institutional environment, the data that are aggregated in HuNI have been heavily curated and intended to convey authority through their closely defined scope, through their controlled vocabularies, ontologies, hierarchical relationships and concepts, and even through their very data models. They present a view of the world that is intended to impose order and structure and that reflects the expertise of a particular academic or curatorial group. There is no room for the messiness and contestation that are integral to the humanities and creative arts. These databases are designed to exist as separate universes that take little, if any, notice of external sources—even similarly authoritative ones such as those belonging to other GLAM institutions.

Bringing the vocabularies and structures of these various data sources together could be an exercise in normalization, reconciliation, and conformity to a single, authoritative structure, as Europeana aims to do. HuNI, however, takes a fundamentally different approach, partly through its method of ingesting and minimally modeling the data and partly through what it enables its users to do with (and to) the data. The incoming data is mapped to a modest model, consisting of only six basic types of entity: Person, Place, Concept, Event, Organization, and Work. For example, while there are nearly eighteen million entities in HuNI, these may include multiple versions of the same person since no reconciliation or merging of entities from different sources has been imposed.

The platform's design reflects the belief that there is value for researchers in knowing how often a record appears in different collections and how the differences and similarities between these records can yield valuable insights. Relationships between entities have not been imported from the source datasets, with one exception: the 43,000 persons, events, and organizations originating from DAAO have generic "associated with" and "participated in" relationships. As a result, HuNI is like a "connect-the-dots" puzzle, where HuNI provides the dots (in the form of records) and leaves users to draw their own connecting lines (and conclusions) by creating a picture of their own devising rather than following a predetermined pattern.

The key feature of HuNI is that users (essentially anyone with a social media or Australian Access Federation account; see AAF, "AAF 10") can create their own links between entities. To create links, they can reuse existing expressions of a relationship (since HuNI offers these as prompts) or they can invent their own. Since multiple relationships can be made between the same two entities, a HuNI user can add a different relationship that contests an existing one—or simply expresses a different perspective or interpretation.

HuNI users can also group entities into public or private collections based on whatever categorizing or organizing principle they choose and share the collections publicly. There are almost 200 public collections in HuNI, ranging from scholarly subjects ("Literary Works About Bushfires") to topics associated with popular culture ("Women Being Haunted by Things" or "Dead Guys Named David"). Users can also add their own entity records through comma-separated value (CSV) uploads. Recent uploads have included more than 1,100 entities from spreadsheets constructed by Canadian students in DH, women's and gender studies, and library and information science. Through working with HuNI, these students learn how data and power are mutually implicated, especially when data is integrated, exchanged, and interoperated.

HuNI's multiplicity of relationships and categorizations does create complexity and messiness. The graphlike visualizations of relationships between entities can quickly blow up into immensely complex networks with many interconnections.

But this reflects the complexity and multiplicity of meanings inherent in the humanities and the social worlds that are their focus. At the same time, by noting the source of each entity, every relationship link, and every collection, HuNI ensures that users are accountable and responsible for their own interpretations and contributions. It is not a reproducible experiment in the scientific sense, but its infrastructure is designed to allow users to re-traverse the associations made by other users and replicate their connections, while at the same time contesting and reinterpreting them.

Over to You

> There is no "the truth," "a truth"—truth is not one thing, or even a system. It is an increasing complexity. The pattern of the carpet is a surface. When we look closely, or when we become weavers, we learn of the tiny multiple threads unseen in the overall pattern, the knots on the underside of the carpet.
>
> —Adrienne Rich

Insistence on the value of replication and reproducibility centers on the success of elevating the purity of process over the vagaries of people. The idea that data and methods exist outside the self is anathema to humanities methods. In the humanities, social and ethical questions are typically paramount. Who has the authority to create public datasets, to control vocabularies and ontologies, and to define which data relationships are valid? What and, most important, who have been eliminated in the singular, universal logics of digital information infrastructure? A humanities-centered view of data acknowledges that information is woven in a relationship to the curator, creator, and keeper. Data sovereignty initiatives suggest that we should also think of information as "belonging to" its subjects. Mukurtu and HuNI are digital information platforms that recognize and attempt to redress the solipsistic curation of knowledge and understanding embedded in collections built on supposedly neutral "authoritative" vocabularies, ontologies, and relationships. By creating spaces in which existing meanings can be challenged, and new or alternative meanings and truths constructed, they offer a powerful opportunity to explore new processes of validation in the humanities and sciences alike by embracing complexity, multiplicity, and contestation.

Finally, we would be remiss if we did not acknowledge that these proposed approaches to contestation are themselves necessarily incomplete and partial in the sense of always being positioned, of being or having a-side. Contestation is critically a form of collaborative and social knowledge-making, of elaborating and adding nuance to existing knowledge with humility, a way of appreciating the manifold ways in which acts of revision, resistance, and resilience can be creative rather than simply reproductive.

NOTES

1. https://huni.net.au/#/search.
2. https://mukurtu.org.
3. https://www.go-fair.org/fair-principles/.
4. https://allea.org.
5. https://plateauportal.libraries.wsu.edu.

BIBLIOGRAPHY

All European Academies (ALLEA). Home page, 2023. https://allea.org.
Australian Access Federation (AAF). "AAF 10." 2020. https://aaf.edu.au/wp-content/uploads/2020/10/AAF_10-year_brochure_collated_For-website.pdf.
Baker, Monya. "1,500 Scientists Lift the Lid on Reproducibility." *Nature* 533 (2016): 452–54. https://doi.org/10.1038/533452a.
Ball, Philip. "Is AI Leading to a Reproducibility Crisis in Science?" *Nature* 624 (2023): 22–25. https://doi.org/10.1038/d41586-023-03817-6.
Begley, C. Glenn, and Lee M. Ellis. "Drug Development: Raise Standards for Preclinical Cancer Research." *Nature* 483 (2012): 531–33. http://doi.org/10.1038/483531a.
Berez-Kroeker, Andrea L., Lauren Gawne, and Susan Smythe Kung, et al. "Reproducible Research in Linguistics: A Position Statement on Data Citation and Attribution in Our Field." *Linguistics* 56, no. 1 (2018): 1–18. https://doi.org/10.1515/ling-2017-0032.
Berlant, Lauren, and Lee Edelman. *Sex, or the Unbearable*. Duke University Press, 2014.
Bernstein, Shelley. "Crowdsourcing in Brooklyn." In *Crowdsourcing Our Cultural Heritage*, edited by Mia Ridge, 17–43. Ashgate, 2014.
Borda, Ann, and Jonathan P. Bowen. "Turing's Sunflowers: Public Research and the Role of Museums." *Proceedings of EVA London 2020: AI and the Arts: Artificial Imagination*. ScienceOpen, 2020. http://dx.doi.org/10.14236/ewic/EVA2020.5.
Bueger, Christian. "Practices, Norms, and the Theory of Contestation." *Polity* 49, no. 1 (2017): 126–31.
Bunge, Mario. "A General Black Box Theory." *Philosophy of Science* 30, no. 4 (1963): 346–58. http://www.jstor.org/stable/186066.
Burrows, Toby, Mike Jones, and Deb Verhoeven. "Selling Our Soul (for Total Control)? Linked Open Data and GLAM." In *Libraries, Archives, and the Digital Humanities*, 1st ed., edited by Isabel Galina Russell and Glen Layne-Worthey, 187–203. Routledge, 2025. https://doi.org/10.4324/9781003327738.
Bush, Rosemary, Andrea Dutton, and Michael Evans, et al. "Perspectives on Data Reproducibility and Replicability in Paleoclimate and Climate Science." *Harvard Data Science Review* 2, no. 4 (2020). https://doi.org/10.1162/99608f92.00cd8f85.
Cecire, Natalia. "Introduction: Theory and the Virtues of Digital Humanities." *Journal of Digital Humanities* 1, no. 1 (2011). http://journalofdigitalhumanities.org/1-1/introduction-theory-and-the-virtues-of-digital-humanities-by-natalia-cecire/.

Christen, Kimberly. "Does Information Really Want to Be Free? Indigenous Knowledge Systems and the Question of Openness." *International Journal of Communication* 6 (2012): 2870–93. https://ijoc.org/index.php/ijoc/article/view/1618/828.

Christen, Kimberly. "Opening Archives: Respectful Repatriation." *American Archivist* 74, no. 1 (2011): 185–210. https://doi.org/10.17723/aarc.74.1.4233nv6nv6428521.

Christen, Kimberly, Alex Merrill, and Michael Wynne. "A Community of Relations: Mukurtu Hubs and Spokes." *D-Lib Magazine* 23, no. 5/6 (2017). https://doi.org/10.1045/may2017-christen.

Christie, Michael. "Aboriginal Knowledge Traditions in Digital Environments." *Australian Journal of Indigenous Education* 34 (2005): 61–66.

De Weerd-Wilson, Donna, and William Gunn. "How Elsevier Is Breaking Down Barriers to Reproducibility." *Elsevier Connect*, Elsevier, January 31, 2017. https://web.archive.org/web/20211029041141/https://www.elsevier.com/connect/archive/how-elsevier-is-breaking-down-barriers-to-reproducibility.

Edelman, Lee. *No Future: Queer Theory and the Death Drive*. Duke University Press, 2004.

Eijnatten, Joris van, Toine Pieters, and Jaap Verheul. "Big Data for Global History: The Transformative Promise of Digital Humanities." *BMGN—Low Countries Historical Review* 128, no. 4 (2013): 55–77. https://doi.org/10.18352/bmgn-lchr.9350.

Einstein, Albert. *The Ultimate Quotable Einstein*. Collected and edited by Alice Calaprice. Princeton University Press, 2011.

Fanelli, Daniele. "Negative Results Are Disappearing from Most Disciplines and Countries." *Scientometrics* 90, no. 3 (2012): 891–904. https://doi.org/10.1007/s11192-011-0494-7.

Fanelli, Daniele. "Opinion: Is Science Really Facing a Reproducibility Crisis, and Do We Need It To?" *Proceedings of the National Academy of Sciences (PNAS)* 115, no. 11 (2018): 2628–31. https://www.pnas.org/doi/10.1073/pnas.1708272114.

Fraser, Hannah, Tim Parker, and Shinichi Nakagawa, et al. "Questionable Research Practices in Ecology and Evolution." *PLOS One* 13, no. 7 (2018): e0200303. https://doi.org/10.1371/journal.pone.0200303.

Fricker, Miranda. *Epistemic Injustice: Power and the Ethics of Knowing*. Oxford University Press, 2007.

GO FAIR. "FAIR Principles." 2023. https://www.go-fair.org/fair-principles/.

Goodman, Steven N., Daniele Fanelli, and John P. A. Ioannidis. "What Does Research Reproducibility Mean?" *Science Translational Medicine* 8, no. 341 (2016): 341–3. https://doi.org/10.1126/scitranslmed.aaf5027.

Harrower, Natalie, Maciej Maryl, and Timea Biro, et al. *Sustainable and FAIR Data Sharing in the Humanities: Recommendations of the ALLEA Working Group E-Humanities*. ALLEA—All European Academies. Digital Repository of Ireland, 2020. https://repository.dri.ie/catalog/tq582c863.

Hicks, Daniel J. "Open Science, the Replication Crisis, and Environmental Public Health." *Accountability in Research* 30, no. 1 (2023): 34–62. https://doi.org/10.1080/08989621.2021.1962713.

Humanities Networked Infrastructure (HuNI). Home page. 2023. https://huni.net.au/#/search.

International Council of Museums (ICOM). "Museum Definition." August 2022. https://icom.museum/en/resources/standards-guidelines/museum-definition/.

Jones, Mike. *Artefacts, Archives, and Documentation in the Relational Museum*. Routledge, 2021.

Kitchin, Rob. "Big Data, New Epistemologies and Paradigm Shifts." *Big Data & Society* 1, no. 1 (2014). https://doi.org/10.1177/2053951714528481.

Koster, Lukas, and Saskia Woutersen-Windhouwer. "FAIR Principles for Library, Archive and Museum Collections: A Proposal for Standards for Reusable Collections." *Code4Lib Journal* 40 (2018). https://journal.code4lib.org/articles/13427.

Larregue, Julien. "Sentencing Social Psychology: Scientific Deviance and the Diffusion of Statistical Rules." *Current Sociology*, August 13, 2022. https://doi.org/10.1177/00113921221117604.

Lash, Timothy L., Lindsay J. Collin, and Miriam E. Van Dyke. "The Replication Crisis in Epidemiology: Snowball, Snow Job, or Winter Solstice?" *Current Epidemiology Reports* 5, no. 2 (2018): 175–83. https://doi.org/10.1007/s40471-018-0148-x.

Leonelli, Sabina. "Rethinking Reproducibility as a Criterion for Research Quality." In *Research in the History of Economic Thought and Methodology. Including a Symposium on Mary Morgan: Curiosity, Imagination, and Surprise* 36B, edited by Luca Fiorito, Scott Scheall, and Carlos Eduardo Suprinyak, 129–46. Emerald Publishing Limited, 2018. https://doi.org/10.1108/S0743-41542018000036B009.

Lynch, Bernadette. "Migrants, Museums and Tackling the Legacies of Prejudice." In *Museums in a Time of Migration: Re-thinking Museums' Roles, Representations, Collections, and Collaborations*, edited by Christina Johnson, and Pieter Bevlander, 225–42. Nordic Academic Press, 2017.

Marwick, Ben. "Computational Reproducibility in Archaeological Research: Basic Principles and a Case Study of Their Implementation." *Journal of Archaeological Method and Theory* 24, no. 2 (2017): 424–50. https://ro.uow.edu.au/smhpapers/4034/.

McGillivray, Barbara, Beatrice Alex, and Sarah Ames, et al. *The Challenges and Prospects of the Intersection of Humanities and Data Science: A White Paper from The Alan Turing Institute*. Alan Turing Institute. Figshare, 2020. https://doi.org/10.6084/m9.figshare.12732164.v5.

Message, Kylie. *Museums and Racism*. Routledge, 2018.

Mukurtu. Home page, 2023. https://mukurtu.org.

Mukurtu CMS. "Our Mission." 2023. https://mukurtu.org/about/.

Munafò, Marcus R., Brian A. Nosek, and Dorothy V. M. Bishop, et al. "A Manifesto for Reproducible Science." *Nature Human Behaviour* 1, no. 0021 (2017). https://doi.org/10.1038/s41562-016-0021.

National Academies of Sciences, Engineering, and Medicine; Policy and Global Affairs. *Reproducibility and Replicability in Science*, National Academies Press, 2019. https://doi.org/10.17226/25303.

Nehamas, Alexander. "A Promise of Happiness: The Place of Beauty in a World of Art." *Tanner Lectures on Human Values* (2001): 189–231. https://tannerlectures.utah.edu/_resources/documents/a-to-z/n/Nehamas_02.pdf.

Nuijten, Michèle B., Chris H. J. Hartgerink, and Marcel A. L. M. van Assen, et al. "The Prevalence of Statistical Reporting Errors in Psychology (1985–2013)." *Behavior Research Methods* 48, no. 4 (2016): 1205–26. https://doi.org/10.3758/s13428-015-0664-2.

Open Science Collaboration. "Estimating the Reproducibility of Psychological Science." *Science* 349, no. 6351 (2015). https://doi.org/10.1126/science.aac4716.

O'Sullivan, James. "The Humanities Have a 'Reproducibility' Problem." *Talking Humanities,* July 9, 2019. https://talkinghumanities.blogs.sas.ac.uk/2019/07/09/the-humanities-have-a-reproducibility-problem/.

Owens, Trevor. "Digital Cultural Heritage and the Crowd." *Curator: The Museum Journal* 56, no. 1 (2013): 121–30. https://doi.org/10.1111/cura.12012.

Pashler, Harold, and Eric-Jan Wagenmakers. "Editors' Introduction to the Special Section on Replicability in Psychological Science: A Crisis of Confidence?" *Perspectives on Psychological Science* 7, no. 6 (2012): 528–30. https://doi.org/10.1177/1745691612465253.

Peels, Rik. "Replicability and Replication in the Humanities." *Research Integrity & Peer Review* 4, no. 2 (2019). https://doi.org/10.1186/s41073-018-0060-4.

Peels, Rik, and Lex Bouter. "The Possibility and Desirability of Replication in the Humanities." *Palgrave Communications* 4, no. 95 (2018): 1–4. https://www.nature.com/articles/s41599-018-0149-x.

Penders, Bart, J. Britt Holbrook, and Sarah de Rijcke. "Rinse and Repeat: Understanding the Value of Replication Across Different Ways of Knowing." *Publications* 7, no. 3 (2019): 52. https://doi.org/10.3390/publications7030052.

Penders, Bart, Sarah de Rijcke, and J. Britt Holbrook. "Science's Moral Economy of Repair: Replication and the Circulation of Reference." *Accountability in Research* 27, no. 2 (2020): 107–13. https://doi.org/10.1080/08989621.2020.1720659.

Plateau Peoples' Web Portal. Home page, 2023. https://plateauportal.libraries.wsu.edu.

Plateau Peoples' Web Portal. 2018. "Woman's Basket Hat." https://plateauportal.wsulibs.wsu.edu/digital-heritage/womans-basket-hat-7.

Plesser, Hans E. "Reproducibility vs. Replicability: A Brief History of a Confused Terminology." *Frontiers in Neuroinformatics* 11, no. 76 (2018). https://doi.org/10.3389/fninf.2017.00076.

Posner, Miriam. "The Radical Potential of the Digital Humanities: The Most Challenging Computing Problem Is the Interrogation of Power." *LSE Impact Blog,* August 12, 2015. https://blogs.lse.ac.uk/impactofsocialsciences/2015/08/12/the-radical-unrealized-potential-of-digital-humanities/.

Rich, Adrienne. *On Lies, Secrets and Silence: Selected Prose, 1966–1978.* W. W. Norton & Co., 1979.

Ridge, Mia. "Crowdsourcing Our Cultural Heritage: Introduction." In *Crowdsourcing Our Cultural Heritage,* edited by Mia Ridge, 1–13. Ashgate, 2014.

Schofield, Tom, Mitchell Whitelaw, and David Kirk. "Research Through Design and Digital Humanities in Practice: What, How and Who in an Archive Research Project." *Digital Scholarship in the Humanities* 32, no. 1 (2017): i103–20. https://doi.org/10.1093/llc/fqx005.

Sentance, Nathan. "My Ancestors Are in Our Memory Institutions, but Their Voices Are Missing." *The Guardian*, March 6, 2018. http://www.theguardian.com/commentisfree/2018/mar/06/my-ancestors-are-in-our-memory-institutions-but-their-voices-are-missing.

Shilton, Katie, and Ramesh Srinivasan. "Participatory Appraisal and Arrangement for Multicultural Archival Collections." *Archivaria* 63, no. 1 (2007): 87–101. https://archivaria.ca/index.php/archivaria/article/view/13129.

Sieg, Katrin. 2021. "Postcolonial Activists and European Museums." In *Reframing Postcolonial Studies,* edited by David D. Kim, 215–47. Palgrave Macmillan, 2021. https://doi.org/10.1007/978-3-030-52726-6_9.

Sikk, Kaarel. "Towards Reproducible Science in the Digital Humanities." *Digital History and Hermeneutics,* May 19, 2020. https://dhh.uni.lu/2020/05/19/towards-reproducible-science-in-the-digital-humanities-how-to-publish-your-data-and-code-alongside-your-research-with-the-help-of-zenodo/.

Simon, Nina. *The Participatory Museum.* Museum 2.0, 2010.

Spino, Eileen. "Woman's Basket Hat." National Museum of the American Indian, 2023. https://americanindian.si.edu/collections-search/object/NMAI_278753.

Suchman, Lucy. "Located Accountabilities in Technology Production." *Scandinavian Journal of Information Systems* 14, no. 2 (2002). http://aisel.aisnet.org/sjis/vol14/iss2/7.

Sui, Daniel, and Peter Kedron. "Reproducibility and Replicability in the Context of the Contested Identities of Geography." *Annals of the American Association of Geographers* 111, no. 5 (2021): 1275–83. https://doi.org/10.1080/24694452.2020.1806024.

Terras, Melissa. "Opening Access to Collections: The Making and Using of Open Digitised Cultural Content." *Online Information Review* 39, no. 5 (2015): 733–52. https://doi.org/10.1108/OIR-06-2015-0193.

UK Reproducibility Network (UKRN). Home page, 2023. https://www.ukrn.org/.

Verhoeven, Deb. "As Luck Would Have It: Serendipity and Solace in Digital Research Infrastructure." *Feminist Media Histories* 2, no. 1 (2016): 7–28. https://doi.org/10.1525/fmh.2016.2.1.7.

Verhoeven, Deb. "Scholarship in a Clopen World." *Pop! Public. Open. Participatory* 4 (2022). https://doi.org/10.54590/pop.2022.002.

Verhoeven, Deb, and Toby Burrows. "Building the Australian Knowledge Graph: HuNI (Humanities Networked Infrastructure)." *Graphs and Technologies in the Humanities: 6th International Conference,* February 3–4, 2022. https://graphentechnologien.hypotheses.org/files/2022/01/Building_the_Australian_Knowledge_Graph_HuNI_Humanities_Networked_etc-Verhoeven_Burrows.pdf.

Wiener, Antje. *A Theory of Contestation.* Springer, 2014. https://doi.org/10.1007/978-3-642-55235-9.

Wilkinson, Mark D., Michel Dumontier, and IJsbrand Jan Aalbersberg, et al. "The FAIR Guiding Principles for Scientific Data Management and Stewardship." *Scientific Data* 3, no. 160018 (2016). https://doi.org/10.1038/sdata.2016.18.

Yunkaporta, Tyson. *Sand Talk: How Indigenous Thinking Can Save the World.* Text Publishing Company, 2019.

PART II][Chapter 13

Scrounging

DARREN WERSHLER

Scrounging is a powerful but overlooked cultural technique that not only helps to ensure the ongoing operation of many labs in many disparate fields, but also can be a way of bringing an entire lab or research space into being. It is distinct both from the formal processes of application and certification that universities and other institutions use and from other informal techniques like gleaning and poaching, which describe forms of subsistence on someone's or something else's resources. The productivity and utility of gleaning and poaching are not in question (de Certeau, 165–76), but the kind of scrounging that I am thinking about operates differently because it shifts material resources from one infrastructural location to another to integrate them into a different research apparatus.

University infrastructure is often orthogonal to the interests and activities of scholars working in materially engaged arts and humanities fields. The university systems and policies that manage many of the practices that define and regulate research spaces, such as the allocation and retrofitting of physical research space, the assignment of course releases for lab management, procurement procedures for equipment and vendor approval, the hiring and payment of personnel, and expense reporting, were established with the sciences in mind, and it can be difficult to make them function in other fields. Even with the explicit support of upper-level administration and (rare) dedicated funding, arts and humanities scholars with research spaces often encounter significant difficulties in simply making university infrastructure support their research because such infrastructure has not been designed to take the arts and humanities into account, and only reluctantly adjusts to their presence. The result is that material media research in the digital humanities (DH) and related fields (e.g., media history, media archaeology, sound studies) frequently positions itself tactically rather than strategically in relation to the production of knowledge, relying on various ad hoc operations to function.

As a "cultural technique," in the way that Bernhard Siegert uses the term, "scrounging" is helpful because it describes "a more or less complex actor network that comprises technological objects as well as operative chains they are part of and that configure or constitute them" (Siegert, 11). As Lori Emerson, Jussi Parikka, and I point out in *The Lab Book,* lab techniques do more than create historical and methodological continuities. They also hop between fields, bring new technologies into existence, and then change in response to those technologies or are abandoned for a variety of reasons. As such, techniques produce hybridity and discontinuity as well as regularity (Wershler, Emerson, and Parikka, 213).

Scrounging concerns the relation of a lab's apparatus (the technologies, objects, practices, and relations inside the lab, which it uses to conduct its investigations)[1] to its infrastructure (in Susan Leigh Star's foundational formulation, the invisible, ready-to-hand system of substrates that serves as the background for other kinds of work).[2] When a given lab's infrastructure does not allow for the necessary expansion or adaptation of its apparatus to begin a new research project, scrounging kicks into action. When thinking about the sort of scrounging techniques that happen in and around media labs and related spaces, it's helpful to consider them as a series of operations that weave together a tangle of institutions, discourses, agents, spaces, objects, imaginaries, and apparatus into a coherent entity that produces knowledge, and then help to ensure that this entity continues to function. Scrounging precedes and produces lab activity, and sometimes even the lab itself.

Popular definitions of *scrounging* explain it as scavenging or foraging.[3] The term is often pejorative (particularly in the British context) because it can imply borrowing something surreptitiously without the intention to return it or pay for it.[4] In the context of contemporary techno-art practice, particularly because of the work of the performance art group Survival Research Laboratories, scrounging produces "obtainium"—any substance that can be easily obtained and incorporated into a current project (Survival Research Laboratories). In this context, scrounging is at least pragmatic and arguably the grounds for resistance and critique. But the term *scrounging* encompasses a range of practices, and, despite U.S. bravado and British scrupulousness alike, it is not synonymous with theft.

The literature on scrounging in the context of lab work is very sparse, and limited in scope to hard science labs, but it implies that the practice can and does occur in labs at all scales and times. In "Scrounging Old Equipment for New Experiments," Toni Feder contends that "not everybody gets giant castoff lasers, but scrounging for used scientific equipment is common" (26). Several of the scientists whom Feder interviews support the contentions of cultural technique theorists by observing that scrounging precedes both labs and lab work:

> "The way you get a new program started is to do some fraction of the work before it's funded," says Jefferson Lab's Fred Dylla. "How do you do that? You either recycle old equipment, or you borrow it, or you find it cheap. You look for equipment you might be able to scrounge, locally through universities, through your network of friends. A good scientist should be constantly on the lookout" (Feder, 26)

Here, scrounging is not only a necessary condition for the creation of a new research program. The ability and predilection to scrounge are also constitutive of the very category of "the good scientist." This is another contention of cultural technique theory: that techniques precede the subjects that use them and play a major role in the formation of those subjects (Siegert, 9).

Feder (28) found "broad agreement" among his interviewees that "personal connections" were the most important aspect of scrounging. Because the interviewees do not elaborate on this point, it's difficult to tell whether they have in mind the long tradition of deceptive communications that early hackers and phreaks referred to as "social engineering" (Jargon File) or something more genteel. One interviewee describes techniques for circumventing official government disposal regulations to secure valuable equipment from publicly funded projects that are winding down (Feder, 27). But there is certainly none of the sense of squeamishness in the British definitions of scrounging among the researchers whom Feder (28) consults: "'It also helps to have no sense of shame,' says [Columbia University physicist Janet] Conrad." There is a broad range of interpersonal communication techniques at play in scrounging, but they are inextricable from it.

Since founding the Residual Media Depot (RMD)[5] at Concordia University in 2016, I have had plenty of experience with scrounging, which was necessary to locate and secure the space, to create RMD's collection of historical and modified video game consoles and signal processing equipment, to maintain it, and then actually to employ it in research. But to broaden my sense of what scrounging is and how it operates in the loosely overlapping fields in which I work (material media history, media archaeology and DH), I spoke with researchers from a number of university media labs in Canada, the United States, and Germany:

- Jason Camlot, director of the AmpLab East (AL) and the Spokenweb Network at Concordia University
- Lori Emerson, director of the Media Archaeology Lab (MAL) at the University of Colorado at Boulder, and libi streigl, the lab manager there
- Rick Prelinger, founder of the Prelinger Archives (PA) and professor at the University of Santa Cruz
- Stefan Höltgen, former curator of the Signal Laboratory (SL) and the Media Archaeological Fundus (MAF) at the Humboldt University of Berlin, at the time of publication working on the cultural history of computing at the University of Bonn
- Marcel O'Gorman, director of the Critical Media Lab (CML) at the University of Waterloo
- Karis Shearer, director of AMP Lab West (ALW) at the University of British Columbia Okanagan, and Emily Murphy, assistant director of Amp Lab West at the University of British Columbia Okanagan and director of the emerging (Re)Media Lab (RL) at the same institution

- Florian Sprenger, director of the Virtual Humanities Lab (VHL) at Ruhr Universität Bochum, and Thomas Nyckel, a research assistant there

Throughout this article, references to these conversations occur using the parenthetical abbreviations given here, such as (AL) or (MAL).[6]

When I asked these researchers about their relationship to scrounging, the concept was deeply familiar to all of them as part of their experience running their labs. Several of them said that it epitomized their careers (PA, CML). In our conversations about scrounging, three concepts (space, temporality, and affect) and two kinds of assemblages (apparatus and personnel networks) appeared consistently as major topics. The sections that follow address each of these topics in order (but with affect delayed until the end, for effect), attempting to point to regularities in the discussions without erasing important differences.

Space

New labs rarely come into the world in their mature form, and perhaps not in their permanent space. But even if they are physically situated in a stand-alone building, labs are almost always contained inside some larger institution. The connective tissue between a given lab and these larger institutions (including, as Alan Liu points out in "Drafts for *Against the Cultural Singularity*," culture itself) is their infrastructure. In her work on research infrastructures, Sheila Anderson (20–21) argues that for a new research project to come into existence, scholars need to learn how to interpret, engage with, and collaborate with the infrastructure of their respective institutions. This is always true, and it's an often-overlooked aspect of scholarly work because it's rarely if ever taught (or valorized), so it's a difficult skill set to pick up unless a given individual learns it in the private sector or has excellent support staff.

Via interpersonal communication or gray literature (i.e., applications and paperwork), access to space and apparatus comes initially via gatekeepers at various points of control (PA): deans, department chairs, vice presidents of research, committees at various levels (including designated space committees and faculty or university research committees), nongovernmental organizations (NGOs), and commercial/private owners and landlords (ALW, CML). For this reason, lab directors who have spent service time in upper administration have a better understanding of how institutions apportion research space and may be more likely to succeed in securing it for their own work when they return to faculty positions (AL).

As Feder's interviewees suggested and my own conversations corroborated, strong interpersonal communication skills and the ability to manage professional connections are crucial to the process of scrounging for lab space. Because many departments don't have formal space policies, lab space can sometimes be obtained by simple conversation with a department chair, or by bartering with a space's current users for access or to trade spaces (MAL). Occasionally, an underutilized room in a departmental footprint can even be "squatted" by the performative act of declaring it a lab (Wershler et al., 32). Of course, even civil conversations to secure lab space, let alone social engineering approaches, can and do lead to interfaculty conflicts over who has the right to use space, and for what.

Managing good relations with colleagues *after* obtaining a research space also requires strong interpersonal skills. Learning how to manage relations with other research units (whether neighboring or in the same shared space) can be crucial to the ongoing success of scrounging as a technique, not to mention to basic civility. Labs raise many special considerations relating to infrastructural needs and institutional policy, both in terms of what labs might produce as unwanted by-products of their research or what environmental and other conditions they have to contend with, especially in loosely regulated spaces or commercial zones (VHL, SL, ALW). Noise, fumes, and electrical hazards move in both directions. On campus, addressing such issues often involves renovations that scrounging alone can't address (AL). Clever administrators interested in upgrading university facilities and generating outside attention will often negotiate with scrounging researchers over the availability of space if there is a good chance that the Canadian Foundation for Innovation (CFI) will fund a renovation of the overall infrastructure (ALW).

By design or necessity, university lab space is increasingly outside the campus footprint. Several of the lab directors have spaces that are partly or even largely outside the university ambit, located above retail space or in strip malls (ALW), inside a corporate innovation hub (CML), or partially in warehouse space owned by a U.S. 501(c)(3) nonprofit organization (PA). Off-campus locations can provide access to resources not available on campus that are necessary for some kinds of research projects, like large amounts of storage space for bulky legacy media such as celluloid film (PA). A space like an incubator may also offer proximity to, and interactions (including scrounging) with, other kinds of organizations not commonly found on campus, such as tech startup companies (CML). The intention behind such spaces is to act as what Peter Galison calls a "trading zone," where practices and professional argots intermingle and change in the interest of solving shared problems. When this occurs, such spaces can provide rich scrounging territory and open up new avenues for research as a result.

Temporality

The temporality of scrounging is arbitrary and variable. Finding random pieces of equipment lying in a hallway feels nearly instantaneous, but scrounging does not always have that sort of speed. Building interpersonal relations with collectors and community experts, as well as with university bureaucrats managing purchases and acquisitions, takes time. One researcher offered to show me an email chain with university procurement services that grew into tens of thousands of words, all in the name of securing from an expert user a nearly impossible-to-find piece of equipment (a disc-cutting lathe). The inability of the university bureaucracy overseeing infrastructure to process the purchase efficiently resulted in the equipment seller being paid many months after the item had changed hands (AL). Another researcher described how obtaining one set of rare materials from a collector took nearly twenty years of discussion, hounding, and contract-drafting (PA). They also suggested that scrounging might be better thought of as an ongoing series of events—rejections and adoptions of objects by various parties over many years—and cited examples in their research of objects that had been acquired and discarded by various parties at least five times (PA).

Changes in a field's standards for operating equipment can also be the occasion for apparatuses becoming available for scrounging. For example, a single piece of digital apparatus may now encompass the functionality of several older pieces of analog or digital equipment (PA), meaning that equipment can be downsized. Or again, some items that are still perfectly usable but not state of the art may wind up not in the trash, but to other labs. As a result, many labs regularly employ various pieces of anachronistic but perfectly functional equipment.

Apparatus

If *apparatus* designates the tangle of material technologies and practices inside a lab that allows it to conduct its research, it is also what made me first think about how scrounging operates, and to what end. Securing and maintaining lab apparatus seem to perpetually involve scrounging, regardless of the research field.

The lab apparatus of the people I spoke with is bewildering in its variety, ranging from computer code and robots (VHL) to Arduino boards (VHL, CML), forklifts (PA), penny-farthing bicycles and resin cows (CML), Edison Phonographs (AL), Scientology E-Meters and Minitel terminals (MAL), and embossing telegraphs and Korsakov machines (MAF). Most of this apparatus is operational, and all of it has figured in research projects at some point. Many odd things end up in the sorts of labs that I am discussing because, as working collections, they use these items at various times as both apparatus (the instrument you study *with* and *through*) and as objects of study themselves. As Lisa Gitelman indicates in *Always Already New*, this dual status of media as subject and object is characteristic of media history and related fields, and it also is part of what makes conducting such research difficult (Gitelman, 2–7). At any point, the arrival of some new piece of apparatus in the lab may occasion the need for even more esoteric supplementary devices to operate it.

The question of unforeseen needs for apparatus has much to do with a lab's imaginary—how it sees itself, especially in relation to how it would like to be seen from outside. In his research on the Chinese typewriter, Thomas Mullaney (22–23) states that "the technolinguistic imaginary" is a vital component of the material infrastructure that makes it possible for cultural objects to function. This imaginary also links a technological object to larger infrastructures and assemblages, and/or to prototypes, clones, models, and analogs, so it's important to take into consideration.

There is always a gap between a lab's early imaginary (what the lab's founders think it needs and can account for in an official startup budget) and what it *actually* needs to conduct research after the money has been spent and the researcher has been in place long enough to understand what is available in their specific institutional context (ALW). This difference creates a primary occasion for scrounging. It's axiomatic that the eventual use of a given piece of media technology is rarely what its inventor intended because various publics imagine it in different articulations (in Stuart Hall's sense that a given technological artifact can be articulated to different discourses and different ideologies and different times and places; see Grossberg, 53). Media labs seem to operate in a similar manner. As a result, scrounging for apparatus occurs both intrainstitutionally and extrainstitutionally.

The implicit question in intrainstitutional scrounging is ownership, which is complex. Some scrounged media is garbage—things for which no one wants to claim ownership or responsibility. One colleague pointed out that one of the reasons that scrounging is a successful technique in the U.S. context is because the United States throws away more media than most nations ever produce (PA). For the researchers I spoke with who are working with code, scrounging is a matter of course, and it always has been. Copying code happens routinely, sometimes following academic and programming citation protocols, but usually without notification if the purpose is noncommercial (VHL). The protocols for scrounging physical equipment, however, are somewhat different. The scientists whom Feder speaks with frequently scrounge decommissioned equipment from various U.S. government agencies and the military (Feder, 27). Sometimes, however, the government retains ownership of such equipment, which can lead to additional complications, like annual equipment audits (Feder, 28). Even when ownership has changed officially as apparatus changes hands, physical markings on the equipment, documentation, or even informal oral histories create a continuity of use that can go back for decades (PA) and become an important part of lab culture.

In Canada, research equipment purchased with federal grants comes from taxpayer dollars. The equipment policies of the Social Sciences and Humanities Research Council (SSHRC) are clear: "All items purchased with SSHRC funds are the property of the university or organization administering the award. Any nondisposable items (including books, research materials and documents) must be formally listed in the university's or organization's inventory." In practice, though, how universities manage apparatus varies widely once equipment has been purchased. Perhaps the most accurate description of the scrounging that occurs inside universities and similar institutions is that it is an informal redistribution of institutionally owned resources.

Within a university, equipment depots (universitywide and department-specific) are fruitful scrounging sites. Many faculty members are unaware that such resources even exist, but long-term support staff know how to access them. Several researchers whom I spoke with mentioned the importance of administrative assistants and other staff to the securing of space, apparatus, and research collections (MAL, AL, ALW). Depot managers know when the value of equipment has been amortized to zero and have developed informal practices for disposing of it (sometimes via email lists, sometimes by word of mouth, sometimes by shipping them to universitywide storage facilities, or sometimes by placing it outside the door). In all cases, getting to know depot managers and personnel at all levels is the key to cheaply and quickly furnishing a lab with apparatus.

Other labs are another excellent source of scroungeable equipment. In many universities, placing superfluous equipment in the hallway outside the door is a sign that it is available for use by whoever feels moved to cart it away. The growing pervasiveness of practices for assigning unique serial numbers to equipment, and for tracking such equipment in databases, ties scrounged equipment via purchasing and accounting departments back to its original status as formally purchased and accounted-for infrastructure more firmly than has ever been the case historically. This fine-grained tracking may well make scrounging practices more difficult and uncommon over time (AL). But even in individual universities, there are different expectations of how equipment will be managed for different units. For example, research equipment may be handled differently (with more or fewer purchasing and tracking restrictions) than teaching equipment. Equipment purchased more than a few years ago may have no digital tracking associated with it, and, in some cases, there is no tracking at all.

Extra-university scrounging complicates the picture further because it is more likely to be a form of donation rather than theft. Some labs are lucky enough to receive so many donations that they don't actually have to go looking for apparatus very often (MAL); but they are the exception rather than the rule. University labs may appeal to private companies, public institutions, and private individuals for apparatus at different times and in different ways. One researcher whom I spoke with is in the habit of walking into the local hardware store and the nearest architectural salvage yard, describing the lab's current project, and asking for donations of everything from hammers and other tools to lumber (CML). In another instance, the researcher funded the construction of an arcade cabinet via a small donation from a local insurance company (CML). A second researcher described combing through Craigslist looking for classified ads from parents selling old video game consoles and computers (MAL).

Because humanities media labs often begin with the specific research interests of a particular scholar in an area that the academy has largely been unable to capture, it's not uncommon for lab apparatus to be seeded from a privately owned collection—often the collection of the researcher themselves (AL). This is only one of the ways that private property becomes intermingled inextricably with institutional property in lab apparatus. The researchers whom I spoke with described a whole spectrum of scenarios, from extremes where the entire lab apparatus is fully privately owned by the director because they have purchased everything themselves (MAF, SL) to fully university-funded apparatus (VHL). But in most cases, the ownership of the apparatus is usually a mix of public and private, especially in the lab's early days.

I would be remiss if I didn't point out the importance to the functioning of many labs of online auctions and third-party sellers like Alibaba/Aliexpress, Etsy, and third-party sellers on Amazon (and associated payment platforms like PayPal). This bears both on establishing working collections in labs and sourcing odd bits of apparatus on an ongoing basis. I am including these sites as a form of scrounging because, despite their reasonably long lifespan and assured role in world culture, they retain a whiff of disreputability that make them difficult to use or justify in an academic context, especially in conjunction with institutional credit cards.

There is now a substantial cultural studies literature about eBay, but there is very little literature on its utility to academic researchers seeking lab equipment, and most such discussion pertains to the sciences (Greenslade, 512; Ledford; Schneider et al.). As searchable digital bazaars, such sites provide lab directors with a huge network of potential sources of apparatus not obtainable elsewhere, or obtainable for a much lower cost than would otherwise be imaginable. Further, research relationships with online sellers can also emerge because they will often have items not listed that might be of use, and they may even make donations to the lab (MAL). Information about unlisted holdings and other useful kinds of information can come out of conversation only after establishing a relationship with online sellers, usually as a result of buying something and providing favorable ratings.

Buying from online sellers can also afford entry points into amateur expert communities that might not otherwise be visible to an academic researcher. One of my colleagues referred to these communities as "cultures of dissemination and distribution" (PA), and another pointed out that these amateur expert communities and the links between them were also an important kind of infrastructure (SL). And it can be the only market for the niche, hand-built, and specialized electronics that are core to emergent practice in subfields like DH and media archaeology. For these reasons, several of my respondents mentioned eBay and other online sellers in our discussions, but they also pointed out that gaining authorization to make purchases with a university credit card through PayPal can take some effort (AL, MAL, VHL).

Personnel

Labs often find themselves searching for people who are particularly skilled in scrounging for apparatus. Such people are often deeply embedded in expert amateur and professional networks on both local and global levels, via online discussion groups on a variety of platforms. They have experience in operating specialized and even obsolete equipment and know where to source it because of their interpersonal connections (AL). This can make them invaluable to any lab, especially when knowledge about how to operate the apparatus in the lab may have all but vanished from the general population. But there is generally no mechanism in university procurement systems to make possible the kinds of apparatus purchases that such individuals can arrange.

Interested individuals or small companies not affiliated formally with the lab or its larger institutions can also act as intermediaries in the process of scrounging (AL, MAL). In this case, the lab personnel search for ways to establish relationships with such individuals, who then scrounge on behalf of the lab. When someone has done some work for a lab and is registered as a company, it is easier to register the company in the university accounting system as a vendor, which then makes it much easier to pay them for equipment that they source from eBay, fairs, flea markets, collectors' meets, and other events designed for collectors and enthusiasts or elsewhere (AL, MAF).

In some cases, individuals acting as second-order scroungers for labs will be collectors of the sort of apparatus that the lab requires; in others, they may be donating materials that have long been in storage or belonged to a relative (MAL, CML). In still other cases, hobbyists, enthusiasts, and philanthropists who are familiar with a lab's mission may begin actively seeking materials to assist in lab work (MAL, CML, ALW). Sometimes a box of stuff just shows up at the lab's door, and the contents may or may not prove to be useful (MAL, MAF, ALW). These relations can be one-off or ongoing, but many labs now list donors, regularly advertise for equipment they need, and publicly elicit donations via events (MAL, ALW).

Affect

The final topic that I want to discuss is affect—the emotional charge that circulates around scrounging as a practice. The deep difference that I noted at the outset between British and U.S. definitions of scrounging is affective in nature. I was thus not surprised to find a wide affective range in the ways that my colleagues talked about scrounging.

On one end of the spectrum, a researcher described their relationship with university accountants—their home institution's infrastructural guardians—in terms of shame. Having to justify the purchase of apparatus that might seem like frivolous toys to people outside their field, they reflected, was a function of an imbalance in power across university disciplines. In the sciences, CFI-funded purchases would normally come from suppliers of laboratory equipment and would have criteria like a warranty, a certain number of years of projected use, and others; in the DH and media arts, there are not always such institutional markers of legitimacy on equipment purchases (AL).

On the other end of the spectrum is the joy that results from feeling personally empowered by securing and then learning how to operate complex and potentially dangerous apparatus (PA). Joy can also come from pushing the envelope of what constitutes academic research within an institutional context (CML).

Somewhere between the affective extremes of shame and satisfaction is frustration. Almost everyone whom I spoke with expressed some degree of frustration with university bureaucracy, but some also expressed frustration with anonymous donors for using their labs as a kind of charity drop-off, leaving boxes of random technology on the lab doorstep for them to sort through (MAL, ALW), or frustration with being unable to source specific pieces of equipment because of postpandemic supply chain issues (VHL). The researchers whom I spoke with also pointed out that the frustration that lab denizens can experience from having to work with randomly donated, semifunctional equipment can be transformed into satisfaction though Marcel O'Gorman's notion of "crapentry" (xi) or crudely hacking things together in a process that requires lab members to spend intense reflective time with materials, reclaiming them and turning them into something new. Because the ultimate goal of crapentry is discursive—to get the students to write about the experience—flaws in fabrication can be forgiven because failure and flaws are part of the process (CML).

Scrounging is orthogonal to university infrastructure. This was my working hypothesis for this chapter, and it was confirmed by the researchers whom I spoke with; one said, "Being invisible to the university is important, in some ways. Not invisible to the people that are interested in us, but invisible to the structure of the university" (MAL). Others spoke about how the labor of carefully piecing together funding for a lab manager's salary was quite essential to them—and to the actual human occupying the lab manager position—but was apparently invisible on "the faculty level or the university level, some level above us" (ALW). Another way of phrasing this, and what I hope this chapter demonstrates, is that scrounging is what Slavoj Žižek (after Donald Rumsfeld) calls an "unknown known"—something that everyone does but many disavow; something that is essential to the functioning of research labs across all fields, but largely undocumented and invisible; something that allows small labs to function economically but that university accounting is often functionally incapable of addressing. In such a case, Žižek argues, the challenge is to recognize and undertake the technique without denying it (Žižek, 95).

This is not the same as saying that scrounging should be formalized somehow. The power of scrounging lies in its ad hoc nature, what one of my colleagues calls a "positive, shared sort of leakiness" (AL). Another observed that they were able to accomplish what they had precisely because of the lack of formal support from the institution, which required them to look into options that would not otherwise have been available. At the same time, they made it clear that they didn't want to doom future generations of material media researchers to doing everything by the seat of their pants (MAL). Scrounging functions at its peak in the spontaneous moments of generosity and brashness that occur when someone recognizes what they actually need to launch or maintain a research trajectory, rather than what they think they are supposed to need, and then figures out how to get it.

What scrounging requires is tolerance. As technologies like radio frequency identification (RFID) tagging allowing for the databasing of all university equipment become more widespread, scrounging is increasingly difficult to do. But it remains utterly necessary for the existence of the sorts of labs that the university wishes to see—labs that have the capacity to document some hitherto undocumented space of knowledge production. If university infrastructure is going to function more efficiently, sometimes its guardians need to look the other way for a moment as a researcher scoops up a box of heterogeneous oddities from the floor of a back corridor and scurries away with it, scarcely believing their luck.

Interviews Conducted for This Chapter

For convenience, the interviews conducted for this chapter are cited by the parenthetical initials prefixed here (e.g., "AL"):

(AL) Jason Camlot (director of AmpLab East and the Spokenweb Network, Concordia University). Zoom conversation with Darren Wershler, August 29, 2022.

(ALW) Karis Shearer (director of AMP Lab West, University of British Columbia Okanagan) and Emily Murphy (assistant director of AMP Lab and director of (Re)Media Lab, University of British Columbia Okanagan). Zoom conversation with Darren Wershler, August 24, 2022.

(CML) Marcel O'Gorman (director of Critical Media Lab, University of Waterloo). Zoom conversation with Darren Wershler, August 23, 2022.

(MAF) Stefan Höltgen (former curator of Media Archaeological Fundus, Humboldt University, Berlin). Zoom conversation with Darren Wershler, August 31, 2022.

(MAL) Lori Emerson (director of Media Archaeology Lab, University of Colorado, Boulder) and libi streigl (lab manager of Media Archaeology Lab). Zoom conversation with Darren Wershler, August 23, 2022.

(SL) Stefan Höltgen (former curator of Signal Laboratory). Zoom conversation with Darren Wershler, August 31, 2022.

(PA) Rick Prelinger (founder of Prelinger Archives). Zoom conversation with Darren Wershler, August 25, 2022.

(RL) Emily Murphy (assistant director of AMP Lab, University of British Columbia Okanagan). Zoom conversation with Darren Wershler, August 24, 2022.

(VHL) Florian Sprenger (director of Virtual Humanities Lab. Ruhr Universität Bochum) and Thomas Nyckel (Research Assistant at Virtual Humanities Lab). Zoom conversation with Darren Wershler, August 5, 2022.

NOTES

1. Wershler, Emerson and Parikka, 80–81.
2. Star, 380.
3. "Scrounge," Merriam-Webster; Dictionary.com; Collins English Dictionary.
4. "Scrounge," Cambridge Dictionary; Collins English Dictionary.
5. See http://residualmedia.net/.
6. Interviews are cited in a separate section of this chapter's bibliography. The Uniform Resource Locators (URLs) of labs, centers, and other entities associated with the interviewees are cited separately in the main part of the bibliography. Interviews are not made available publicly because to allow interviewees to speak freely, they were conducted under the express condition that video recordings and transcripts would not be shared in full.

BIBLIOGRAPHY

AMP Lab, University of British Columbia Okanagan. Home page, n.d. https://amplab.ok.ubc.ca/.

Anderson, Sheila. "What Are Research Infrastructures?" *International Journal of Humanities and Arts Computing* 7, nos. 1–2 (2013): 4–23.

Canadian Foundation for Innovation (CFI). Home page, 2023. https://www.innovation.ca/.

Critical Media Lab, University of Waterloo. Home page, 2023. https://criticalmedia.uwaterloo.ca/crimelab/.

de Certeau, Michel. *The Practice of Everyday Life.* Translated by Steven Rendall. University of California Press, 1984.

Feder, Toni. "Scrounging Old Equipment for New Experiments." *Physics Today* 54, no. 7 (2001): 26–28.

Filecoin Foundation for the Decentralized Web. "FFDW Works with Prelinger Archives to Make Rare Historic Films More Accessible Using the Decentralized Web." Medium.com, August 24, 2022. https://medium.com/@FFDWeb/ffdw-works-with-prelinger-archives-to-make-rare-historic-films-more-accessible-using-the-558e4dbec990.

Galison, Peter. "Trading Zone: Coordinating Action and Belief." (1997, abridged 1998). In *The Science Studies Reader,* edited by Mario Biagioli, 137–60. Routledge, 1999.

Gitelman, Lisa. *Always Already New: Media, History, and the Data of Culture.* MIT Press, 2006.

Greenslade, Thomas. B., Jr. "Hold on to Your Heritage." *The Physics Teacher* 42 (2004): 512.

Grossberg, Larry. "On Postmodernism and Articulation: An Interview with Stuart Hall." *Journal of Communication Inquiry* 10 (1986): 45–60.

Jargon File. Version 4.4.7. Edited by Eric Raymond. S.v., "Social Engineering." http://www.catb.org/jargon/html/S/social-engineering.html.

Ledford, Heidi. "Garage Biotech: Life Hackers." *Nature* 467 (2010): 650–52.

Liu, Alan. "Drafts for *Against the Cultural Singularity.*" 2016. https://hcommons.org/deposits/item/mla:699/.

Media Archaeological Fundus. Home page, 2023. https://www.musikundmedien.hu-berlin.de/de/medienwissenschaft/medientheorien/fundus/media-archaeological-fundus.

Media Archaeology Lab (MAL), University of Colorado, Boulder. Home page, 2024. https://www.mediaarchaeologylab.com/.

Mullaney, Thomas S. *The Chinese Typewriter: A History.* MIT Press, 2017.

O'Gorman, Marcel. *Making Media Theory: Thinking Critically with Technology.* Bloomsbury Academic, 2021.

Prelinger Archives. Home page, 2024. http://www.panix.com/~footage/.

(Re)Media Lab, University of British Columbia Okanagan. Home page, November 12, 2020. https://fccs.ok.ubc.ca/2020/11/12/remedia-lab/.

Residual Media Depot. Home page, 2024. http://residualmedia.net/.

Schneider, Ethan, Paul J. Schenarts, Valerie Shostrom, et al. "'I Got It on Ebay!' Cost-Effective Approach to Surgical Skills Laboratories." *Journal of Surgical Research* 207 (January 2017): 190–97.

"Scrounge." Cambridge Dictionary. https://dictionary.cambridge.org/dictionary/english/scrounge.

"Scrounge." Collins English Dictionary. https://www.collinsdictionary.com/dictionary/english/scrounge.

"Scrounge." Dictionary.com. https://www.dictionary.com/browse/scrounge.

"Scrounge." Merriam-Webster. https://www.merriam-webster.com/dictionary/scrounge.

Siegert, Bernhard. *Cultural Techniques: Grids, Filters, Doors, and Other Articulations of the Real.* Translated by Geoffrey Winthrop-Young. Fordham University Press, 2015.

Signal Laboratory (*Signallabor*). Home page, 2022. https://www.musikundmedien.hu-berlin.de/de/medienwissenschaft/medientheorien/signallabor.

Social Sciences and Humanities Research Council (SSHRC). "Grant Holder's Guide." September 6, 2013. https://www.sshrc-crsh.gc.ca/funding-financement/using-utiliser/grant_regulations-reglement_subventions/strat_grants-subventions_strat-eng.aspx.

Spoken Web. Home page ("Sounding Literature"), 2024. https://spokenweb.ca/.

Star, Susan Leigh. "The Ethnography of Infrastructure." *American Behavioral Scientist* 43, no. 3 (1999): 377–91.

Survival Research Laboratories. *The Will to Provoke: An Account of Fantastic Schemes for Initiating Social Improvements.* VHS. Directed by Jonathan Reiss. Survival Research Laboratories. 43 minutes. Color. 1988.

Virtual Humanities Lab (VHL), Ruhr Universität Bochum. Home page, n.d. https://vhl.blogs.ruhr-uni-bochum.de/.

Wershler, Darren, Lori Emerson, and Jussi Parikka. *The Lab Book: Situated Practices in Media Studies.* University of Minnesota Press, 2022.

Williams, Raymond. "Dominant, Residual, and Emergent." In Raymond Williams, *Marxism and Literature,* 121–27. Oxford University Press, 1977.

Young, Vershawn Ashanti. "'Nah, We Straight': An Argument Against Code Switching." *Jac* 29.1–2 (2009): 49–76.

Žižek, Slavoj. *Organs Without Bodies: On Deleuze and Consequences.* Routledge, 2004.

PART III

(RE)ENVISIONING DIGITAL HUMANITIES INFRASTRUCTURE

Resisting BYOI (Bring Your Own Infrastructure) in Digital Humanities Learning Spaces

KUSH PATEL, ASHLEY CARANTO MORFORD, AND ARUN JACOB
(PEDAGOGY OF THE DIGITALLY OPPRESSED COLLECTIVE)

We have crafted a fictional creative writing piece—a Twine narrative (discussed later in this chapter)—based on our own lived experiences but set within fictional institutions and using fictional technological platforms. For this creative piece, we have chosen email communication as the form because this mode of engagement resonates across all three institutional contexts in our fiction and has been widely used to bridge distances amplified by the pandemic. Structuring a Twine narrative around this form of writing allows us to illuminate both the material infrastructure of the institutions under discussion and congruities of managerial problem-solving using educational technologies across countries. The choose-your-own-adventure, nonlinear storytelling of Twine juxtaposes a set of distinct institutional-infrastructural contexts; provokes an understanding of their entanglements; and invites the reader to reflect on how technosolutionist responses to the pandemic might become (and indeed already are being) reenacted and exacerbated across interconnected crises that affect postsecondary institutions and society more broadly.

Our Twine narrative is set at the beginning of the global Covid pandemic and consists of a series of email chains. Readers follow the experiences of faculty, staff, students, and precarious workers at postsecondary institutions in India, the United States, and Canada as they navigate changes to educational processes and organize their activities through online infrastructures. The locations of the central characters highlight the material overlaps of infrastructure "not just as a 'thing', a 'system' or an 'output', but as a complex social and technological process that enables—or disallows—particular kinds of action" (Graham and McFarlane, 1). The first email is written from an emerging not-for-profit school of design in India, offering undergraduate-level education. The second is written from a long-established small, private liberal arts college offering undergraduate and graduate studies in an urban center in the so-called United States.[1] The third email features a Canadian community college in an urban setting. As the combined emails explain, for precarious workers' digital learning to be productive, their media technology hardware, software, and wetware[2] have to work properly, seamlessly, and synchronously. Our story then shifts into imagining a digital otherwise, wherein these same characters have all the infrastructural and institutional support they need to organize, teach, and learn in nourishing, affirming ways online.

In the accompanying reflection that follows the Twine story, we provide a rationale and analysis of the creative piece involving three concepts: responsibilization, crisortunity, and digital infrastructural otherwises. With each of these concepts, grounded in specific sociotechnical realities, we reflect on the possibility of reimagining infrastructure to enable more care-filled and just teaching, learning, and organizing. By moving past the common institutional expectation that people with precarious lives provide their own infrastructures (BYOI: bring your own infrastructure) to support critical pedagogical commitments, we argue that BYOI: exacerbates structural oppressions by stratifying and stultifying learning and organizing; keeps us dependent on network connectivity via subscriptions to proprietary service providers; and pressures students and faculty to view the pandemic as an opportunity to be realized with "business as usual." Working against BYOI, the crafting of digital infrastructural otherwises ensures that teachers, learners, and organizers have the infrastructural and institutional support they need to organize, teach, and learn in nourishing, affirming ways online.

Twine Narrative

Our Twine narrative is available online as a downloadable Hypertext Markup Language (HTML) file for playing the story (and also as the Twine source-code file).[3] We also transcribe the fictional emails in it as plain text here, though of course sacrificing the nonlinear form of the Twine narrative.

PLAIN-TEXT TRANSCRIPTION

Location: An Emerging Not-for-Profit Undergraduate School of Design in India

> To: Staff, Faculty, and Students
> Subject: New Work Arrangement
> Date: March 2020
>
> Dear students, staff, and faculty members,
>
> This is an unprecedented situation in our country and the whole world. Let us individually be responsible in this pandemic and amplify the government's

"Break The Chain" campaign to contain the spread of COVID-19 virus across the state.

However, to maintain academic excellence and conclude the semester on a timely basis, we will move to a "work from home" mode, starting tomorrow. Each of you is expected to follow the academic calendar, meet academic deadlines, and observe the prescribed expectations for student work and year-end faculty reports. All "work from home" assignments will be critical for students' academic success. All students must also note that your faculty members will be available online and over the call at any time for answering questions or concerns that emerge.

The Academic Chair will be the key contact person for all your academic matters during this time. As a reminder, let us not compromise the high quality of education and learning at our institution.

With best wishes for a successful semester,
Principal

* * *

To: Reply All
Subject: New Work Arrangement

Dear Principal, and dear faculty members, students, and administrative colleagues,

Thank you for your warm welcome. These uncertain times of the pandemic will require us to be remote and online until it is safe for us to return to the physical campus. We will follow the state and national COVID-19 guidelines for higher educational institutions and campus residences, and take decisions about reopening as a collective.

While the institution's subscription to Google educational workspace, including your access to Google Meet and Google Drive will continue without interruption, I do not expect our immediate move to "work from home" mode to be as smooth, or remain as uninterrupted. We may find ourselves logging in from spaces busy or small; internet connectivity erratic or slow; homes with people distant or close; stresses personal and familial; and questions about maintaining a good academic standing both contested and uncertain. We may also find ourselves sharing and planning for far more responsibilities at home than usual, caring for our loved ones, and trying to keep focus. As a community, let us post reminders to each other about what is at stake in the context of pandemic learning and teaching lest we fail in adjusting or our adjustments fail us.

Starting tomorrow, the faculty cohort will shift away from weekly administrative meetings to protect time for quiet and work, and to balance out a continuous range of engagements that we have otherwise been participating in or scheduling with our colleagues and students this semester. The priority for me is our mental and physical health, so please take care of your energy levels and reach out to me with any concerns that you may have as a student, faculty colleague, or staff member at this institution.

In community,
Academic Chair

* * *

To: Reply All
Subject: New Work Arrangement
Date: April 2020

Dear Colleague,

As a wellness coordinator, and following our revised academic plan, I have been meeting with the undergraduate student groups regarding their well-being and educational challenges in these remote and online environments. Through phone and video calls, students (cc'd) have opened up about their hopes and struggles with me, and a lot of them find it difficult to access online spaces or even stay online for extended periods of studio and seminar work due to inconsistent and even absent internet connectivity across the state, or within limited data plans and financial securities. From joining on mobile phones at roadside tea stalls to connecting from terraces or family members' homes where there is a greater possibility for a stable internet, these students are navigating academic expectations under extreme duress. As we move through this extended remote semester, it would be useful for us to recalibrate synchronous and asynchronous project engagements and pedagogical formats. I am copying everyone on this email with students' permission and following your invitation to continue to post community reminders regarding teaching and learning in pandemic times. Thank you for your consideration.

Best wishes,
Wellness Coordinator

* * *

To: Reply All
Subject: New Work Arrangement

Dear Wellness Coordinator,

Thank you so much for reaching out to the students at all academic levels and representing their concerns with consent on this shared email thread. I have taken note of these challenges and reached out to the Principal (cc'd) and members of the management team for advice regarding pooling resources for purchasing adequate internet plans for any and all students, faculty colleagues, and staff members seeking reliable network connectivity.

My thanks are also to the students for sharing their challenges with us directly and across these different fora. With limited IT support and administrative strength, we are currently also looking into the possibility of sharing internet connectivity remotely.

In the meantime, and to all my faculty colleagues, please make edits to your respective course plans and continue to center the question of academic excellence away from its pre-pandemic framings. Pedagogy of wellness was not just the theme of our last faculty colloquium, but also an ethic in practice around which we will structure our commitments moving forward. Thank you all for your cooperation.

Best wishes,
Academic Chair

Location: A Long-Established, Small, Private Liberal Arts College Within an Urban Center of the United States

To: Department Faculty, Staff, and Graduate Students
Subject: Urgent Updates
Date: March 2020

Dear Teaching Assistants:

In light of the pandemic, the campus and this department will be closing to in-person gatherings and we will be shifting our classes online effective immediately. You will be leading your seminars via Room. Please promptly make a Room account and please note that a Basic account will not suffice for teaching purposes. Be aware too that, in terms of etiquette, it is expected that, as the seminar lead, you have your camera on at all times during the seminar, and please share slides with students via Room's Share Screen feature.

Thank you and happy teaching.

Best,
Graduate Department Chair

* * *

To: Reply All
Re: Urgent Updates
Date: March 2020

Hi Graduate Department Chair,

Thanks for this update. I am appreciative that the department is making moves to try to keep our community safe during this pandemic by halting in-person classes and gatherings.

However, I do have some concerns and considerations I wanted to raise. What is our department—and the institution more broadly—doing to structurally support graduate student teaching assistants in successfully continuing to teach during this shift to online learning? For instance, my seminars have 30 students, but I do not have an internet connection sufficient to hold a 30-person Room call while having my camera on and sharing my screen, nor do I currently have the financial ability to pay for such a connection. Given that high-speed internet is now integral to my role as a teaching assistant, I believe that the department should provide this infrastructure for its instructors. Furthermore, there has been no information about how the institution and department will help us access licensed Room accounts. Are teaching assistants expected to pay for our own licensed Room accounts? Thanks for any insight you can provide.

Best,
PhD Candidate and Teaching Assistant

* * *

To: Reply All
Re: Re: Urgent Updates
Date: March 2020

Dear PhD Candidate and Teaching Assistant,

Thanks for letting me know your concerns.

With regards to your concerns about a Room account, please be aware that, if you sign up for a Room account using your institutional email address, the account will be licensed. The institution is generously providing this Room access to all of its members during the course of this pandemic.

Now, in terms of your concerns about internet connection—internet connection is widely accessible throughout the city, and it seems to me that, since you have been able to respond to my initial email update, you do have ready internet access or, at the very least, you know of someone whose internet you can use.

However, if you feel that you are not currently able to fulfill your teaching obligations due to the circumstances of the pandemic, please know that we can find someone who can take over your seminars for the remainder of the semester. Let me know as soon as possible if you would like us to seek a replacement teaching assistant. I hope that this offer helps alleviate any undue stress.

Stay safe!

Best,
Graduate Department Chair

* * *

To: PhD Candidate and Teaching Assistant
Re: Re: Urgent Updates
Date: March 2020

Dear PhD Candidate and Teaching Assistant,

I am so sorry that the department is once again neglecting and, quite frankly, undermining the needs of its graduate students and most precarious workers. Please let me offer a small gesture of support given my privilege as a full professor and my role as your supervisor. I am happy to pay for adequate internet connection for your apartment for the remainder of this term, so that you may successfully fulfill your online teaching duties this semester.

Best,
Full Professor and Doctoral Supervisor

Location: A Canadian Community College in an Urban Setting

To: Staff, Faculty, and Students
Subject: Covid Pandemic Updates
Date: March 2020

Hello Colleagues,

I hope this email finds you well.

As you all know, the world has been transformed beyond recognition by Covid-19.

It is clear that digital technology will be essential to teaching, research and administration under these truly uncertain conditions. We at the college feel it is urgent that this online pivot is done as smoothly and efficiently as possible. Students and learning are at the heart of what we do and it is imperative that we keep the college fully operational and functioning at a distance. Our institution has many faculty and staff who are experts on the use of digital technologies for educational purposes, and we will be leveraging their knowledge and expertise to roll out state of the art educational technology solutions to ensure the organization runs smoothly and seamlessly. We will be using industry standard technologies like Room, which will enable operations of the college to continue at a distance during the lockdown, and we will be offering webinars to our students and staff on how to ensure they are able to access this software.

We would like to ensure faculty readiness to make sure that we have mobilized all possible models of socially distanced teaching and learning. We would also like to make sure that our faculty are supported in the best possible way to develop, transition, and implement online teaching as they rapidly pivot to remote learning solutions and distance education. We are striving to make sure the student experience is not compromised in any way as the academic term is delivered through the digitally mediated educational technology solutions. We rest assured that our experienced faculty will be able to help assuage whatever minimal impact the digital migration will have on students' learning experience.

Our Centre for Teaching and Learning are developing strategies and techniques for creating community in a socially distanced university as we speak and will be in touch with you shortly to consult with you on ensuring that your courses are ready to be pivoted for online delivery in the next little while and the course will be completed on schedule with minimal impact of social distancing on our teaching, learning and research culture.

Looking forward to hearing from all of you.

Best,
Associate Dean of Liberal Studies

* * *

To: Colleague
Re: Covid Pandemic Updates
Date: March 2020

Hi Colleague,

I hope this email finds you well in these truly distressing times. I wanted to reach out to you because I find the Associate Dean's email rather concerning. I don't think that email really quite gets to the matter of faculty readiness and institutional support. It is quite disheartening how so many of our real life concerns have been flattened out in that email. I don't quite get what kind of support and help we are getting here. I feel there's a tethering of our home internet connection, personal computer and our work that is happening here that is not being addressed at all. Let me elaborate: In order for the instructor's virtual classes to go right, the hardware has to work right, the software has to work right, the network connection has to work right, if and only if all these media architectures nod their heads in harmony, can the virtual instructor deliver their classes.

Our reliability as instructors (Can you hear me now??) and trustworthiness (Oops! Sorry, I have a faulty connection!) are tethered to the kind of internet connections we have at home. So rather than the virtual classroom championing underprivileged students, it will stratify our classrooms evermore. We as virtual instructors are both yoked to the tech that we have access to, so what that means is we are only as good as our internet connection, our laptop, and the software on our machine. Moreover, what is troubling is the likelihood that the state of exception becomes the new norm. i.e. post-Covid-19, the Covid-austerity measures will in all likelihood be implemented, at which time, virtual instructors will find the supports that are available as stop-gap measures will be clawed back, with the virtual class option dangled as a cost-effective option for hyflex (hybrid and flexible) learning, one in which sessional instructors like myself will be expected to do all the work and receive little to no institutional help.

Cheers,
Sessional Instructor

* * *

To: Union Steward
Re: Covid Pandemic Updates
Date: March 2020

Dear Union Steward,

I hope this email finds you well in these truly distressing times.

I wanted to bring to your attention as our union steward some of the problematic things that I noted in our Associate Dean's email to the faculty regarding the pivot to online instruction, firstly the use of proctoring software to surveil student exams at home and secondly the adoption of teaching material from e-textbook vendors writ large.

    The potential merger of the world's two publishing giants would make them the largest publishing house on the planet. The sharks are circling the waters. The mergers and acquisitions that are happening in the sector suggest that these players no longer just offer textbooks. They include several goodies to go with the combo meals that they sell. Lecture slide decks, animations and videos, multiple-choice question banks and most recently a large publishing house has partnered with a proctoring company to offer remote exam invigilation services if you're subscribed to their platform. The shift here is that the book publishing house no longer offers just a book—they have so many other add-on features to sweeten the deal. e.g. AI can deal with the drudgery of grading now according to the new publishing platform, which is only available to those schools and professors that are locked into the platform ecosystem that the publishing house dictates. You are locked into the walled garden of products and services as devised and directed from them.

    Moreover, the growing amount of student data amassed through digital courseware will make the large textbook publishing houses attractive targets for consumer credit bureaus, recruiting agencies, and data vendors, who would find access to alternative data about students highly valuable.

Cheers,
Sessional Instructor

Commentary

As this creative narrative illustrates, we are living in a society, structure, and period rife with—indeed, driven by—responsibilization, "the process whereby subjects are rendered individually responsible for a task which previously would have been the duty of another—usually a state agency—or would not have been recognized as a responsibility at all" (O'Malley, 276). The process and practice of responsibilization constitute a legacy and aspect of capitalist colonialism. Historically and in an ongoing fashion, colonization has sought to destroy collective ways of being by forcing people into nuclear families and other isolated and individualized structures of being, actively separating colonized communities into fractured units as a means of attempting to prevent anticolonial resistance. As Indigenous queer studies scholar Chris Finley (Colville Confederated Tribes) asserts, individualized structures of being such as "the nuclear family need [. . .] to be thought of as [. . .] colonial system[s] of violence," given how such a structuring has been used as "both an ideological and a physical tool of [. . .] colonialism" (32).

The pandemic made the pervasiveness and harms of responsibilization ever clearer—pervasiveness and harms that seep into, interweave with, and can be both confronted and exacerbated by digital spaces and "infrastructural subjectivation."[4] As our lives shifted online in the early stages of the pandemic, digital infrastructural responsibilization accelerated, when educational institutions demanded that individuals provide and pay for their own digital infrastructures. Subsequently, and ironically, the capitalist push to pre-Covid "normal" undermined online options and infrastructural possibilities for teaching, learning, and working: responsibilization was rhetorically framed by colonial institutions as the freedom to individually choose how to negotiate health risks in unsafe and unjust physical conditions.

The vision of freedom forwarded by Black and Queer of color feminists provides a useful lens to interpret these issues, collective at its root and positioned against the individualistic and fallacious conceptualization of freedom at the core of responsibilization—a supposed freedom that strips others of their agency and safety. As Black feminist scholar-organizer Robyn Maynard writes, "There are two different visions of freedom at play here. One is the freedom to evade, to deny one's responsibilities to a collective social body; the other forwards a freedom that is relational, holds up freedom as collective safety" (76). It is this second freedom, this understanding that freedom is relational and collective, that will help to move our infrastructural approaches away from perpetuating colonial harms, and that must guide our actions and approaches to digital infrastructures in the here and now and for the future. Such a vision of freedom offers the possibility and foundation for building and fostering digital infrastructures of collective care and community freedom, grounded in disability justice organizer Leah Lakshmi Piepzna-Samarasinha's conceptualizations of care work as "collective responsibility" (33) and "collective access" (32), and as "a deep possibility model, not a one-size-fits-all solution for everyone who needs care" (46).

As the Covid-19 pandemic continues in its effects, even as the emergency has been declared officially over by the World Health Organization (2023), scholars and practitioners across virtually all domains are thinking critically about how to continue their work via digital and virtual means. What does a robust and useful technological response, aware and attentive to the biases and messages of digital media, look like in the context of a crisis like the Covid-19 pandemic? As Barbie Zelizer (890) asks, is "a crisis that ensues following long periods of neglect as sudden as we make it out to be? Why is disruption more deserving of attention than what precedes it?"

This creative narrative centers on "crisortunity," a neologism coined by *The Simpsons*[5] and since expanded on by critical media scholars such as Tanner Mirrlees. Crisortunity, or a crisis situation that also presents the opportunity for someone to gain something, unifies our narratives. Our creative responses are linked by an attentiveness to the uneven precarities and vulnerabilities so often symptomatic of institutional responses to crises, as well as the production, circulation, and management of information. Techno-fixes often fail to fix; instead, they reinstate unequal and inequitable relations in the name of repair. In contrast, when communities organize to respond to crises on their own, tensions arise between attempts for self-organization and anticapitalist modes of creating community and the infrastructure required to adequately support their labor. Our narrative thus explores the implications of qualifying something as a "crisis," with a discrete beginning and end, and what it might mean to offer fixes to something that is broken, rather than curating, managing, repairing, or caring for something that is not yet irreparable (Gàl). As our creative provocations lay themselves bare, "crisortunity" is a generative concept for highlighting what institutions see as the bugs, and what are the features; in other words, when a crisis becomes critical enough to justify intervention, and for whom. Is the platform crashing,[6] or is it only glitching for those few who are accustomed to a seamless interface and the social defaults of an inclusive classroom (Barnett et al.)? What additional challenges are introduced when the response to a crisis is primarily technological?

If the experiences of our students and faculty colleagues across both geographical and infrastructural difference highlight shared concerns of institutional apathy, how might these experiences also help us to critique more materially what feminist writer and scholar Sara Ahmed calls our "relationship to institutional worlds" (12) and what the self-described "Black, lesbian, mother, warrior, poet"[7] Audre Lorde describes as "the master's tools" (120)? If the correspondence between academic and administrative staff translates historically unequal and inequitable digital pedagogy engagements, making it easier for only certain populations to access and enter "the online classroom," what alternative worlds might we conceive of that would be capable of building community infrastructures enabling of more equitable learning and teaching goals? How, where, and with whom might we build such classrooms? And what critical relationships to digital pedagogy infrastructural work might these classrooms embody?

Thinking of Black feminist author-activist bell hooks's call to "work for justice" (xiii), and alongside Sara Ahmed's figure of the "stranger" (3) to explain the urgency of constructing new worlds that structurally uproot the experiences of academic and digital "unbelonging," Black Latina scholar and academic Lorgia García Peña discusses "freedom spaces," including the Freedom University, which she cofounded in the state of Georgia with human rights activists, undocumented students, and allies.[8] At their core, both freedom spaces and the Freedom University constitute a "third space" (Peña, 61) of academic freedom, where the justice-oriented work of teaching, learning, healing, and community-building may be facilitated, and where no student is left to learn and survive on their own (either materially or infrastructurally), even during pandemic emergencies. "To not ask the university," García Peña writes, "to 'love us back,' to not demand the university—a neoliberal, colonizing, racializing institution—provide that which is against its own nature, but rather to take its resources and structures and repurpose them to create freedom spaces, freedom schools, and liberation movements within and through its violent exclusion" (19–20). How might our infrastructural reimaginings make possible and sustain the construction of freedom schools around digital community infrastructures along the campus-community spectrum?

Within our own digital pedagogical experiences, we have witnessed a range of community movements and initiatives that begin to help address this question. The Detroit Community Technology Project[9] on "digital stewardship," for example, asks what it means to install, advocate for, and build local capacities for accessible and affordable internet networks in a city where around 40 percent of homes do not have an internet connection.[10] While colonial nation-states continue to stifle Indigenous community access to internet infrastructures throughout so-called North America—even as these nation-states dig up and extract from Indigenous lands to create digital infrastructures for settlers—Indigenous nations have been building internet and digital network connections (Duarte) and crafting robust online and social media presences (Caranto Morford and Ansloos). Their goal is to support network and cultural sovereignty and to help foster Indigenous learning spaces rooted in Indigenous worldviews. In India, the two-decades-long work of Janastu[11] around "software commons" demonstrates their success in developing Wi-Fi mesh networks with communities across rural Karnataka and "building," in their words, "technological solutions for indigenous communities and nomadic tribes in the region." While functioning as information and communications technology (ICT), each of these initiatives also serves as an infrastructure for technology coownership and exploration, enabling community-centered and community-led survival and flourishing outside corporate machineries. How might we bridge these worlds of collective liberation and infrastructure re-building? How might we proceed from BYOI: Bring Your Own Infrastructure to BYOI: Build Your Otherwise Infrastructure in digital humanities (DH) learning spaces?

In an interview with Peter James Hudson on "The Geographies of Blackness and Anti-Blackness," Katherine McKittrick, a Black anticolonial feminist critical geographer, offers a radical critique of discourse and assumptions related to safe classrooms, wondering whether we can or should ever achieve safe classrooms, given that many of the participants—students, faculty, and nonacademic community members—experience unsafe and precarious lives.[12] Being accountable to these everyday and structural contexts in which learning comes into being is central to our project of otherwise infrastructures; a project where we see students, faculty, and nonacademic community members as partners in coconstructing the freedom spaces that we desire.

Filipino queer studies scholar Robert Diaz and Black studies scholar Rinaldo Walcott root the ongoing emergence of otherwise spaces and infrastructures in the lived, intergenerational, and embodied experiences of marginalized communities. The lived experiences and community-based knowledges of Black, Indigenous, people of color, immigrant and refugee, 2SLGBTQIA+, disabled and neurodivergent, and class- and caste-oppressed communities, and the dreams, expressions, and actions that emerge from these experiences, provide "the necessary fuel for imagining better futures that do not rely on the finite grammars of the present" (Diaz, xix). Those perspectives provide the fuel to imagine into being infrastructural otherwises that exist across and traverse digital and analog spaces, places, and experiences. Working and dialoguing with the writings of Cuban American academic José Esteban Muñoz (1), Filipino queer studies scholar JP Catungal posits that "imagining the future otherwise is a queer refusal of the unacceptable present" (34). Overtly positioning the otherwise as a refusal of the colonial present, Black and Indigenous scholars and community organizers on the Black-Indigenous solidarities podcast The Henceforward recognize the otherwise, or the elsewhere, as the "places we yearn for" and also recognize that these are places and infrastructures that marginalized communities work to bring into being every single day in small- and large-scale ways (Habtom).

Against the responsibilization-driven, crisortunity-centered, and structurally estranged worlds of conventional digital pedagogy environments, we expect the otherwise infrastructure of digital learning to be simultaneously occupying higher educational spaces; building new worlds from those very resources (labs, centers, and pedagogical initiatives); and also extending beyond the limits and confines of academic spaces to collaborate with and ultimately emerge from and be deeply rooted within community spaces. We return, too, to Leah Lakshmi Piepzna-Samarasinha's work in disability justice to emphasize that digital infrastructural otherwises must reflect and embrace "a deep possibility model, not a one-size-fits-all solution for" digital access (46). Rather, thinking with design scholar Sasha Costanza-Chock's writings on design justice, we assert the need to think collectively and produce infrastructural designs that use community knowledge. Success should not be linked to profit generation, and technical prowess should not be antithetical to humanistic and community-based ethics and values. We remain cognizant of the complexities and contradictions in working through these relationships, diverse needs, and possibilities, but it is to the messy, ongoing, mutually accountable, widely citing, and always consensual practices of unlearning, relearning, and colearning in the context of critical digital pedagogy and otherwise infrastructure building that we remain committed. Our combined creative and academic narrative highlights infrastructure as a technoefficient and technodeterministic conduit distinct from any social makings of infrastructure—and particularly educational infrastructure in a DH classroom—in embodied, material, and situated terms.

Emergency online instruction, we argue, initiated an epistemic shift in how postsecondary education operates. The Covid-19 pandemic set the stage for a rewiring of society and a reordering of things. While this transformation could have been oriented toward building and sustaining digital infrastructural otherwises, the crisis instead added catalysts to the epistemic rupture that was already happening in the education sector. Catastrophe capitalists and technosolutionists are pushing ethically compromised products and services, and those in positions of power and privilege are using the moment—this crisortunity—to push existing policy positions. For example, remote proctoring policies, which surveil students through cameras, microphones, and various other data harvesting and tracking techniques, were instituted at the height of the pandemic and have now become normative practices (Camara, 2020). We emphasize that online education is not the issue. Indeed, online learning spaces can and do provide enhanced access and are necessary for helping to keep ourselves and one another safe while learning together. However, how institutions have created and facilitated online education during crises has perpetuated, recongealed, reconfigured, and exacerbated the colonial-capitalist affinities, allegiances, and alliances that have been embedded in mainstream education for far too long.

As instructors, when we necessarily shifted to online learning to protect our communities, we were committed to providing students with a nourishing, supportive, accessible, and meaningful learning environment, especially as all of us struggled to navigate and survive the stress, uncertainties, and grief of the unfolding pandemic. However, as we worked to offer this type of learning environment, we were far too often left on our own, without the proper infrastructural support from our institutions to ensure that we could develop a holistically accessible and caring online learning space for us and our students. Whereas emergency online instruction initiated an epistemic shift in how postsecondary education is being imparted, we were responsibilized to maintain normative standards of academic excellence by doubling as a virtual help desk to aid students with navigating instructional software issues, as well as moonlighting as instructional designers to ensure that course designs met an institution's expectations of effective delivery during the pandemic's height. That is, as an online instructor, one does not merely perform the necessary pedagogical tasks, being a content matter expert, mentor, coach, and so on. Here, one takes on the added responsibilities of being a commercial content moderator when using chat rooms, discussion boards, or video conferencing software; a podcaster and/or vlogger, if one produces asynchronous teaching and learning material using audio-video technologies; a virtual help desk to help students address instructional software issues and on occasion deal with student hardware and software issues; and even an instructional designer to ensure that the course design meets universal design for learning principles. By entering the job with the expectation of having all these competencies, fluencies, and literacies, an online instructor is also reified into the infrastructures of subjectivation and responsibilization.

Online instruction is not only about delivering content over a new medium: there is a technodeterminist undercurrent to the move, an ideological temperament that the instructors themselves must espouse both within and across geographical specificities. Therefore, when a course development cycle begins to look like a software development cycle, or when course development involves coming up with software solutions for the various pedagogical activities that are performed in the platformed virtual teaching and learning spaces, we pay more attention to software instead of putting care and focus into holistically developing the course, as well as its content and community. Since the pandemic started, we are seeing how instrumental reasoning and the techno-logics of the software developer are taking hold in mainstream academia. The purveyors of technologies and media infrastructures (i.e., the enterprise resource management platform, the learning management systems, the knowledge discovery systems) are exerting their power and authority in this reordering of things and reorganization of the institution.

To foster digital infrastructural otherwises, we must resist BYOI, which involves capitalist-colonial priorities and induction into the effective and efficient neoliberal machinery of the higher education system. Communities are building their own internets and virtual teaching, learning, and organizing spaces; students and academics are cocreating antioppression reading lists within historically white and upper-caste academia; and campus-community connections are coming together to facilitate virtual teach-ins and digitally circulated mutual aid campaigns to provide ongoing support to students, contingent staff, faculty peers, and community members who experienced precarity amid the pandemic and afterward.

This community- and collectivist-centered otherwise work reminds us of and embodies Black organizer-educator Mariame Kaba's assertion that hope is more than a mere emotion; rather, it is "a discipline," an ongoing and grounded philosophy or practice that is "practiced everyday, that people actually practice [. . .] all the time" (26–28). As we engage in the creation of more just digital teaching, learning, and organizing spaces, and as we work with digital infrastructures to develop more just teaching, learning, and organizing practices, we must learn from this radical community and collectivist work and embrace grounded practices of hope. Indeed, we might see hope as a discipline that we embody as we enter and engage with digital infrastructures. And we might also see hope as an infrastructure itself, cobuilt with and within communities, and emergent, functional, and developed across and in connection with lands, waters, and cyberspace, moving us from BYOI: as Bring Your Own Infrastructure to BYOI: as Build Your Otherwise Infrastructure.

NOTES

1. We use the term "so-called" to problematize and challenge the legitimacy and supremacy of settler colonial nation-states. The phrase is deployed in the context of decolonizing theory presented in works such as Mignolo and Walsh, and this is a practice that we have witnessed and learned from certain Indigenous studies and community-organizing spaces we have been part of in North America.

2. *Wetware* refers to the human labor that keeps platforms and infrastructures operational. See Winthrop-Young.

3. The HTML file for playing the Twine narrative in a web browser and the .twee file containing the source code for the Twine may be downloaded from https://dhdebates.gc.cuny.edu/projects/critical-infrastructure-studies-and-digital-humanities/resources?tag=chapter%2014.

4. Langlois and Elmer (238) explain how the new political economy of subjectivation relies on distributed impersonal infrastructures made of impersonal subjects homing in on the idea of "infrastructural subjectivation," wherein a set of relationships are established within the media platform to aggregate the human user and nonuser machinic traits as ascertained from the hardware and software are concatenated into one informationalized unit—the data point.

5. This portmanteau, created by one of the characters on the popular American animated series *The Simpsons,* season 6, episode 11, "Fear of Flying," conveys this idea (that every crisis provides an opportunity) well. See Benjamin Zimmer.

6. Plantin and Punathambekar (165) point out how we are living through the infrastructuralization of digital platforms, as well as how a critical infrastructural lens enables us to think through/about digital platforms vis-a-vis questions of scale, labor, cultural practices, policies and regulations, and other issues.

7. This positioning of the self is one that Lorde would offer continuously during public talks and appearances.

8. See Garcia Peña (2022). The goal of Freedom University was to serve the needs of undocumented students who had been excluded from established public universities in Georgia. García Peña's work builds on the 1960s work of education, belonging, and empowerment in connection with the civil rights movement in the United States, but also with the teaching and justice work of bell hooks, Arthur Schomburg, and Gloria Anzaldúa, among others.

9. See https://detroitcommunitytech.org/.

10. See the Teaching Community Technology Handbook (2017) at https://detroitcommunitytech.org/teachcommtech, which brings together all the projects that the members of this collective have used and built in practicing community technology in Detroit from 2008–2015.

11. See https://open.janastu.org/.

12. See McKittrick (238). See also our engagement with McKittrick's provocation in Friend, Caranto Morford, Patel, Jacob et al.; Patel, Caranto Morford, Jacob; and Patel, Jament, and Mathew.

BIBLIOGRAPHY

Ahmed, Sara. *On Being Included: Racism and Diversity in Institutional Life.* Duke University Press, 2012.

Barnett, Fiona, Zach Blas, Micha Cárdenas, et al. "QueerOS: A User's Manual." In *Debates in the Digital Humanities,* 50–59. University of Minnesota Press, 2016.

Camara, Wayne. "Never Let a Crisis Go to Waste: Large-Scale Assessment and the Response to COVID-19." *Educational Measurement: Issues and Practice* 39, no. 3 (2020): 10–18. https://doi.org/10.1111/emip.12358.

Caranto Morford, Ashley, and Jeffrey Ansloos. "Indigenous Sovereignty in Digital Territory: A Qualitative Study on Land-Based Relations with #NativeTwitter." *AlterNative: An International Journal of Indigenous Peoples* 17, no. 2 (2021): 293–305.

Caranto Morford, Ashley, Kush Patel, and Arun Jacob. "Pedagogy of the Digitally Oppressed: Anti-Colonial DH Pedagogy as Care Work." Paper presented at the Canadian Society of Digital Humanities/Société Canadienne des Humanités Numériques Annual Conference Making the Net Work, May 30–June 3, 2021.

Catungal, John Paul. "Toward Queer(er) Futures: Proliferating the 'Sexual' in Filipinx Canadian Sexuality Studies." In *Diasporic Intimacies: Queer Filipinos and Canadian Imaginaries,* edited by Robert Diaz, Marissa Largo, and Fritz Pino, 23–40. Northwestern University Press, 2017.

Costanza-Chock, Sasha. *Design Justice: Community-Led Practices to Build the Worlds We Need.* MIT Press, 2020.

Detroit Community Technology Project. "Technology Rooted in Community Needs." Accessed September 15, 2022. https://detroitcommunitytech.org/.

Diaz, Robert. "Introduction: The 'Stuff' of Queer Horizons and Other Utopic Pursuits." In *Diasporic Intimacies: Queer Filipinos and Canadian Imaginaries,* edited by Robert Diaz, Marissa Largo, and Fritz Pino, xv–xxxvi. Northwestern University Press, 2017.

Duarte, Marisa Elena. *Network Sovereignty: Building the Internet Across Indian Country.* University of Washington Press, 2017.

Finley, Chris. "Decolonizing the Queer Native Body (and Recovering the Native Bull-Dyke): Bringing 'Sexy Back' and Out of Native Studies' Closet." In *Queer Indigenous Studies: Critical Interventions in Theory, Politics, and Literature,* edited by Qwo-Li Driskill, Chris Finley, Brian Joseph Gilley, et al, 31–42. University of Arizona Press, 2011.

Friend, Chris, Ashley Caranto Morford, Arun Jacob, Kush Patel, et al. "Care." Teacher of the Ear Podcast: Hybrid Pedagogy Podcast (podcast), November 10, 2021. https://hybridpedagogy.org/care/.

Gàl, Réka Patrícia. "Climate Change, COVID-19, and the Space Cabin: A Politics of Care in the Shadow of Space Colonization." *Mezosfera.* October 2020. http://mezosfera.org/climate-change-covid-19-and-the-space-cabin-a-politics-of-care-in-the-shadow-of-space-colonization/.

Galison, Peter. "The Ontology of the Enemy: Norbert Wiener and the Cybernetic Vision." *Critical Inquiry* 21, no. 1 (1994): 228–66.

García Peña, Lorgia. *Community As Rebellion: A Syllabus for Surviving Academia as a Woman of Color.* Haymarket Books, 2022.

Graham, Stephen, and Colin McFarlane, ed. *Infrastructural Lives: Urban Infrastructure in Context.* Routledge, 2015.

Habtom, Sefanit. "Meditating on the Elsewhere." *The Henceforward* (podcast), November 12, 2018. https://luminarypodcasts.com/listen/indian-&-cowboy/the-henceforward/episode-26–meditating-on-the-elsewhere/254f6684-d3cd-4ad7-8bb8-bbd6c80aa3d8.

hooks, bell. *Teaching Community: A Pedagogy of Hope.* Routledge, 2003.

Janastu. About. Accessed September 15, 2022. https://open.janastu.org/about.

Kaba, Mariame. *We Do This 'Til We Free Us.* Haymarket Books, 2021.

Kittler, Friedrich A. *Gramophone, Film, Typewriter.* Translated by Geoffrey Winthrop-Young. Stanford University Press, 1999.

Langlois, Ganaele, and Greg Elmer. "Impersonal Subjectivation from Platforms to Infrastructures." *Media Culture & Society* 41, no. 2 (2019): 236–51. https://doi.org/10.1177/0163443718818374.

Lorde, Audre. *Sister Outsider: Essays and Speeches.* 1984. Crossing Press, 2007.

Maynard, Robyn, and Leanne Betasamosake Simpson. *Rehearsals for Living.* Haymarket Books, 2022.

McKittrick, Katherine. "The Geographies of Blackness and Anti-Blackness: An Interview with Katherine McKittrick." *C.L.R. James Journal* 20, no. 1–2 (2014): 233–40.

McLuhan, Marshall. *Understanding Media: The Extensions of Man.* MIT Press, 1994.

Mignolo, Walter, and Catherine E. Walsh, eds. *On Decoloniality: Concepts, Analytics, Praxis.* Duke University Press, 2018.

Mirrlees, Tanner. "Ghoulish EdTech Innovations." *The Scarlet Standard,* October 25, 2020, https://socialistproject.ca/podcast/ghoulish-edtech-innovations/.

Muñoz, José Esteban. *Cruising Utopia: The Then and There of Queer Futurity.* New York University Press, 2009.

O'Malley, Pat. "Responsibilisation." In *The SAGE Dictionary of Policing,* edited by Alison Wakefield and Jenny Fleming, 276–78. SAGE, 2009.

Packer, Jeremy, and Peter Galison. "Abstract Materialism: Peter Galison Discusses Foucault, Kittler, and the History of Science and Technology." *International Journal of Communication* 10, no. 1 (2016): 3160–73.

Patel, Kush, Johnson Jament, and Merin Mathew. "Framing Survival: Questions of Safe Space in Design Pedagogies." *Journal of Architectural Education* 76, no. 2 (2022): 190–192. https://doi.org/10.1080/10464883.2022.2097545.

Patel, Kush, Ashley Caranto Morford, and Arun Jacob. "Workshops in Anti-Colonial Digital Humanities: Towards Building Relationships with Critical University and Community Movements." In *Lessons Learned: Digital Humanities Workshops*, edited by Laura Estill and Jennifer Guiliano, 117-127. Routledge, 2023.

Piepzna-Samarasinha, Leah Lakshmi. *Care Work: Dreaming Disability Justice.* Arsenal Pulp Press, 2018.

Plantin, Jean-Christophe, and Aswin Punathambekar. "Digital Media Infrastructures: Pipes, Platforms, and Politics." *Media, Culture & Society* 41, no. 2 (2019): 163–74. https://doi.org/10.1177/0163443718818376.

Walcott, Rinaldo. *Black Like Who? Writing Black Canada.* Insomniac Press, 2003.

"WHO Director-General's Opening Remarks at the Media Briefing—5 May 2023." World Health Organization. Accessed 20 July 2023. https://www.who.int/director-general/speeches/detail/who-director-general-s-opening-remarks-at-the-media-briefing---5-may-2023.

Winthrop-Young, Geoffrey. "Hardware/Software/Wetware." In *Critical Terms in Media Studies,* edited by William John Thomas Mitchell and Mark Hansen, 186–214. University of Chicago Press, 2010.

Zelizer, Barbie. "Terms of Choice: Uncertainty, Journalism, and Crisis." *Journal of Communication* 65, no. 5 (2015): 888–908.

Zimmer, Benjamin. "Crisis = Danger + Opportunity: The Plot Thickens." *Language Log.* March 27, 2007. Accessed July 20, 2023, http://itre.cis.upenn.edu/~myl/languagelog/archives/004343.html.

PART III][Chapter 15

Making Infrastructure Writable

LUCIE KOLB

We face digital interfaces whether we use the browser search window, ChatGPT, or library and archival catalogs. The interface is not merely a tool to access data; it also provides a "platform for interpretative work in knowledge production" (Drucker, 92). This platform strengthens some things, slows others down, or stops them by making them harder to use. It also defines what questions we can ask, often also what answers we get while rendering invisible, how the results became available, and what decisions regarding categorizations, descriptions, and classifications led to their use. As a form of "thinking infrastructure," such interfaces create bureaucracy, channeling our behavior and thinking without us noticing that it is there (Kornberger et al.). Interfaces are powerful and deeply political technological-intellectual tools that often appear to be everything but that. Their internal logic and politics are usually hidden in the name of "user-friendly design" (Rossenova). However, "user-friendliness" applies only to those at home in a body, a discipline, and a world (Ahmed, 19). For those of us who live and work transversally, they fail to be useful.

Making Interfaces Tangible and Writable

I want to raise the question of how we negotiate such a "thinking infrastructure" in a way that lets us see and access it.[1] This would be a way that helps us be mindful of how our actions and reflections are being shaped by it and creates the possibility for change—for writability.

I want to tackle these questions by discussing two interventions in the search interface in the library that help users understand the context of their search query: "Feminist Search Tools" and "Infrastructural Manoeuvres." I will discuss how their interventions, modifications, or supplementation strategies work toward the possibility of a "writable library infrastructure"—a discursive and performative space exposing the ways that our actions and reflections are shaped by infrastructure and creating the possibility for a more emancipatory practice.

Feminist Search Tools

Feminist Search Tools[2] is an ongoing artistic research project manifesting in workshops and experimental tool-making. The project is organized by members of the collectives Read-In (Sven Engels, Annette Krauss, and Laura Pardo) and Hackers & Designers (André Fincato, Anja Groten), Ola Hassanain and Aggeliki Diakrousi, and Alice Strete. Departing from a shared interest in hierarchies of knowledge that inhabit our bookshelves, reading practices, and library search movements, Feminist Search Tools challenges knowledge systems, articulates ways to operate within them, and explores different ways of retrieving knowledge.

THE ROLE OF CATEGORIES

Starting with the question, "Why are the authors of the books I read so white, so male, and so eurocentric?" the collective developed their own search interface for catalog items published between 2006 and 2016 in the Utrecht University Library (see Figure 15.1). The group activated the interface in the context of workshops that aim to discuss the biases around literature search and support collective processes of creating awareness toward biases that might be implicit and inherent in specific search movements in the Utrecht University library catalog.

The Feminist Search Tools interface is not designed as an alternative to the library catalog search interface. Rather, it is a supplementary tool: instead of providing an

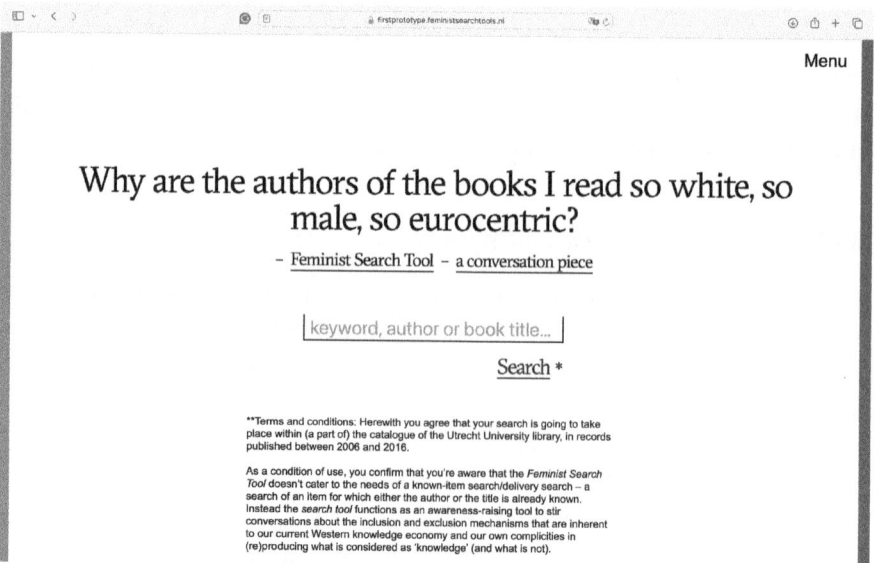

Figure 15.1. The Feminist Search Tool's alternative search interface for the Utrecht University library catalog (Aggeliki Diakrousi, Sven Engels, Anja Groten, and Annette Krauss), 2024. Prototype created in 2017.

item search or delivery search, Feminist Search Tools speaks to a more contextual approach. It asks, "In which context is our search embedded within the library catalog? What kind of books can we find in the library?" The interface showcases the catalog's searchable categories, such as "original language," "translated work," and "place of publishing." Furthermore, it articulates the lack of categories, such as "gender" and "race," that are critical for answering the starting question, "Why are the authors of the books I read so white, male, and Eurocentric?"

Without these categories, this question cannot be answered directly. Although categories such as "place of publishing" can determine how many books in the catalog were published in Europe, it is questionable whether we can make a deduction based on this knowledge as to whether a book is Eurocentric or not. Based on the existing categories of the library catalog, the question as to how many books were authored by people who identify as nonmale and nonwhite cannot be evaluated at all.

The developers of Feminist Search Tools needed to engage differently with publications and available information to answer the initial question. They did so by retrieving information on the gender of the author from the free and open knowledge base Wikidata, which encompasses gender categories such as "female," "male," "transgender-male," "transgender-female," and "unknown"; and from Gender API, a commercial software assigning binary gender categories based on names. By supplementing the available fields with data from Wikidata and Gender API, the tool links different datasets. On the one hand, Feminist Search Tools demonstrates that the gender category is missing and points out that we cannot search and find books based on the gender of the author in library catalogs. On the other, it demonstrates the potential of enriching existing library systems and standards, such as the machine-readable description standard for libraries widely used in Europe, the cataloging standard MARC 21, which the Utrecht University library uses, with additional information from other datasets.

While the project reflects the lack of a wide spectrum of gender identification in the datasets (Groten, 198) and problematizes that it is impossible to reconstruct how gender categories are assigned in the case of Gender API, it does not consider the discussions of the context and history of the Wikidata datasets and their inherent biases (Zhang and Terveen).

STAYING WITH THE TROUBLE OF CATALOGING

Feminist Search Tools offer tools as conversation pieces and an occasion to speak about issues hands-on and make sense of the catalog by sorting and categorizing things differently or otherwise. In her PhD thesis, Anja Groten, a group member and designer, reflected on the process of collective tool building and how it created conditions in which "tools are not presumed as an inevitable outcome but as ongoing and discursive" (157). According to the collective, the tools are not supposed to resolve the problems; they are part of the group's "unlearning" method, which works toward reinvestigating assumptions, prejudices, and histories and raising awareness about the search processes and how particular power dynamics are reproduced (Kolb and Weinmayr, "A Syllabus (Session 4): Read-ability").

Therefore, the tools' usage heavily relies on the discursive context of the workshop format. Embedded in a discussion context, the tools work toward making the catalog writable, in that they help us see our role in the collaborative process (i.e., literature search), where our decisions intersect with those of librarians, coders, and algorithms. Working with the tools and conversing with one another help to see our implication in reproducing particular power dynamics and proposes strategies of unlearning as constantly trying to find ways to search otherwise.

As one of the developers of Feminist Search Tools, Annette Krauss said, "What does it do to me if I am over-represented? If I'm constantly mirrored in the cataloging system, in the world, I live? I have not been equipped with many tools to look in this social-political-psychic mirroring that I'm constantly confronted with" (Kolb and Weinmayr, "A Syllabus (Session 4): Read-ability"). The processes proposed by Feminist Search Tools work toward filling this gap by providing tools that stay with the trouble of cataloging.

Infrastructural Manoeuvres

For Infrastructural Manoeuvres, librarian and designer Anita Burato collaborated with computer programmer and researcher Martino Morandi to modify the library catalog of the Gerrit Rietveld Academy and Sandberg Institute (Rietveld/Sandberg Library) in Amsterdam (see Figure 15.2). Presented in the Infrastructural Manoeuvres search interface, the library catalog radically questions its nature and provides possibilities for different layers of usage and engagement by the library users.[3] These modifications introduce a doorway into conversations about standard-making bodies in library systems. The "manoeuvres" aim at creating cracks in library infrastructure as a truth-telling mechanism by raising questions regarding authority, responsibility, and accountability in the representation of knowledge in the catalog and by making porous categories that were assumed to be stable and fixed.

Accessing the digital library catalog of the Gerrit Rietveld Academy and Sandberg Institute, on the left side of the browser page, we see a long snake of field numbers such as "002 klu 1," "020.1 ine 1," and "090 rao 7," corresponding to the respective MARC 21 fields. Clicking on one of these field numbers shows the typical metadata available about the corresponding book, such as title, publisher, and keywords, supplemented with their respective field numbers. On the right side of the browser page, a menu provides options for the user to modify the catalog. The first modification that we can choose is to switch off the librarian's view displaying the MARC 21 field numbers.

By switching the field numbers off, we see the display of an entry that we would typically get when using any other library catalog. By default, Infrastructural

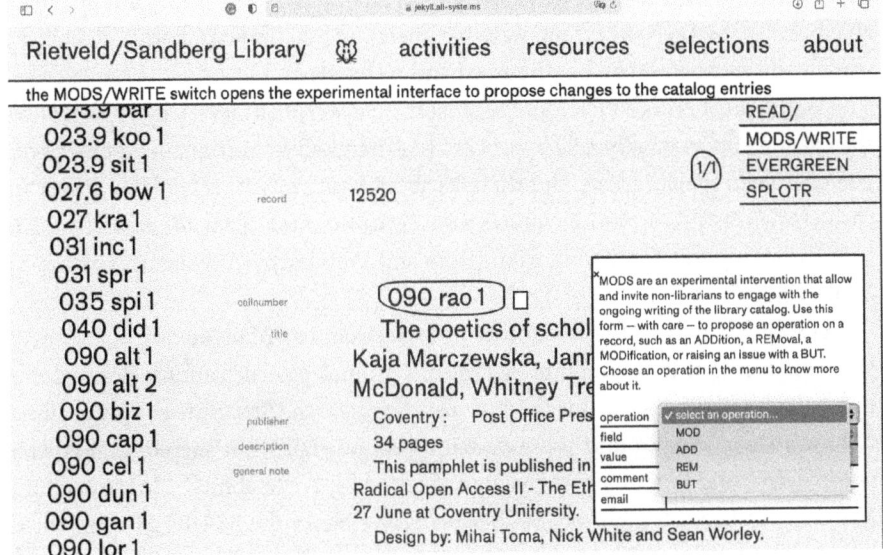

Figure 15.2. Experimental interface of the Rietveld/Sandberg Library catalog by Infrastructural Manoeuvres (Anita Burato, Martino Morandi), 2023.

Manoeuvres displays the view of the librarians for the user. This breaks down the barrier between the professional librarians' view of the catalog and the users' limited view, aiming to question this separation and the responsibilities and decision-making processes accompanying them. The second modification option is linked to the MODS tool, which offers options for the user to modify catalog entries. The tool includes four options: "addition," "subtraction," "modification," and "but," as in "objection." The description reads: "Your proposal will be incorporated into the website and remain linked to this record, but librarians will need some time to integrate the proposal into the record itself. Please take care of motivating your proposed changes in the 'comment' field and of providing a contact email so that librarians can get in touch with you. Your email address will not be visible" (Rietveld /Sandberg Library).

MAKING RECORDS WRITABLE

Listed and displayed in the catalog, the proposals become part of the catalog. For Burato and Morandi, they serve the purpose of "conversations on the record" (Kolb and Weinmayr, "A Syllabus (Session 5): Write-ability"). They track and record the modifications to the catalog records and present them as cultural objects. Collecting and publishing these modifications, the project asks who is writing these records and with what agenda. Conceived as an ongoing project, Infrastructural Manoeuvres aims to foreground the role and possibilities of a library's technical infrastructure, opening it up to reflection and experimentation. Burato and Morandi are experimenting with making the catalog (and its implicit limitations) not just readable but also, to a certain extent, writable, meaning that library records—which are usually untouchable—can be discussed and negotiated.

Writability here is educational because it creates critical literacy by showing how the catalog comes together. This process is also linked to introducing a new pedagogical role to the librarians. Burato talks about how in an art school, librarians are cashiers. Infrastructural Manoeuvres aims to carve out a space for education and a new teaching role for the librarian (Kolb and Weinmayr, "A Syllabus (Session 5): Write-ability").

Another aspect of the writability of Infrastructural Manoeuvres is sensitivity to history, in that it documents discussions around categorizations. Books move through the library, says Burato, but also through the structures through time. "The conversations around these decisions are normally not shared; that is what we do" (Kolb and Weinmayr, "A Syllabus (Session 5): Write-ability"). Infrastructural Manoeuvres keeps the history of changes accessible in the MARC 21 "666" field, where deleted categories are stored. For Burato and Morandi, it is not about overwriting the catalog's errors by implementing proposals but rather about creating an educational space. The collected proposals are a testament to processes of collective care about the catalog, providing context and connection. The proposals show and map some of the histories of categories, demonstrate how they might be linked, and carve out the meanings embedded there. Opening a discussion around categories in the catalog lets us see how certain words keep imaginaries in place and how others introduce reimagination. Sharing proposals, discussions, and decisions and being able to "write" the catalog create possibilities of reimagining and reordering the reality of the thinking infrastructure.

Learning from . . .

Both Feminist Search Tools and Infrastructural Manoeuvres use prototyping as a discursive tool to initiate continuous rethinking and reconfiguring processes of technological-intellectual infrastructure. They create "recursive publics," in which users are engaged in maintaining and modifying the technical, legal, and practical means of the respective interfaces (Kelty, 256).

The projects rehearse the interface as a thinking infrastructure. Rehearsal is a tool that can be used to play with categories, probe rules, take on different roles, and, by doing so, become aware of them. It is a way to center change, center modification, of our patterns, proceedings, and protocols. By rehearsing, the projects find ways to make tangible the decisions, histories, and conditions inscribed in the tools that we use. Moreover, in the case of Feminist Search Tools, the project demonstrates how missing categories slow down questions such as "Why are the authors of the books I read so white, so male, and so eurocentric?" Neither Feminist Search Tools nor Infrastructural Manoeuvres provides a simple fix, a solution about what needs to be adjusted for us to be able to answer such questions. They rather complicate things: they have designed interfaces that reflect the sociotechnical context of the users and "act more like mirrors" (Rossenova). These interfaces create awareness of the infrastructural work and procedures behind it, demonstrating that this awareness is needed to understand where intervention is necessary.

The projects also demonstrate how critical infrastructure practices need to consider different aspects such as design, social, legal, and economic processes, as well as processes of classification and maintenance. Further, they show how entangled technical and social processes are, as the ways that the communities around these projects use the respective tools to inform the design of the interface and are, to a certain extent, mirrored in it.

I want to highlight the methodology that these projects developed, combing competencies and factors from design, coding, and critical thinking for hands-on approaches to reconfiguring infrastructure. Working on a structure while simultaneously questioning it and linking a critical activity with an engagement with its infrastructure provide the foundation for critical infrastructural practices. Such practices aim at learning from those of us who do not feel quite at home "in a body, a discipline, a world," as Sara Ahmed (19) has put it. Through their interventions into existing library catalogs, they map out the cracks, the things disappearing in the unindexed abyss, or those not even disappearing but never becoming visible. Moreover, in these cracks, we should dig deeper and collectively map transversal infrastructures that render tangible the ways that they guide, shape, and inform us and help us notice it, and help us see how certain results became available to us and what led to their use.

NOTES

1. This contribution draws on research conducted in "Teaching the Radical Catalog," a collaborative artistic research project that the author conceived with Eva Weinmayr for the exhibition "Reading the Library" (2021) at the Art Library of the Sitterwerk Foundation, a cultural institution in St. Gallen, Switzerland. Weinmayr and I contributed to the exhibition by developing the online syllabus "Teaching the Radical Catalogue." Taking City University of New York (CUNY) librarian Emily Drabinski's homonymous text as an inspiration, the syllabus explores ways to teach library users how books are cataloged and categorized. It talks about the people, processes, and infrastructures involved in how a book ends up in a catalog. It demonstrates what decisions, standards, and conditions led to a book being searchable with specific keywords and information on the author and publisher. In her text "Teaching the Radical Catalog," Drabinski points to the fact that by teaching how to use a catalog, libraries perpetuate the dominant story told by the classification, which carries traces of "all the intentional and unintentional racism, sexism, and classism of the workers who create them" (198). For library users to have critical literacy and to know why they find specific books and not others, Drabinski proposes to teach

the catalog mechanisms instead of uncritically teaching the catalog (204). The online syllabus assembles a series of projects by artists, designers, activists, coders, and librarians who have developed experimental tools that work toward "teaching the catalog" in Drabinski's sense, among those Feminist Search Tools and Infrastructural Manoeuvres.

2. See https://feministsearchtools.nl.

3. For the catalog of the Rietveld/Sandberg Library, see https://library.rietveldacademie.nl. For the catalog's entry about Infrastructural Manoeuvres, which provides the catalog's interface, see https://library.rietveldacademie.nl/projects/infrastructural-manoeuvres.html.

BIBLIOGRAPHY

Ahmed, Sara. *What's the Use? On the Uses of Use.* Duke University Press, 2019. https://doi.org/10.1515/9781478007210-002.

Drabinski, Emily. "Teaching the Radical Catalog." In *Radical Cataloging: Essays at the Front,* edited by K. R. Roberto. Jefferson, 198–205. McFarland, 2008.

Drucker, Johanna. *Visualization and Interpretation: Humanistic Approaches to Display.* MIT Press, 2020.

Feminist Search Tools. Home page, 2023. https://feministsearchtools.nl.

Groten, Anja. *Figuring Things Out Together: On the Relationship Between Design and Collective Practice.* PhD diss., 2022. https://scholarlypublications.universiteitleiden.nl/handle/1887/3487176.

Kelty, Christopher. *Two Bits: The Cultural Significance of Free Software.* Duke University Press, 2008.

Kolb, Lucie, and Eva Weinmayr. "A Syllabus (Session 4): Read-ability. Bibliothek Wyborada, Lucie Kolb and Eva Weinmayr in conversation with Feminist Search Tools." Accessed September 1, 2022. https://www.sitterwerk.ch/En/Journal/642/A_Syllabus_Session_4.

Kolb, Lucie, and Eva Weinmayr. "A Syllabus (Session 5): Write-ability. Lucie Kolb and Eva Weinmayr in conversation with Infrastructural Manoeuvres." Accessed September 1, 2022. https://www.sitterwerk.ch/En/Journal/643/A_Syllabus_Session_5.

Kornberger, Martin, Geoffrey C. Bowker, Julia Elyachar, et al., eds. *Thinking Infrastructures.* Emerald Publishing, 2019.

Rietveld/Sandberg Library. Home page, 2023. https://library.rietveldacademie.nl.

Rietveld/Sandberg Library. Infrastructural Manoeuvres, n.d. https://library.rietveldacademie.nl/projects/infrastructural-manoeuvres.html.

Rossenova, Lozana, and Marialaura Ghidini. "Designing Interfaces to Support Contextual Interpretation and Open Conversations with Users." *Curating Online,* 2021. https://www.curating.online/interview/lozana-rossenova/.

Sayers, Jentery, ed. *Making Things and Drawing Boundaries.* University of Minnesota Press, 2017.

Zhang, Charles Chuankai, and Loren Terveen. "Quantifying the Gap: A Case Study of Wikidata Gender Disparities." In *17th International Symposium on Open Collaboration* (OpenSym 2021), September 15–17, 2021, Spain. ACM. https://doi.org/10.1145/3479986.3479992.

PART III][Chapter 16

Online Feminist Publishing and Content Creation as Feminist Infrastructure in India

PUTHIYA PURAYIL SNEHA AND SAUMYAA NAIDU

The role of infrastructure has been central to work in the digital humanities (DH), particularly in the contexts of the Global South and the majority world (Alam), where digital divides continue to remain widely prevalent even today. Technology access and use are crucial to the development of scholarship, practice, and pedagogy in DH as a field premised on the "digital." Much debate around DH, particularly with regard to the affordances of digital technologies for research and practice in the humanities, has also focused on the need for sustainable infrastructural interventions, be it in the form of tools, platforms, software, resources, capacities, or funding; or thinking about the humanistic and relational aspects of such infrastructure (McGrail, Nieves, and Senier; del Rio Riande; T and Menon; Pawlicka-Deger). Over the last decade, an emerging but substantive body of critique on DH, especially relating to its Anglo-centric origin stories, has also illustrated how the discussion of infrastructure-building predates the field and is in fact intertwined with complex, colonial histories of knowledge production (Fiormonte; Risam). Importantly, several recent efforts to "decolonize DH" have highlighted visible knowledge gaps in theory and practice and the need to create a diverse, multilingual, and accessible field of research and practice (Aiyegbusi; Ricaurte).

Feminist work is particularly relevant as an important precursor to discussions of infrastructure within feminist, postcolonial, and global DH. Concepts such as the "feminist internet"[1] and "forms of feminist infrastructures"[2] offer a way to understand how the engagement with digital spaces has been both empowering and challenging for structurally marginalized communities, where often systemic forms of injustice perpetuate within modern and neoliberal frameworks (see Feminist Principles of the Internet; Zanolli et al.; Malhotra, Hussen, and Fossatti). Building a feminist internet, therefore, involves creating structural changes in access to information, technologies, and public spaces; capacity-building; and the development of networks of solidarity and care. Such actions, however, are also affected by persistent general infrastructural challenges in the access to and use of the internet and digital technologies.

The downloadable and zoomable "Bridging Intersectional Gaps in Digital Infrastructures in India" infographic visualization that we have developed as a large PDF file, hereafter referred to as "visualization" (Figure 16.1), is an effort to explore some of these challenges at the intersections of feminist work, DH, and critical infrastructure studies.[3] The visualization represents key observations from an ongoing study that we undertook of digital feminist infrastructures in India through a review of literature, and interviews with creators, editors, and other participants of online feminist publishing and content-creation spaces to understand how these digital spaces engage with and inform the contemporary discourse on feminism, gender, and sexuality.[4] As broken out in the section of the visualization labeled "A Snapshot of Publications and Content Creation Platforms and Creators Interviewed," our study comprises interviews with those involved in an independent feminist publishing house (A); intersectional feminist platforms sharing information for specific groups such as youth and women with disabilities (B and C); a multimedia project that shares information and stories on intimacy and desire (D); a women-run digital media network that practices feminist rural journalism (E); online magazines on sexual rights and reproductive health in the Global South (F), and caste, gender, politics, and environment (I); a social media account that runs crowdsourced discussions on gender, sexuality, and feminism (G); an alternative social media network for legal, health, and safety support and resources (H); an Instagram handle run by an artist and illustrator sharing people's experiences and stories through visual art (J); individual writers, researchers, and activists working on caste, gender, and sexuality (K).

In our interviews, we delve into our research participants' motivations, content materials, challenges, choice of media, reach, experience with censorship, platform aesthetics, and perspectives on feminism and feminist infrastructure. We also conducted two workshops on unpacking what feminist infrastructures are or could look like, with one workshop focused on how feminist principles can be used in design (Costanza-Chock; Bardzell). These exercises revealed key themes related to feminist infrastructures, including that of inclusivity, safety, access, and policy intervention. Importantly, by seeking to understand how the growth of digital infrastructures mediates contemporary feminist work in India, we intend for our study also to examine how such growth engages with related fields, such as critical infrastructure studies and DH. Of course, while the focus of our research (and our visualization representing it) is primarily on digital publishing and content creation, we are clear that digital and print media coexist. Our overall engagement with infrastructures extends to the transition from print to digital media, with its myriad gaps and affordances.

The term *feminist infrastructures* is still relatively new in academic, technology, and policy discourse, especially in India and other parts of the majority world. However, the concept offers a critical entry point into intersectional knowledge gaps in

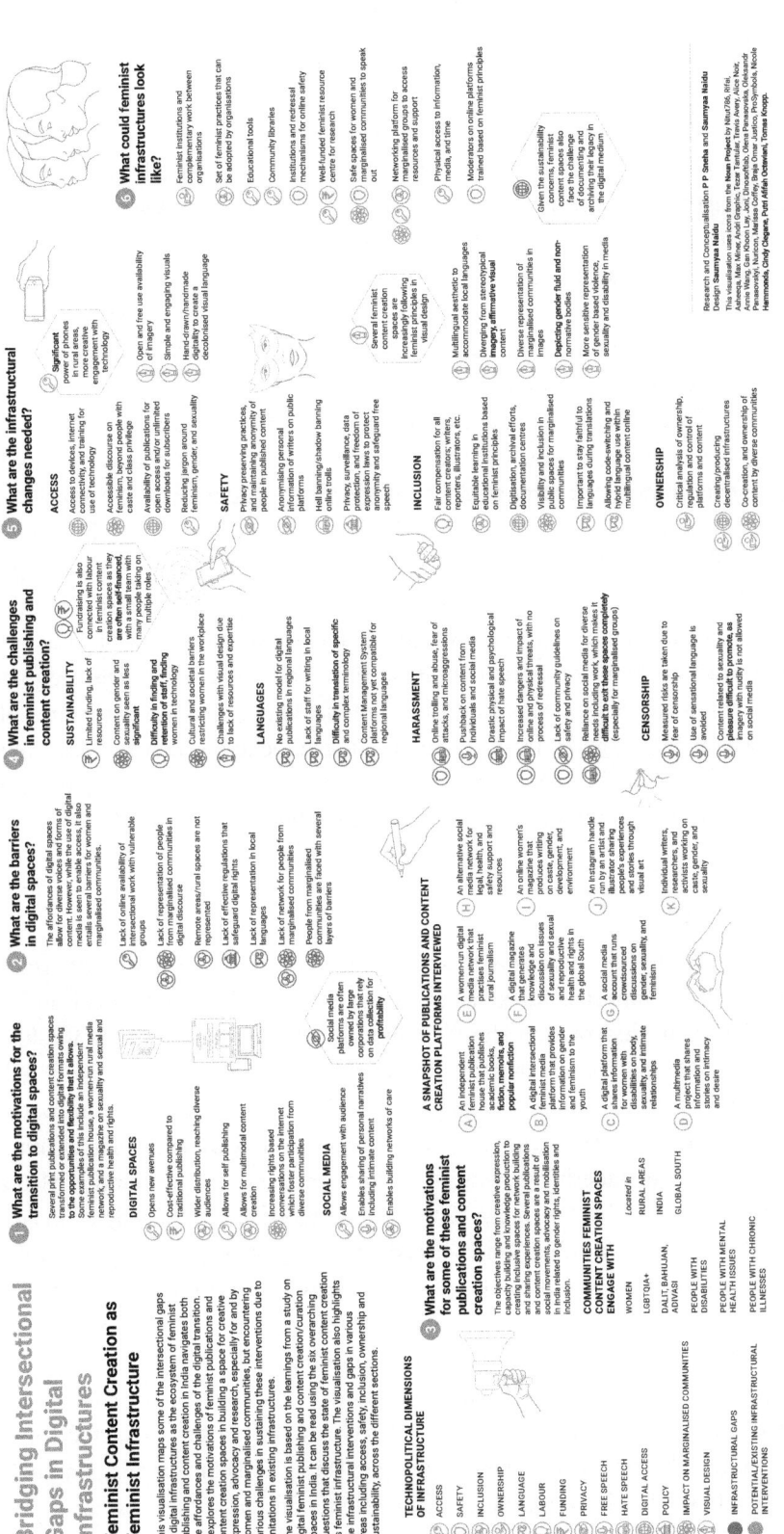

Figure 16.1. "Bridging Intersectional Gaps in Digital Infrastructures in India" (infographic visualization created for this chapter). (For a full-size visualization, see the PDF version at https://dhdebates.gc.cuny.edu/projects/critical-infrastructure-studies-and-digital-humanities/resources?tag=chapter%2016).

fields such as DH, enabling examination of the theoretical and political underpinnings of infrastructural design and development. Digital spaces have been crucial for access to feminist work, and also to its outreach. Still, they have also raised questions concerning what constitutes a feminist internet—especially in light of broader challenges in the access to and use of infrastructure in society with which a feminist internet intersects. Even as infrastructures encompass anything from built environments and networks to affective and sensorial experiences of desire and promise (Mrázek; Larkin), they often become evident only through gaps or in a moment of breakdown (Star; Bowker; del Rio Riande). Most recently, the Covid-19 pandemic has revealed infrastructural inequities in India (as elsewhere), whether in terms of access to health care and resources, challenges in remote working and learning, risks to the safety of migrant and gig labor, and lapses in essential services like mass public vaccination mediated through digital platforms. Lockdowns and restricted mobility have also led to an increase of invisible and disproportionately burdened care work (Deshpande) and a higher rate of gender-based violence at home (Kapoor).

Especially over the last decade, these challenges have driven an increased focus on studying infrastructures through a feminist lens by drawing on concepts and practices from broader feminist critiques of science and technology across disciplines. Growing debates on the relational and global infrastructures of DH (Pawlicka-Deger) and infrastructure-building in the Indian context (Nishant Shah; Shanmugapriya T and Menon) also indicate what is at stake in broader areas of information technology policy and practice in the majority world.

Technopolitics of Infrastructure

Drawing on literature and insights from our study, we take *feminist infrastructures* here to mean infrastructures that are inclusive, accessible, multimodal, and developed with an ethics of care. This is an understanding of infrastructure as inherently political and determined by structures of power and privilege. The gendered production of infrastructures (Siemiatycki, Enright, and Valverd) and its impact on the visibility, access to resources, and safety of marginalized and vulnerable communities has been an underlying theme of feminist work and writing in India (Mehta and Eapen; Gurumurthy; Surepally; Kumari). The affordances of the digital in facilitating the creation of a public sphere in India that allows creative expression, activism, and building networks of solidarity and care has been a primary reason that many content creators and publishers working on gender, sexuality, and feminism have turned to digital media (see section 3 of Figure 16.1). As section 1 also indicates, the transition from print to digital and multimodal forms of content creation in social media has been a significant aspect of this change in India as well, gradually shifting content beyond text as the primary publishing and content creation medium.[5] This shift is related to important issues of limitations in the formats, languages, and accessibility of content, which are also not entirely resolved in the digital medium.

A related development in India is the growth of public discourse on the internet and in mainstream media about rights, access, and identities and the availability of online content (especially on social media); that is, about a diversity of voices that were earlier less heard in the public sphere due to systemic and intersectional forms of social exclusion across gender, sexuality, religion, caste, and disability. These former absences in public discourse, however, were not new. At the same time, the engagement of women in nationalist and anticolonial movements, including social movements related to caste, language, and ethnicity, has become increasingly prominent in recent years in ways that inform Indian feminist writing—particularly on access, identity, and marginalization (Pawar and Moon; Roy; Schöningh, Maya, and Eismann). Debates around intersectionality, therefore, continue to be an important aspect of the discussion in India on antisubordination frameworks in feminist writing and feminist practice (Abhishek Shah). Many emerging content-production platforms also increasingly incorporate feminist principles into their visual design (Costanza-Chock; Bardzell), thereby offering a critical perspective on the form, diversity of representation, and accessibility of existing digital infrastructures. (See points connected to the visual design icon—under "Technopolitical Dimensions of Infrastructure"—in Figure 16.1.)

Despite the many affordances of digital platforms that have aided the growth of feminist publishing and content production, however, constraints remain. Safety continues to be a primary concern because of the urgent need for methods and tools to address online gender-based violence, discrimination, and hate speech. Also needed are policies to address data protection, surveillance, and censorship, along with measures to safeguard privacy and anonymity. (See points under "Safety" in section 5 of Figure 16.1.) Uneven access to capacity-building in some of these areas—not just in terms of technical skills or languages of access, but also of legal expertise—is also a significant problem. The growth of community infrastructures such as feminist servers, anticaste and queer archives, and alternative social media networks has been an essential intervention in this regard.[6] Many community infrastructures have been set up with limited funding, resources, and technical capacities. Open-access platforms, tools, and software are crucial aspects of such community initiatives, enabling a minimal-computing approach to infrastructure-building that is adapted for the challenges imposed by prohibitive costs of proprietary technologies (especially as it relates to accessibility and digital security). However, open-access tools and software often entail their own challenges in ease of use, safety, and sustainability. The very terminology required to navigate their policies and best practices (a terminology that is technical, legal, and difficult to translate conceptually across languages) is an impediment for which approachable solutions are needed that would much aid in feminist infrastructure building. (See under "Languages" in section 4 of Figure 16.1.)

Many of these affordances and constraints of digital platforms bearing on feminist infrastructure also need to be contextualized in a changing environment in

India of media-platform ownership and regulation (see "Ownership" in section 5 of Figure 16.1), the growing digital economy, and increasing digital surveillance. These factors shape the growth of feminist infrastructures and affect the nature and forms of the content that they convey about gender, sexuality, and feminism.

On Infrastructure and Content

The number and variety of feminist content-production platforms using digital media for the purposes listed in sections 1 and 3 of Figure 16.1 indicate how infrastructure significantly influences content related to feminism, gender, and sexuality. As also occurs in the general case of the publishing industry and commercial interests affecting work on these issues (Butalia and Menon), infrastructure that amplifies and alters the spaces of activism, research, and policy-making opens up new possibilities for focus and intervention—and thus for the overall advance of the women's movement in India as it has coevolved over the last few decades with social movements addressing identities, languages, and exclusion. Social media platforms and self-publishing initiatives, for example, have enabled the emergence of more work previously considered niche, such as on sexuality and disability.[7] Also, more publishing venues for positive narratives about bodies, desire, and pleasure and for the creation of affective networks of care and solidarity are now mediated through some of these platforms.[8]

At base, such new discursive spaces for content creation are motivated by the need to critique colonial infrastructures of knowledge production that once determined not just what content was produced but how it could be distributed through networks of publication, translation, archiving, and now digitization. The question that we ask based on our study and visualization is how the development of critical digital infrastructures can help advance work that, while it acknowledges continuing constraints on knowledge production and sharing, advances openness, inclusivity, and accessibility.

Digital Infrastructures as Sustainable Communities

DH in a global context can contribute to work on critical infrastructure studies and, specifically, feminist infrastructures through shared values of openness, accessibility, multimodality, and inclusion—especially in the context of efforts to decolonize DH by drawing attention to the uneven and inequitable development of digital infrastructures in different parts of the world. While the early focus of DH in India has largely been on academic research (because the field evolved primarily in the university setting), Indian DH has increasingly turned to broader, socially engaged questions of access, ownership and regulation of the internet, and digital infrastructures. In the process, however, DH in India reveals itself to be open to critical questions about its own resource-intensive infrastructure. Subject to limitations in access and digital literacy across regions in a way that affects what content can be digitized and whom that content can reach, DH in India does not just revolve around classic DH problems regarding, for example, corpora, tools, and skills, but also around the infrastructural conditions for those capacities. These are issues bearing on intersectional forms of marginalization (including linguistic and sociocultural ones) both within and outside academic research. The issues are critical for the development of feminist infrastructures that, creatively and innovatively, might advance access to information, an affordable internet and devices, digital literacy, capacity-building, safety, resources, and funding, and, importantly, the centering of the voices and ownership of the communities affected by systemic forms of marginalization and disparity. (See "Access," "Safety," "Inclusion," and "Ownership" in section 5 of Figure 16.1.) The outcomes to be hoped for from feminist infrastructures in India, assisted by a DH that addresses critical infrastructure problems, will be the feminist infrastructural institutions, organizations, practices, tools, community libraries, mechanisms, and platforms set forth in section 6, on "What could feminist infrastructures look like?" Additional future goals include creating sustainability for online feminist publishing and content production in India by doing the documenting and archiving work needed to build a legacy for such infrastructure.

Such is a vision, which is what our visualization really is, of what digital feminist infrastructures in India—engaging with critical infrastructure studies and global DH to develop open and inclusive technologies spanning across geopolitical and disciplinary boundaries—can be.

NOTES

1. Although there is no available definition of this concept, one of the earliest instances of the use of the term was the first "Imagine a Feminist Internet" meeting called by the Association for Progressive Communications (APC) in 2014, and subsequent efforts to build a "framework to articulate how we imagine the internet as a public space that is safe, engaging, open, free, and conducive to feminist movements." See Fascendini.

2. While still to be defined, this term denotes several infrastructural aspects that have been imperative to the growth and sustenance of feminist networks and practice. See Toupin and Hache.

3. Our full-size visualization is downloadable as a PDF file at https://dhdebates.gc.cuny.edu/projects/critical-infrastructure-studies-and-digital-humanities/resources?tag=chapter%2016

4. This study was conducted by Puthiya Purayil Sneha and Saumyaa Naidu at the Centre for Internet and Society, India. It includes a literature review and twelve semistructured interviews with founders, editors, and contributors of feminist publications and content creation platforms in India. The authors also conducted two workshops with a diverse set of participants working in the areas of arts and design and urban infrastructures to understand the current perception of a feminist internet or feminist infrastructures among

the public and what they could be. The study commenced in the latter half of 2018, and while the research was completed, the final report was significantly delayed owing to the Covid-19 pandemic and other institutional challenges; it was published in March 2024 and can be seen here: https://cis-india.org/raw/understanding-feminist-structures.

5. As an illustrative example of this, see Genderlog India and work by Indu Lalitha Harikumar on Instagram.

6. For work on feminist servers, see the Community Owned Wireless Knowledge Infrastructure (COWKI) and Co-Creating Local Knowledge Network (CLKN) projects in Karnataka, India (Ray Murray et al.). An example of an anticaste archive is Dalit Art Archives. An example of an alternative social media network is Smashboard.

7. For example, see a website created in India, Sexuality and Disability.

8. For example, see the Agents of Ishq website, created in India. This is a "multimedia project about sex, love and desire with the aim of creating an inclusive, pleasure-positive and celebratory language for sex and sexuality," thus opening a space to engage with topics that may still be considered taboo, especially in mainstream media. The reference here is also to networks of care that emerge with increasing online feminist content, often mediated through social media. See Toffoletti et al.

BIBLIOGRAPHY

Agents of Ishq. Home page, 2023. https://agentsofishq.com.

Aiyegbusi, Babalola Titilola. "Decolonizing Digital Humanities: Africa in Perspective." In *Bodies of Information: Intersectional Feminism and Digital Humanities,* edited by Elizabeth Losh and Jacqueline Wernimont, 434–46. University of Minnesota Press, 2018.

Alam, Shahidul. "Majority World: Challenging the West's Rhetoric of Democracy." *Amerasia Journal,* 34, no. 1, 88–98, DOI: 10.17953/amer.34.1.l3176027k4q614v5.

Bardzell, Shaowen. "Feminist HCI: Taking Stock and Outlining an Agenda for Design." In *Proceedings of the SIGCHI Conference on Human Factors in Computing Systems (CHI'10), New York, 2010,* 1301–10. Association for Computing Machinery, 2010. https://doi.org/10.1145/1753326.1753521.

Bowker, Geoffrey C. "Sustainable Knowledge Infrastructures." In *The Promise of Infrastructure,* edited by Nikhil Anand, Akhil Gupta, and Hannah Appel, 203–22. Duke University Press, 2018. https://doi.org/10.1515/9781478002031-010.

Butalia, Urvashi, and Ritu Menon. *Making a Difference: Feminist Publishing in the South.* Bellagio Publishing Network, 1996.

Costanza-Chock, Sasha. "Design Justice: Towards an Intersectional Feminist Framework for Design Theory and Practice." *Proceedings of the Design Research Society* (2018). https://ssrn.com/abstract=3189696.

Dalit Art Archives. Home page, 2023. https://www.instagram.com/dalitartarchive/?hl=en.

del Rio Riande, Gimena. "Digital Humanities and Visible and Invisible Infrastructure." In *Global Debates in the Digital Humanities,* edited by Domenico Fiormonte, Sukanta Chaudhuri, and Paola Ricaurte, 247–58. University of Minnesota Press, 2022.

Deshpande, Ashwini. "The Covid-19 Pandemic and Gendered Division of Paid Work, Domestic Chores, and Leisure: Evidence from India's First Wave." *Economia Politica* 39 (2022): 75–100. https://doi.org/10.1007/s40888-021-00235-7.

Fascendini, Flavia. "Imagine a Feminist Internet: The Conversation Is On!" Association for Progressive Communications (APC) and Hivos, July 17, 2015. https://www.apc.org/en/news/imagine-feminist-internet-conversation.

Feminist Principles of the Internet. Home page, 2023. https://feministinternet.org.

Fiormonte, Domenico. "Digital Humanities and the Geopolitics of Knowledge." *Digital Studies/Le champ numérique* 7, no. 1 (2017): 1–18. https://doi.org/10.16995/dscn.274.

Genderlog India (@genderlogindia). Twitter, 2023. https://twitter.com/genderlogindia.

Gurumurthy, Anita. "A History of Feminist Engagement with Development and Digital Technologies." Association for Progressive Communications (APC). *Issue Paper 1* (2017). https://ssrn.com/abstract=3873762.

Indu Lalitha Harikumar (@induviduality). *Instagram*, 2023. https://www.instagram.com/induviduality/.

Kapoor, Anuradha. "An Ongoing Pandemic: Domestic Violence During COVID-19." *Economic and Political Weekly* 56, no. 17 (2021). https://www.epw.in/journal/2021/17/review-womens-studies/ongoing-pandemic.html.

Kumari, Harshita. "Is Public Commute Gender Inclusive? Access, Safety, And Utility of Our Transport Infrastructure." *Feminism in India,* June 6, 2022. https://feminisminindia.com/2022/06/06/is-public-commute-gender-inclusive-transport-infrastructure/.

Larkin, Brian. "The Politics and Poetics of Infrastructure." *Annual Review of Anthropology* 42, no. 1 (2013): 327–43. https://doi.org/10.1146/annurev-anthro-092412-155522.

Malhotra, Namita Aavriti, Tigist Shewarega Hussen, and Mariana Fossatti. "How to Build a Feminist Internet and Why It Matters." *APRIA Journal* 4, no. 4 (2022): 3–22. https://www.ingentaconnect.com/content/artez/apria/2022/00000004/00000004/art00001.

Mehta, Aasha Kapur, and Mridul Eapen, "Gendering the Twelfth Plan: A Feminist Perspective," *Economic and Political Weekly,* 47, no. 17 (2012). https://www.epw.in/journal/2012/17/review-womens-studies-review-issues-specials/gendering-twelfth-plan-feminist.

McGrail, Anne B., Angel David Nieves, and Siobhan Senier (eds.). *People, Practice, Power: Digital Humanities Outside the Center.* University of Minnesota Press, 2021.

Mrázek, Rudolf. *Engineers of Happy Land: Technology and Nationalism in a Colony.* Princeton University Press, 2002.

Pawar, Urmila, and Meenakshi Moon, *We Also Made History: Women in the Ambedkarite Movement.* Zubaan Books, 2014.

Pawlicka-Deger, Urszula. "Infrastructuring Digital Humanities: On Relational Infrastructure and Global Reconfiguration of the Field." *Digital Scholarship in the Humanities* 37, no. 2 (2022): 534–50, https://doi.org/10.1093/llc/fqab086.

Ray Murray, Padmini, Naveen Bagalkot, Siddhant Shinde, et al. "A 'Feminist' Server to Help People Own Their Own Data." *The Bastion,* August 12, 2022. https://thebastion.co.in/politics-and/tech/a-feminist-server-to-help-people-own-their-own-data/.

Ricaurte, Paola. "Data Epistemologies, The Coloniality of Power, and Resistance." *Television & New Media* 20, no. 4 (2019): 350–65. https://doi.org/10.1177/1527476419831640.

Risam, Roopika. *New Digital Worlds: Postcolonial Digital Humanities in Theory, Praxis, and Pedagogy.* Northwestern University Press, 2018.

Roy, Srila. *Changing The Subject: Feminist and Queer Politics in Neoliberal India.* Duke University Press, 2022.

Schöningh, Ingo, Maya, and Sonja Eismann, eds., *Movements and Moments: Indigenous Feminisms in the Global South.* Zubaan Books, 2022.

Sexuality and Disability. Home page, 2023. https://sexualityanddisability.org.

Siemiatycki, Matti, Theresa Enright, and Mariana Valverd. "The Gendered Production of Infrastructure." *Progress in Human Geography* 44, no. 2 (2020): 297–314. https://doi.org/10.1177/0309132519828458.

Shah, Abhishek (curator). "Is 'Intersectionality' a Useful Analytical Framework for Feminists in India?" Discussion map. *Economic & Political. EPW Engage.* 2023.

Shah, Nishant. "Beyond Infrastructure: Re-humanizing Digital Humanities in India." In *Between Humanities and the Digital,* edited by Patrik Svensson and David Theo Goldberg, 95–107. MIT Press, 2015.

Smashboard. Home page, 2023. https://site.smashboard.org/about.

Star, Susan Leigh. "The Ethnography of Infrastructure." *American Behavioural Scientist* 43, no. 3 (1999). https://doi.org/10.1177/00027649921955326.

Surepally, Sujatha. "Pandemic in the Time of Dalit Feminism." *Economic and Political Weekly* 56, no. 28 (2021). https://www.epw.in/journal/2021/28/alternative-standpoint/pandemic-time-dalit-feminism.html.

T, Shanmugapriya, and Nirmala Menon. "Infrastructure and Social Interaction: Situated Research Practices in Digital Humanities in India." *Digital Humanities Quarterly* 14, no. 3 (2020). https://www.digitalhumanities.org/dhq/vol/14/3/000471/000471.html.

Toffoletti, Kim, Holly Thorpe, and Rebecca Olive, et al. "A Feminist Embodied Ethics of Social Media Use: Corporeal Vulnerability and Relational Care Practices." *New Media & Society,* 27, no. 1. 2023. https://doi.org/10.1177/14614448231171560.

Toupin, Sophie, and Alexandra Hache. "Feminist Autonomous Infrastructures." Global Information Society Watch: Sexual Rights on the Internet. Association for Progressive Communications (APC) and Hivos. 2015. https://www.giswatch.org/en/internet-rights/feminist-autonomous-infrastructures.

Zanolli, Bruna, Carla Jancz, Cristiana Gonzalez, et al. "Feminist Infrastructures and Community Networks." Report. Global Information Society Watch. 2018. https://giswatch.org/en/infrastructure/feminist-infrastructures-and-community-networks.

PART III][Chapter 17

Digital Humanities from Below
Speculating on Solidarity Infrastructure

MATTHEW N. HANNAH AND MIRIAM POSNER

In the 1966 issue of the *Times Literary Supplement*, Marxist historian E. P. Thompson advocated an analysis of history "from below." Rather than emphasize the "great men" model of history—from above—Thompson posited that historical research should focus on the commoner, the everyday, the masses. Such an emphasis on "below" proffers a Marxist critique of history, elevating the laboring masses as subjects of history over the powerful, rich, and important. But his model of history was also infrastructural. It required a conceptual restructuring of societal models, attending to base over superstructure, emphasizing below over above. In this chapter, we adapt Thompson's theory of history to advance "digital humanities from below," critiquing current infrastructural topologies of the field, which still rely on marquee projects, elite institutions, and academic stars as prime drivers of new research. Instead, we advocate for a "solidarity infrastructure," offering a speculative infrastructural space that addresses the needs of information workers as a laboring class, redistributes resources from top to bottom, and employs an implicit commitment to openness.

Solidarity infrastructure suggests critiques of late capitalism and the higher-education industrial complex from within digital humanities (DH). We offer a view from the United States but acknowledge the increasingly global nature of these conditions, and of local responses. Similarly, we focus on academic DH to confront issues in that sector but also hope that rehabilitation of academic DH can foster cross-sector solidarity with colleagues laboring elsewhere. Such a speculative infrastructure might seem unlikely, but it maps the topography of current practices while revealing possible futures. This kind of speculation is visible in recent approaches to the study of history, as described by Gavriel Rosenfeld: "I hope to demonstrate that alternative histories lend themselves quite well to being studied as documents of memory. By examining accounts of what never happened, we can better understand the memory of what did" (90). But we can also examine nonexistent current situations as a way to imagine alternatives to what is. Laura Bear (1) describes the magic of "speculation": "Speculation is akin to practices of divination or magic because it aims to reveal a hidden order of human and non-human ethical powers that explain the past, present and future and make it possible to act." Reimagining speculation beyond finance capital's use of the term creates a space for action by revealing the possible.

Within the field of critical infrastructure studies, speculation offers ways to imagine new modes of organizing and structuring our cities, our societies, our ecosystems, and our futures. "Where extrapolation is grounded in probabilistic reasoning," claims Steven Shaviro, "speculation is rather concerned with possibilities, no matter how extreme and improbable they may be" (1). Speculative infrastructure allows us to imagine new possibilities for the way that things are structured in our world, generating new possible futures (Chattopadhyay; Ziser; Badami). Unlike speculation occasioned by the economic conditions that have exacerbated so many of our most intractable problems, speculative infrastructures provide a glimmer of hope and a trenchant critique of the present, making it possible to act.

Speculation leads us to imagine another possible DH, an alternative universe, not entirely unlike our own, in which organizing structures are built from below. While this proletarian energy has animated the field in particular moments, the concretization over the past twenty years has tended toward increasing institutionalization and big-name initiatives rather than rhizomatic solidarity. And while there have been important and sustained critiques of DH and labor (Flanders; Keralis; Boyles et al; Pawlicka-Deger; Smithies, Ffrench, and Ciula), there have been few developments at an infrastructural level to mitigate neoliberalism within the field or suggest a political economy of DH (Hannah). Situated as we are in the United States, which has witnessed one of the worst declines in public education due to neoliberal economic austerity, we propose a speculative infrastructure for DH from below.

Three Spheres of Solidarity Infrastructure

IDEOLOGY

We need to reconceptualize our relationship to our own labor. We know that labor in DH is delightfully heterogeneous, occurring in various modalities and at various locations within and beyond academia, but we must accept that academic DH distills a particular set of labor issues. Such labor extends beyond the technical to organizational, affective, administrative, intellectual, and physical labor, which forms the base of program development and maintenance in DH. Glaringly, most DH labor in the United States is not performed by tenure-track faculty. However, most of the leadership in academic DH still is. We must imagine ourselves as a laboring class, across all ranks, disciplines, and DH communities, and organize according to class interests.

Cue immediate bad-faith dismissal of our labor as some form of "privileged labor" because we don't work in factories. This characterization misses the point. All who labor are bound together because all of us, together, make higher education function. And all who labor at universities and colleges—including custodial staff, graduate and undergraduate student workers, and contingent and tenured faculty—stand to benefit from the recognition and valorization of that labor. Rather than debating "hack" versus "yack," digital humanists should be arguing for solidarity contra disunity. Indeed, solidarity among academic workers (and from there across to workers in other DH communities) is the only way we can effect real, lasting change. As the wave of labor actions across higher education in recent years has shown, such solidarity can make a huge impact on material conditions.

Class relations are complex, as Marx understood so long ago, and they defy easy categorization into "proletariat" and "bourgeoisie." It would be foolish to deny that some academic labor is more prestigious, as anyone who has worked in a staff role can attest. It would be absurd to claim, in light of conditions faced by students and workers of color, that we all benefit from the same degree of racial privilege. But the fact that institutions value some labor and people more than others does not erase the basic structural antagonism between administration and university or college labor. How can we establish solidarity while recognizing that our experiences differ in critical ways?

Solidarity infrastructure means developing bottom-up power, in which tenure-track faculty feel more solidarity with staff than with management. Perhaps we can take a cue from theorist McKenzie Wark, who glimpses new laboring relations. Wark argues that the class relation in the twenty-first century is that between information workers and information owners: "Together we form a class, a class as yet to hack itself into existence as itself–and for itself" (para. 013). As we build DH initiatives, we must remember that such efforts are mediated by the owners of our production, the institution. Hacking our way into a recognition of the class relation that structures DH requires a recognition that our work may be coopted by or rely on neoliberal academic initiatives.

To develop a consciousness of ourselves as labor that produces surplus for the higher education system, we need sustained scholarly engagement with neoliberal ideology as such. It remains remarkable to us that the explosion of critical theory in DH has not yet produced a Marxist DH. How can we expect to expose the contradictions inherent within the capitalist institutions of higher education—between knowledge as a public good and the profitability of education, between academic labor and academic austerity, between respect for intellectual pursuit and cynical economic calculation—if we don't understand our work within capitalism? As Slavoj Žižek puts it: "Now is the time to think." Now is the time to think about the role of solidarity in our discipline.

LABOR

If we understood ourselves as a laboring class, what is the nature of our work? As a hacker class, how do we contribute our labor to the function and sustenance of the neoliberal university or college? Who owns the information we produce, and how do we organize ourselves to ensure our rights as workers, thereby improving working conditions across the institution? How can we spread awareness that the material conditions of higher education within 21st-century capitalism affect the lived experience of everyone who works for our institutions?

With a few notable exceptions, DH in the United States has avoided such questions, focusing on the symptoms rather than the virus. A solidarity infrastructure arising out of class consciousness recognizes that, while a tenure-track professor at an Ivy League university is not the same as a public librarian at a regional library, a lecturer at a small liberal-arts college, or a graduate student at a regional university, we are all subject to forms of class oppression, simply because we are workers from whose labor the institution extracts value. These vectors of oppression are different, but still operative and visible if we choose to see them. By recognizing the differences and highlighting the similarities, we can advance solidarity across institutions, disciplines, and classes to support the most marginalized, while inculcating class consciousness among the most privileged.

Solidarity infrastructure would enable us to redistribute privileges toward the most marginalized while establishing a network of cooperation and mutual aid. Solidarity infrastructure would be built on organizing from the bottom up as follows:

- Start academic unions on every campus, using our digital skills to organize.
- Commit to solidarity with all campus workers, including with service workers, whose unionization efforts so often serve as an example to the rest of us.
- Establish a DH worker's caucus to raise the concerns of constituencies with the community and share local grievances with the entire field.
- Establish a repository of resources for students, postdocs, and nontenure-track faculty.
- Establish fund-raising mechanisms for labor strikes and actions, for the material support of precarious colleagues, and for student resources.

HEGEMONY

Antonio Gramsci sought to understand the role that power can play in maintaining its legitimacy, calling this infrastructural relationship "hegemony." For Gramsci, hegemony was "cultural, moral, and ideological" leadership of a group over the subaltern (Forgacs 423). But hegemony can be maintained only through the consent of those who are led. How are we offering that consent? How can we refuse or withhold it? As we have seen all too clearly, institutional values do not necessarily align with personal and ethical values. In far too many instances, universities ignore the values of their workers, choosing to retain a problematic coach or professor, choosing to cut benefits or salaries of workers, choosing to destroy entire academic ecosystems to save money rather than redistribute resources. How can we use our labor to resist the hegemony of academic administrations and boards of trustees?

In addition, we should reimagine the role that our professional organizations can and should play. DH professional organizations either can shore up the hegemony of neoliberal educational institutions by feigning neutrality or they can operate in solidarity with workers everywhere through focused political education, organizing, and solidarity efforts. For too long, our professional organizations have remained silent as our comrades struggle within an increasingly cruel and austere academic infrastructure, an infrastructure that is falling apart due to the contractions of capitalism. We are witnessing our friends and colleagues being consumed by the machinery of the neoliberal academic market, with little notice from the organizations that we fund. Rather than attempt to remain neutral as higher education collapses—as though neutrality were possible—our organizations could:

- Strategize with existing labor organizations such as the American Federation of Teachers. Colleagues at colleges and universities in the United Kingdom do this as a matter of course, as members of the University and College Union (UCU).
- Offer training and events at conferences around workers' rights and solidarity, such as what occurred at the American Comparative Literature Association (ACLA) in 2023.
- Employ the existing organizational framework to support the most vulnerable among us, especially in an age of increased assaults on higher education by far-right ideologues.
- Redirect a portion of member dues to support organizing efforts at campuses around the country.
- Lobby political leaders to improve conditions at universities through expanded state funding.
- Speak publicly in support of striking faculty, staff, and students.

Fostering a solidarity infrastructure means recognizing the intersections between our work as laborers and other forms of marginalization and oppression. As is said often on social media, the university will not love you back. While we have seen neoliberal institutions respond to oppression and marginalization of specific identities through administrative diversity, equity, and inclusion (DEI) initiatives, we argue that such efforts must recognize the economic disparities which affect such identities at disproportionate rates. If we want a healthy educational system, we must recognize the economic and material conditions in which we work and how they impede broader efforts to build a more just academy. Higher education is broken because it has been transformed from a public good into a capitalist knowledge industry. But as capitalism continues to collapse, so does our current model of higher education. Unfathomable debt, increasing austerity, and attacks on academic freedom are eroding the very raison d'être of education in the United States, and we can no longer afford to ignore the economic and material realities in which DH exists. We can take a cue from the progress toward solidarity already happening around the world and advance a solidarity infrastructure that means getting our hands dirty to make material conditions better for all.

BIBLIOGRAPHY

American Comparative Literature Association (@ACLAorg). "At This Year's Conference in Chicago." Twitter, January 23, 2023, https://twitter.com/ACLAorg/status/1617609674748227585.

Badami, Nandita. "Solarpunking Speculative Futures." Theorizing the Contemporary, *Fieldsights,* December 18 (2018). https://culanth.org/fieldsights/solarpunking-speculative-futures.

Bear, Laura. "Speculation: A Political Economy of Technologies of Imagination." *Economy and Society,* 49, no. 1 (2020), 1–15. https://doi.org/10.1080/03085147.2020.1715604.

Boyles, Christina, Anne Cong-Huyen, Carrie Johnston, et al. "Precarious Labor and the Digital Humanities." *American Quarterly* 70, no. 3 (2018): 693–700. https://doi.org/10.1353/aq.2018.0054.

Chattopadhyay, Bodhisattva. "Speculative Futures of Global South Infrastructures." In *Urban Infrastructuring: Reconfigurations, Transformations and Sustainability in the Global South,* edited by Deljana Iossifova, Alexandros Gasparatos, Stylianos Zavos, et al., 297–308. Springer Nature, 2022. https://doi.org/10.1007/978-981-16-8352-7_18.

Flanders, Julia. "Time, Labor, and 'Alternate Careers.'" In *Debates in the Digital Humanities,* edited by Matthew K. Gold, 292–308. University of Minnesota Press, 2012.

Forgacs, David. "Glossary of Key Terms." In *An Antonio Gramsci Reader*, edited by David Forgacs, 420–31. Schocken, 1988.

Hannah, Matthew. "Toward a Political Economy of the Digital Humanities." In *Debates in the Digital Humanities,* edited by Matthew K. Gold and Lauren Klein, 3–26. University of Minnesota Press, 2023.

Keralis, Spencer. "Disrupting Labor in the Digital Humanities; or, The Classroom Is Not Your Crowd." In *Disrupting the Digital Humanities,* edited by Dorothy Kim and Jesse Stommel, 273–94. Punctum Books, 2018. https://doi.org/10.2307/j.ctv19cwdqv.

Pawlicka-Deger, Urszula. "Digital Humanities Needs Equality Between Humanists and Technicians." *Times Higher Education,* July 9, 2022, https://www.timeshighereducation.com/blog/digital-humanities-needs-equality-between-humanists-and-technicians.

Rosenfeld, Gavriel. "Why Do We Ask 'What If?' Reflections on the Function of Alternate History." *History and Theory* 41, no. 4 (2002): 90–103. https://doi.org/10.1111/1468-2303.00222.

Shaviro, Steven. "Defining Speculation." *ALIENOCENE: Journal of the First Outernational.* December 23, 2019. https://alienocene.com/2019/12/23/defining-speculation/.

Smithies, James, Patrick Ffrench, and Arianna Ciula. "*Droit de cité*: The Digital Lab as Digital Milieu." In *Digital Humanities and Laboratories: Perspectives on Knowledge, Infrastructure and Culture,* edited by Urszula Pawlicka-Deger and Christopher Thomson, 52–66. Routledge, 2023.

Thompson, Edward Palmer. "History from Below." 1966. In *The Essential E. P. Thompson,* edited by Dorothy Thompson, 481–89. New Press, 2001.

Wark, McKenzie. *A Hacker Manifesto.* Harvard University Press, 2004.

Ziser, Michael. "Living with Speculative Infrastructures: Reading Our Present Dilemmas in Science Fiction's Past." *Boom* 3 no. 4 (2013): 27–34. https://doi.org/10.1525/boom.2013.3.4.27.

Žižek, Slavoj. "Don't Act. Just Think." *Big Think,* 2012. https://youtu.be/IgR6uaVqWsQ.

PART III][Chapter 18

Imagining a Future of Multimedia E-books

SYLVIA K. MILLER

Scholarly books in digital or "electronic" form ("e-books") should routinely include multimedia content. Multimedia e-books not only will publish today's multimodal scholarship more accurately and appropriately than traditional books, but also will more easily engage and excite audiences, including scholars, researchers, students, and the general public.

Scholars are increasingly frustrated as the multiple forms of work that they find useful and inspiring are sidelined. Although they continue to pursue multimodal scholarship anyway, the richness of this scholarship is being stripped away during the publishing process, providing an impoverished experience of authors' work for the readers of today and tomorrow—not least tenure and promotion committees, which focus on the book as the most important form of scholarship. The limited nature of book publishing is one reason that the multilayered complexity of today's scholarship often goes unappreciated and uncredited in the process of faculty advancement. This lack of scholarly credit, in turn, discourages academic institutions from supporting the multimodal scholarship that faculty and graduate students want to carry out, putting barriers in the way of development and sustainability so that digital scholarship too often vanishes into old servers and outdated software, leaving little or no trace of the work for future audiences to study and emulate (Miller, "'Horrors of Recoding': Reflections on the 2019 UNCG Scholarly Communications Symposium"), or it requires continued editorial attention and technological investment indefinitely into the future (Hansen, Milewicz, Mangiafico, et al.). I see the publication of multimedia e-books as a potential key to breaking this unfortunate cycle.

This chapter is a highly limited manifesto, focused on incorporating multimedia into the established book-publishing infrastructure in the humanities and the social sciences (reflecting my particular experience). I outline a series of juxtapositions setting the current scenario ("Now"), admittedly in simplified form for concision, against an "imagined near-future."

In my proposed model, each multimedia e-book publication would provide an important snapshot of scholarly work that occurred in multiple simultaneous modes, and the publication would serve as an archival record of that multimodal work. At the same time, the imprimatur of the scholarly publisher would present to the academic world a convincing statement of value that carries with it the potential to legitimatize the nontextual forms of scholarship that are currently regarded in many departments and institutions as superfluous. The larger effect, therefore, would be to encourage disciplinary and institutional support for multiple forms of scholarship.

This proposal is based on my experience in scholarly publishing, including more than twenty-five years in the publishing of e-books, digital encyclopedias, and grant-funded digital editions and e-publishing experiments. (See the glossary at the end of this chapter for definitions of the key terms discussed here.) For updates on the current state of publishing technology, I consulted a number of publishing professionals, some of whom are quoted herein and all of whom are acknowledged in the list of interviews accompanying this chapter. These conversations were carried out in a journalistic spirit, and they revealed information that was new to me and might indeed be new to others, leading to the exciting conclusion that normalization of multimedia e-book publishing might be closer to reality than I had expected when I started this project. I also mention enhanced e-books published under the Mellon-funded Publishing the Long Civil Rights Movement project (2008–2012), which I directed at the University of North Carolina Press (UNCP).[1] (See Figure 18.1, from a video explaining one such e-book.)[2]

Figure 18.1. From video by the Publishing the Long Civil Rights Movement team demonstrating UNCP's enhanced e-book version of Katherine Mellen Charron's *Freedom's Teacher: The Life of Septima Clark* (2010). Video copyright 2012 University of North Carolina Press; reproduced courtesy of University of North Carolina Press.

Inside the Publishing Process

What is the editing and publishing process that produces books now? And what might it become if we imagine our way into the future?

EDITORIAL PLANNING AND GUIDELINES

Now, editors advise authors on the number and type of illustrations, according to their own editorial experience, knowledge of the available budget, and curatorial sense of the right balance suited to the content of each book. They refer authors to company guidelines on acceptable formats and limits for photographs and line drawings; such guidelines form what we might term the *author-editor toolkit*. However, if an author asks whether the publishing company is interested in publishing other aspects of a multimodal project, editors usually say "No."

In an imagined near-future, editors would universally develop the experience and judgment to say "Yes." New production guidelines would support them. Even better, editors would proactively ask authors to tell them about their *multimodal project*, not just their book.

In addition to audio and video materials such as interviews and documentary footage, the author and editor could discuss how to represent other components of a multimodal project in the form of short videos. While not the same as the original experience, whether digital or physical, such videos might become the most long-lasting, archival representation of ephemeral project elements.

Publishers also would develop ways to embed digital items other than audio and video that are common to many multimodal scholarly projects. John Sherer, director of UNCP, envisioned a multimedia toolkit of about five elements, including an interactive map. "Let's start with a few things that people would use pretty often," he suggested. These would become part of the standard author-editor toolkit.

The following list suggests items to be added to publishers' existing guidelines for manuscript preparation in order to produce multimedia e-books routinely:

- Suggested types of video content (content types such as interviews, music, exhibit tours, and demos of digital humanities projects)
- Limits on the number and size of multimedia excerpts
- Preferred and accepted file types for multimedia excerpts
- Manuscript callout style, including how to distinguish callouts for the print book from callouts for the multimedia e-book
- Sample permission-request letters and general advice on fair use for audio and video
- Accessibility guidelines, including alt-text and transcripts
- Instructions for preparing the captions manuscript, including how to distinguish captions for the print book from callouts for the multimedia e-book

- Any revised guidelines on including outbound links in the text and captions
- Submission instructions for multimedia excerpts, including labels/file names
- What to expect when checking "page proofs" of multimedia e-books

PEER REVIEW

Now, peer reviewers usually review illustrations (including photographs, line drawings, maps, and charts), along with the text of a book manuscript. However, other elements of the scholarship that are not in traditional narrative-plus-image form are not submitted to the reviewers, who indeed may not even be aware of them. In an imagined near-future, the manuscript to be reviewed would include the multimedia elements, already selected and excerpted for publication. Editors would devise specific questions for the reviewers about whether the multimedia elements are well chosen and contextualized, and whether they are integral to the scholarship and enrich the narrative. Reviewers would not need to be experts in digital scholarship, but rather, as usual, they would review the manuscript in the role of a disciplinary or area expert and a potential reader of the published book.

COPYRIGHT PERMISSIONS

Now, publishers require authors to obtain written permission to include images and long quotations. In an imagined near-future, they would similarly require permission for multimedia content. Releases signed by subjects interviewed by the author during the research process would include multimedia publishing rights. For multimedia from other sources, the editor and the author would discuss proper interpretations of fair use, just as they do now. Appropriate sample permission-request letters would be provided by the publisher.

MANUSCRIPT PREPARATION

Now, authors insert illustration callouts in the manuscript and provide a list with captions. In separate files, they provide illustrations per the publisher's guidelines. In an imagined near-future, they would do the same with multimedia elements.

DESIGN AND DIGITAL PRODUCTION

Now, the interior design for a book includes all the standard elements, such as chapter titles and picture captions. Production editors supervise copyediting, typesetting, and assembly of all the book's elements. In an imagined near-future, the design would also include a standard visual design and function for audio/video controls, or possibly an icon to signal the inclusion of multimedia. Video might be represented by a still image in the print book, while audio could be represented by transcripts, building in accessibility. Multimedia might be added to the list of elements that the copy editor would be asked to tag and check. A digital production editor would help cut audio and video excerpts and even out sound quality—a new skill requirement. There would be a moment at which production of the multimedia e-book would diverge from the print book; for efficiency, the later in the production process they diverge, the better.[3]

EMBEDDING VERSUS LINKING

Now, there are no normalized standards for including multimedia in e-books. In an imagined near-future, publishers would work out the ideal length (run time) of multimedia elements (1) from an editorial perspective, considering audience engagement, and (2) from a technological perspective, considering file size and load time. In our UNCP model, in the captions, we offered outbound links to archives where the full files could be found.[4] If the archival file became inaccessible or the link failed (the infamous "link rot" that plagues publications of all kinds), at least the excerpt embedded in the e-book would be unaffected as an archival record. An embedded MP3 or MP4 file is remarkably stable; audio and video in the e-books that we produced at UNCP in 2012 still function.

If the author originated the multimedia material, they would deposit it with an appropriate archival partner. If the archive planned to make the material available online, a new publishing challenge would be aligning the publisher's and the archive's production schedules so that the permalinks could be included in the book. New grant-funded publishing platforms developed by university presses offer storage and linking solutions to support multimedia in e-books; this chapter is published on one called Manifold. Publishers may not be aware, however, that it is possible to publish an e-book with multimedia files embedded directly in common e-book file types (PDF and EPUB). One unified file seems more efficient to store and distribute than a linking system, but publishers should experiment with the various possibilities until they find the balance between embedding and linking that works best for them and best suits the content of their books. In other words, the author's content, the editorial perspective, and usability for audiences should be the guides. A proliferation of systems for publishing multimedia e-books might be useful for experimentation, but eventually, it would be efficient to settle on a simple and ubiquitous infrastructure that works routinely for all publishers large and small.

DISTRIBUTION AND DISCOVERABILITY

Now, publishers launch a text with static images into a powerful stream of established dissemination and discovery mechanisms, while leaving the other components of the scholarship behind. Most readers think that publishing e-books is easy and cheap compared to print books and have no idea that varying metadata and file-format requirements make e-book distribution complicated and costly for publishers, who are naturally wary of adding more complexity in the form of multimedia e-books. Publishers have so many files to keep track of in transmitting e-books to multiple vendors and platforms that they have to rely on digital asset management systems. One reason that grant-funded systems to support multimedia in e-books have limited reach is that they come with a commitment to publish free of charge (open access), making it difficult to sustain these platforms financially and to include the books in standard commercial wholesaler databases and digital aggregations for libraries. Publishers work hard to avoid having their multimedia e-books, which they publish in relatively small numbers, sidelined and undiscoverable. Libraries struggle to combine many separate publishing platforms and aggregations into a single usable catalog. Where multimedia publications are included, readers have to click through to access multimedia that are stored separately (sometimes several clicks are needed, which is a usability challenge).

In an imagined near-future, all the standard outlets for e-books would accommodate multimedia. Multimedia e-books would be discoverable and accessible, along with all other e-books.

Now, publishers are not aware that some of the most common publishing platforms already accommodate multimedia. In an imagined near-future, they would collectively review each of the platforms and distribution mechanisms that they use and put pressure on the ones that do not yet accommodate multimedia content. The following is an initial review of some of the most common distribution mechanisms for scholarly books and their current capacity to include multimedia.

Kindle and Kobo. Now, the Kindle reader for tablets and smartphones can play multimedia embedded within an e-book. However, on a computer screen (whether desktop or laptop), the media is "not supported" (see Figure 18.2) The Kobo reader has the same oddly persistent discrepancy (more than a decade after we produced model multimedia e-books at UNCP, it is still "not supported"!), which inappropriately relegates multimedia e-books to the category of apps and games. In an imagined near-future, multimedia e-books would be responsive; that is, they would function fully on all devices. Clearly Amazon and Rakuten Kobo could fix this overnight—if only they felt significant pressure from publishers to do so.[5] Publishers, put on that pressure!

Library Aggregators. Now, the Adobe Portable Document Format (PDF), originally invented to create flat image files to reduce file sizes and freeze layouts, has functional characteristics, apparently unbeknownst to most publishers, who are not using this capability. In an imagined near-future, publishers would take advantage of the fact that JSTOR, Project MUSE, and Oxford Scholarship Online, nonprofit publishing platforms for scholarly books that are ubiquitous in academic libraries, now seamlessly ingest audio and video embedded in PDF e-book files.

Now, of the approximately 80,000 e-books in Project MUSE, only "a handful" of e-books include embedded multimedia. The number is hard to confirm because

Audio content not supported by this device.
Excerpt from interview with Victoria Gray Adams, April 22, 2002, by Katherine Mellen Charron. Courtesy of the author.
View a transcript of this audio clip.
Return to List of Enhancements.

This was Septima Clark's civil rights movement: direct action meant registering to vote, but also combining education, protest, and advocacy to force the state to become more responsive to the black community. Her approach incorporated the young people of the Student Nonviolent Coordinating Committee, but they were not its focus. Building an educational foundation required slow, methodical work and yielded unpredictable results. Black youth invigorated the freedom struggle but often rejected the patience Clark's educational process required. Similarly, citizenship education relied on assistance from white allies, and the program even trained poor whites in interracial sessions before the end of the 1960s, but Clark's strategy first and foremost targeted the needs of black adults. Clark's civil rights movement was not captured in newspaper photographs or broadcast on television, though many of the anonymous faces in the crowds belonged to those who had passed through citizenship education classes. Most important, her movement was sustained, to various degrees in different communities, by grassroots Citizenship School teachers, a corps of go-to women who used their skills to mobilize their neighbors, reconcile conflicts, and answer community welfare needs.

Figure 18.2. "Audio content not supported by this device." This is the explanatory error message offered by the Amazon Kindle app on my computer where, on a smartphone or tablet, audio content is available to be played. Kobo books have the same problem. Publishers should pressure these companies to correct this odd discrepancy so that audiovisual content in e-books sold individually will play on all devices.

"these features flow right through if they are integrated in a PDF," said Phillip Hearn, publisher relations manager at Project MUSE. "An influx of more of this content would not cause any issues on our end, as our system is built to accommodate them without requiring any additional intervention from our production team. Our tech team might need to make adjustments to accommodate file size to optimize download speeds and hosting capabilities, but this already takes place routinely with large files." JSTOR has the same capability, but it appears to have a modest number of multimedia titles as well.[6] Oxford University Press's multipublisher e-book platform, Oxford Scholarship Online, is merging with Oxford Academic, combining e-books and journals in one platform. "You can have as much audio and video as you want on that platform," said Tanya Laplante, senior product manager at Oxford University Press.

It's likely that if these platforms can accommodate multimedia, other scholarly aggregation systems already can, too. Publishers should gather this information from all their distribution platforms and demand multimedia across the board.

Preservation and Access: One Half of Sustainability

At its basic level, preservation should provide information to support the possible future revivification of online experiences that become defunct ("dark archiving"), but libraries demand more; they increasingly rely on publishers not only to preserve publications but also to make them accessible and functional indefinitely, and publishers have responded by committing to digital archives and bearing the associated monetary cost. However, these archives may not preserve multimedia. In an imagined near-future, multimedia would be included in preservation. Publishers are already working to address this challenge.[7]

Markets and Business Models: The Other Half of Sustainability

Now, there is an unfortunate assumption among publishers that there is no market for multimedia e-books because early examples fell flat after initial excitement about them in the early twenty-first century. In an imagined near-future, publishers would remember that early "enhanced e-books" were not published for what trade publishers sometimes dismiss as "the academia segment," and therefore provided no evidence whatever concerning the scholarly publishing marketplace. In an imagined near-future, publishers would recognize that there is a great need for scholarly multimedia e-books. Some already have. "Any publisher in the business of current and future scholarship has to support this [multimedia publishing] if they want to sign authors, because authors demand it," declared Laplante. User statistics so far indicate that "the use of media in chapters creates more engagement and discoverability," she added.[8]

Now, grant-funded multimedia publishing platforms focus on publications that are open access only, which are a challenge to sustain financially in the long term.[9] In an imagined near-future, both funders and customers (individual readers and librarians) would respect publishers' need to charge a fair price for their publications. Multimedia publishing platforms would be business-model agnostic. People who produce books would be fairly compensated for their labor and expertise, and the costs of maintaining and improving the publishing infrastructure would be recognized and supported. Publishers would continue to seek new business models to sustain open access for worldwide equity, and sources of funding for scholarly publishing would continue to expand in number and shift where and how they enter the publishing ecosystem. Multimedia e-books, along with other kinds of publications, would benefit from these changes but would not be forced to take the lead in being made available free of charge.

Now, as always, achieving scale is important from a business standpoint. For a university press, "our goal is to reach the widest possible audience with limited resources; that makes it hard to justify putting resources into developing something new that that can't be disseminated widely yet," explained Ellen Bush, digital initiatives and database director of UNCP, echoing other interviewees. In an imagined near-future, this barrier will be swept away.

The Future Is Now

This highly limited manifesto focuses on the publication of multimedia e-books at scale because a limited format is practical and makes scale feasible. Taking advantage of the scale of already established infrastructure creates time and financial savings, as well as broad distribution. As Charles Watkinson, director of Michigan University Press, put it, "Why *wouldn't* we use our existing infrastructure?"

Now appears to be a moment in which technology has developed just enough that it is possible to establish a practice of routine multimedia scholarly book publishing. Rather than "relying on those who are going to display the e-book to tell us what to do,"[10] publishers should demand this capacity from every publishing platform that does not already offer it.

The proposed next-step evolution of e-book infrastructure is a new expression of the ever-unfolding magic of the codex in time and space. That such technological enhancement produces books that talk and sing like those in fairy tales or science fiction might indicate that they will engage and fascinate readers, and inspire authors and artists, in the generations to come. Experiments with new kinds of storytelling and audience engagement within digital publications are underway that are more ambitious and creative than the limited multimedia e-book format described here and will lead to greater digital-publishing innovations in the future.[11] Meanwhile, however, it's important to implement at scale a future that we have already invented, and indeed have at our fingertips ready to launch, if only publishers can collectively muster the will.

Glossary

Archival In the context of this chapter, describing a publication that is accessible and readable for future generations.

E-book. A book in electronic form, typically made available to readers via software that allows the text, at the most basic level of functionality, to be scrolled through and read. For the purposes of this chapter, I make no distinction between "electronic" and "digital."

Enhanced e-book. A term for a multimedia e-book that was popular in the early 2000s, when Amazon first made it possible to include video in an e-book.

Interactive map. A digital map offering features such as sliders or clickable icons that the reader can navigate at will. For example, one might see the development of a town over time, follow a travel route, or access archival documents or oral histories related to particular spots on the map.

Monograph. In academia, a longform scholarly narrative told predominantly in words (70,000 of them, more or less). In the publishing model that I am focused on in this chapter, the monograph is the unifying center of a constellation of project components or, similarly, it is an interface to a body of information.

Multimedia e-book. An e-book that allows the reader to seamlessly access and interact with content in audio and audiovisual forms from the book's pages.

Multimodal project. A scholarly project that includes multiple components and forms of work, such as documentary films, interactive maps, physical and digital exhibits, digital archival collections drawing from multiple physical archives, databases, interviews and oral histories in audio or video form, three-dimensional (3D) renderings or other visualizations of data, augmented reality or virtual reality experiences or reconstructions, events or performances, and a longform narrative. All are connected; no single component fully represents the project. I focus on digital modes of scholarship in this chapter; other modes, such as performances or other forms of publicly engaged scholarship, could also be well represented via video in a multimedia e-book.

Publishing infrastructure. In this discussion, I focus on the typical production process within publishing houses, in which technologies and workflows turn manuscripts into books, and on some central and widely shared means by which scholarly books reach readers, collectively known in publishing as *distribution*.

Publishing platform. A computer interface that allows readers to search, find, and read digital publications. Usually, it connects to a system that stores and serves up the publications (server plus user interface).

NOTES

1. The model described in this chapter is best exemplified by Charron. Other enhanced e-books that we produced under the Mellon grant were Castro and Ferris.

2. The actual 3 minute, 4 second video, titled "Septima Clark Biography: Demo of Enhanced e\E-book" (posted March 27, 2012, by UNCP), is viewable on YouTube at https://www.youtube.com/watch?v=qTXmqn8VasU&t=5s. (For archival reasons, the video will also be downloadable from the digital version of this book.) As this was the first video produced by the Publishing the Long Civil Rights Movement project, we can only hope that its uneven sound quality and homemade feeling gave it a certain charm. Sylvia Miller storyboarded and directed; Seth Kotch of the Southern Oral History Program narrated; and digital production specialist Kenneth Reed assembled and edited the video.

3. For more details on the production of Charron, see Miller, "Producing the *Freedom's Teacher* Enhanced Ebook." Note, however, that this was a retrospective project, begun after the print book was published; in this chapter, I attempt to apply lessons learned to an envisioned revised production process.

4. Allison Belan, director for strategic innovation and services at Duke University Press, suggested QR codes for the print book, but to me, these seem like slightly more

trouble for the reader than Uniform Resource Locators (URLs), while just as likely to be affected by link rot.

5. Significantly, Amazon has made an effort to respond to calls for the Kindle to accommodate EPUB files. A number of easily findable blog posts review this development (e.g., Cunningham).

6. Quotation from an interview with Cristina Mezuk, books licensing editor, JSTOR | ITHAKA.

7. The Embedding Preservability project by New York University (NYU) placed preservation specialists called "embedders" at select university presses to advise on adjustments for improving preservability (see Elliott).

8. It should be noted that in Tanya Laplante's view, "this" included additional advances not covered in this chapter, such as offering authors a choice of image-based rather than text abstracts in journals.

9. See Ricci for a helpful overview of open access models.

10. Quotation from an interview with John Sherer, Spangler Family director, UNCP. The exact quote that I noted was "Presses—and UNC is as guilty of this as other publishers—rely on those who are going to display the e-book to tell us what to do."

11. See Levy and McKee for some examples.

INTERVIEWS

All the following interviews were conducted virtually on Zoom except where otherwise noted.

Belan, Allison, director for strategic innovation and services, Duke University Press. August 29, 2022.

Bush, Ellen, digital initiatives and database director, University of North Carolina Press (UNCP). September 16, 2022.

Hearn, Phillip, publisher relations manager, and Kelley Squazzo, director of publisher relations and content development, Project MUSE, Johns Hopkins University Press. Email exchange, August 18–September 15, 2022.

Laplante, Tanya, senior product manager—platform, academic, Oxford University Press US. September 8, 2022.

Levy, Allison, digital scholarship editor, Digital Publications Initiative, Brown University. August 24, 2022.

Ohe, Kevin, director of academic publishing, Digital Resources Division, Bloomsbury Publishing, Inc. April 22, 2022.

Mezuk, Cristina, books licensing editor, JSTOR | ITHAKA. August 24, 2022.

Sherer, John, Spangler Family director, University of North Carolina Press (UNCP). August 25, 2022.

Watkinson, Charles, director, University of Michigan Press; associate university librarian, University of Michigan Library; and president, Association of University Presses (2022–2023). August 15, 2022.

BIBLIOGRAPHY

Castro, Sal, and Mario T. García. *Blowout! Sal Castro and the Chicano Struggle for Educational Justice* [Enhanced e-book]. University of North Carolina Press, 2011.

Charron, Katherine Mellen. *Freedom's Teacher: The Life of Septima Clark* [Enhanced e-book]. University of North Carolina Press, 2012. Video demo: https://www.youtube.com/watch?v=qTXmqn8VasU&t=5s.

Cunningham, Andrew. "Kindle E-readers Finally (Kind of) Support ePub Books." *Ars Technica* blog post, May 3, 2022. https://arstechnica.com/gadgets/2022/05/kindle-e-readers-finally-kind-of-support-epub-books/.

Elliott, Michael A. "The Future of the Monograph in the Digital Era: A Report to the Andrew W. Mellon Foundation." *Journal of Electronic Publishing* 18, no. 4 (Fall 2015). https://doi.org/10.3998/3336451.0018.407.

Ferris, William. *Give My Poor Heart Ease: Voices of the Mississippi Blues* [Enhanced e-book]. University of North Carolina Press, 2012.

Hansen, David, Liz Milewicz, and Paolo Mangiafico, et al. *A Framework for Library Support of Expansive Digital Publishing*. Report. Duke University Libraries, 2019. https://doi.org/10.21428/680f3353.

Levy, Allison, and Sarah McKee. *Multimodal Digital Monographs: Content, Collaboration, Community*. PubPub, 2022. DOI: 10.21428/36a3e2c8.e1215c8e. https://multimodal-digital-monographs.pubpub.org/.

Miller, Sylvia K. "'Horrors of Recoding': Reflections on the 2019 UNCG Scholarly Communications Symposium." *Publishing at the Crossroads* (MLA Humanities Commons blog), December 30, 2019. https://publishingcrossroads.hcommons.org/2019/12/30/horrors_of_recoding/.

Miller, Sylvia K. "Producing the *Freedom's Teacher* Enhanced Ebook." *Publishing the Long Civil Rights Movement* (blog), April 30, 2012. https://wayback.archive-it.org/3491/20170224164244/https://lcrm.lib.unc.edu/blog/index.php/2012/04/30/producing-the-freedoms-teacher-enhanced-e-book/#more-3241.

Ricci, Laura. "Every Book, Every Format, All at Once: Exploring the Multiverse of eBook Open Access Models." *Against the Grain* 34 (2022). https://issuu.com/against-the-grain/docs/atg_july22_special_report/s/16467256.

PART III][Chapter 19

Subjective Functions
How Should Humanistic Research Be Quantified?

KYLE BOOTEN

Objective Functions

Humanists, like all scholars, are the objects of machine reading. Google Scholar keeps track of the simplest bibliometric data: how many times a paper or book has been cited. A researcher's profile on the platform will list both "h-index" and "i10-index"[1]—two metrics that attempt to quantify something like "sustained excellence"—while plotting as a bar chart one's citations per year. Bibliometric quantification advances a coherent but one-dimensional (1D) understanding of scholarly activity as a game in which the goal is to accrue the most citations. While humanists tend to play this game less aggressively than scientists, who have been known to form "citation farms" to gratuitously cite each other's work (Van Noorden and Chawla), this does not mean that they are excluded from it. Since Google Scholar and related clearinghouses are important places for academics (including humanists) to see and be seen, making citation-based metrics so highly visible exerts a social pressure, just as the quantified "likes" and "friends" of other social media platforms inherently nudge their users to look for more such entities.

It is not hard to imagine the ways that these metrics may yield subtly perverse incentives. Another field-shifting paper, you wager, will likely bump up your h-index (the "largest number h such that h publications have at least h citations"[2]); hence you politely decline to contribute to a dear mentor's Festschrift. Realizing that, alas, your niche intellectual obsession is *too* niche, you bend—almost to the point of breaking—your latest manuscript to touch on a more popular topic. The rise of "altmetrics," which capture references to academic work on social media platforms (Piwowar), may encourage forays into public scholarship—or spending too much time on Twitter/X cultivating a sufficiently broad and chummy network.

The term *objective functions* is used in applied mathematics, and more recently machine learning, to describe an outcome to be optimized. This optimization is either a matter of making the output of the function smaller (e.g., minimizing the number of miles that the Amazon delivery truck must drive) or larger (e.g., maximizing the linguistic diversity of the chatbot's utterances [Li et al.]). Academic metrics such as citation count and h-index are also "objective functions," in that they may become targets of behavior, and in the additional sense that they presume to measure an objective, almost Newtonian quality: scholarly *impact*. This concept has roots in twentieth-century military logistics, emerging to help planners focus on defining goals while paying less attention to the typical way in which things were done (Cottle): the "specific operations whose relations to the accomplishment of basic ends could only be evaluated subjectively" (Wood and Dantzig, 196). But doesn't being a humanist have something to do with attending carefully not just to the ends but also to the means, styles, and gestures of thought? Oblique allusion, obfuscatory digression, conceptual ellipsis, bizarre puns—the humanist reserves the right to all these and more, h-index be damned.

The question of how humanistic research is quantified is important even for those scholars who pay no heed to their own or their colleagues' personal metrics. For many of us, it would now be difficult to imagine completing any sort of serious research (or even cobbling together a syllabus) without the aid of Google Scholar. And since this platform uses citation-based metrics to rank search results, the 1D quantification of scholarship becomes a core feature of the infrastructure of humanistic thought itself. When we cite an article (or neglect to), we guide the algorithms that will help resurface some texts while leaving others obscured, in part determining what we will read and not read, think and not think.

As a way of thinking beyond the conceptual blandness of standard bibliometrics (as well as most "alt" ones), I offer in this discussion four sketches of ways that humanistic research might be quantified—ways that are more in keeping with the complexity and diversity of humanistic habits and dispositions. These *subjective functions* differ from most bibliometric objective functions in two ways: (1) they aspire to subjectivity rather than objectivity, encoding obviously opinionated, controversial, nonuniversal notions of what virtues scholars should manifest; and (2) they attend not to impact but to the gesture, not to the ends but to the means.[3] While my sketches of these subjective functions are merely speculative, I endeavor to place them on firm technical foundations, although I acknowledge that others would no doubt have different ideas about how best to engineer them.

Subjective Functions

MATCHMAKER SCORE

Take a paper with a title such as "The Languorous Abstraction of James Schuyler and Fairfield Porter." Whatever it ends up arguing, this paper's most basic rhetorical effect is to yoke together two figures by implying that it indeed makes sense to analyze them together. And not just sense but *interest*. Some conjunctions are more unexpected than others. A Google Scholar search of the combined terms "Frantz Fanon" and "Edward Said" returns around 19,600 results; "Frantz Fanon" and "Julia Kristeva" only around 3,800; "Frantz Fanon" and "Lev Vygotsky," fewer than 1,000. An intuition: the more infrequently two authors or texts are discussed in combination, the more potential for surprise—one facet of interestingness—it will have.

How might one calculate a *Matchmaker Score* that describes how unexpected are the combinations of authors or texts discussed by an academic publication? I have already suggested the basis for one metric: the number of preexisting publications that contain a particular name and another name, relative to other names. Examining actual citations, rather than mere in-text references, would be more reliable. Based on the total occurrences of item A, the total occurrences of item B, and the number of co-occurrences of items A and B, *pointwise mutual information* (PMI) describes the strength of the association between items A and B. Calculating PMI scores has been used to surface "serendipitous" combinations within datasets—that is, those with lower PMI values (Jenders et al.).

However, one might wish to exclude those papers that are substantially about author/text A but only tangentially refer to author/text B, or even those that discuss both but do not truly "put them into conversation." Computationally identifying this sort of discursive entanglement is a harder problem. One approach would be to make sure that a publication contains both names a certain number of times within a span of a certain number of characters. Thus, a paper's Matchmaker Score could be operationalized as the minimum PMI between any two authors or texts within a publication (*minimum PMI* here meaning something like "least typically associated"), ignoring those pairs that do not seem to be truly discussed in tandem.

PERSISTENCE SCORE

Is it not better to love deeply, or to worry deeply, than to flit from fixation to fixation? Although the rhythms of scholarly publication seem to be ever-hastening, the academic career still allows one to return to a problem again and again over years or even decades, perhaps never quite satisfied with any solution. To measure this sort of relentless attention to a text or question, we might define a *Persistence Score*. A naive approach to defining such a metric is straightforward enough: a scholar who discusses a certain idea or text—for instance, Anne Bradstreet's poem "Contemplations"—in three papers would earn a higher Persistence Score than a peer who discusses this poem in only one paper. Yet there are important decisions to be made: What if scholar A writes about Bradstreet's poem in five papers over seven years, and scholar B in three papers over fifteen years? Which is the more persistent? The calculation of a Persistence Score could require separate coefficients for the *number of publications referencing* and the *number of days between the first and last references*.

Simply counting repeating references would provide no way of distinguishing between the true target of this proposed metric—the scholar who really does return to Bradstreet's poem in different seasons of life, struggling to see it afresh—from the scholar who keeps handy a clever aperçu, repeating it more or less unchanged whenever it seems fitting. Telling apart these two scholars would require something more complicated: plagiarism detection. For an algorithm to detect plagiarism, it must deal with the fact that plagiarists often make superficial changes (e.g., substituting a word for its synonym) to hide the offending act from a mere string search. The scholar who reheats their own reading of Bradstreet's poem may indeed introduce subtle changes. A typical approach within the field of automatic plagiarism detection is to estimate the semantic distance between two chunks of text using techniques that do not depend on the chunks sharing any of the same words; these include latent semantic indexing (or latent semantic analysis) and word embeddings (Foltýnek, Meuschke, and Gipp, 19–20). These same techniques could be used not just to exclude subsequent readings of Bradstreet's poem that are too similar to one of the author's previous readings but also to reward those recent readings that are strikingly different from those earlier ones. That would be the sign of an author who is stretching to make sense of the familiar text from an unfamiliar vantage.

CONTRARINESS SCORE

A *Contrariness Score* would help identify those scholars who routinely disagree with opinions that their peers laud, or at least take for granted. An essential prerequisite of such a metric is the ability to distinguish between positive ("As Malabou lucidly argues...") versus negative ("What Malabou's account misunderstands...") versus neutral ("For a different opinion, see Malabou (1994)..."). Fortunately, researchers in natural language processing (NLP) have made significant progress in the task of sentiment analysis for citations (Yousif, 2019), and large language models (LLMs) could also be well suited to this task.[4] It might make sense for the Contrariness Score to consider not just a citation's emotional polarity and intensity but also its duration—a barbed quip encased in parentheses versus an extended critique. *Intensity of sentiment about cited text* and *percentage of publication devoted to cited text* could each have its own coefficient. A paper could be given a *Contrarianism Point* for railing on a text that most other scholars tend to discuss in a positive or neutral manner. A scholar's Contrariness Score might be the total number of points accrued across all their publications.

CONCEPTUAL SIMPLICITY SCORE

A common (and indeed at this point quite exhausted) stereotype of certain veins of academic writing is that it is made nearly indecipherable by the thick impasto of jargon, with terms such as *the abject* and *le Nom-du-Père* daubed on every sentence, never to be explained or returned to. Humanists might want to highlight those works that do the opposite. By *opposite* here, I do not mean a piece of writing that is totally free of all terms of art, but rather one that takes a single such term, or perhaps two or three, and returns to the term or terms again and again so that, one hopes, they become that much clearer. The first step toward calculating such a *Conceptual Simplicity Score* would be to surface those words from a publication that could fairly be deemed jargon. One might begin by looking for those that are contextually marked as academic argot:

> ... what Lacan refers to as *"le Nom-du-Père"* ...
> ... reworking Kristeva's notion of *"the abject"* ...

Or one might also attempt a more statistical approach, considering as jargon those words (or brief sequences of words) that (1) appear very infrequently in a general corpus of English and (2) appear preponderantly in one academic subfield relative to academic writing in general. To achieve a high Conceptual Simplicity Score, a paper would need to use one or a few pieces of jargon multiple times throughout a text while also avoiding or minimizing the use of any other jargon. A paper's score could be increased should it include a linguistic marker that suggests a striving for clarity:

> ... Lacan's concept of *"le Nom-du-Père,"* by which he means ...

From Sketches to Infrastructure?

Just as citation counts determine what articles surface most readily on Google Scholar, subjective functions could be integrated into infrastructures of thought that allow more sophisticated and humane ways of navigating the vast archive of scholarship. For instance:

- Confused by a piece of jargon, a student or newcomer to a subfield could search for that esoteric term and filter results based on Conceptual Simplicity Score, assuming that a high-scoring paper could serve as a not-too-steep on-ramp.
- Constructing an introductory syllabus on a given topic, one might search for relevant papers that possess both very low and very high Contrariness Scores.
- A senior scholar might prioritize new publications with high Matchmaker Scores, thinking that these are the most likely to cast an unfamiliar light on a familiar field.

But subjective functions, if embedded into scholarly infrastructure, would also help scholars to better understand their own styles of thinking and to better

Subjective Functions [295]

represent these styles to their peers (as well as to administrators and other evaluators). Subjective functions might be translated into badgelike visualizations that make different scholarly dispositions easily apperceptible. What follows are a few sketches of such visualizations.

APOSEMATIC PORCUPINE

Figures 19.1 and 19.2 present a visualization of a scholar's Contrarianism Points. Each quill represents one such point, with the length of each representing the intensity of the critique.[5] A widely praised text with which the scholar has frequently taken issue can be assigned the same color—or value, in the case of grayscale.[6] Compared to Figure 19.1, Figure 19.2 indicates fewer Contrarianism Points, each earned by attacking a different target. The shorter quills suggest that these critiques are less intense.

TIME LOCKET

Figures 19.3 and 19.4 offer a simple way of depicting the Persistence Score (or, at least, some version of it). Time is represented as a circle and publications as points on that circle. A shape is created by connecting adjacent points with chords; this shape gives a rough but immediate sense of how much of the scholar's career has been spent turning and returning to the text in question. In this case, the area of the shape within Figure 19.3 is larger than the one within Figure 19.4, indicating greater persistence.[7]

GRADIENT GOBLET

Figures 19.5 and 19.6 illustrate the Matchmaker Score. Each of a scholar's top ten papers in terms of this score is depicted as a horizontal line between two black

Figure 19.1. An "aposematic porcupine" illustrating contrarianism.

Figure 19.2. Another porcupine, this one less aggressive.

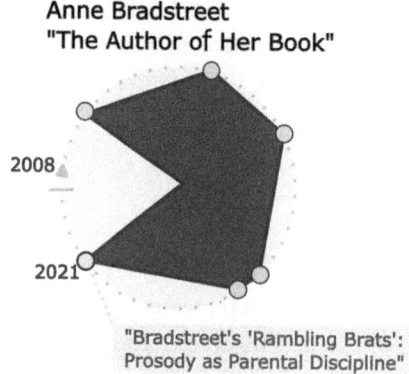

Figure 19.3. A "time locket" visualization of an author's references to a text over time. The author's most recent reference has been selected and the title of the text revealed.

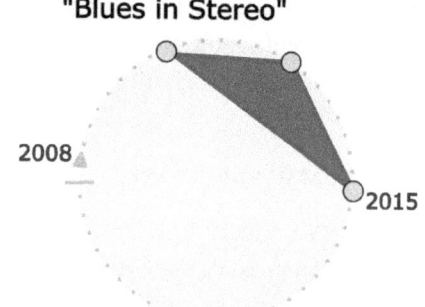

Figure 19.4. Another locket.

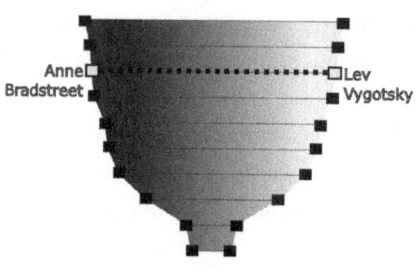

Figure 19.5. A "Gradient Goblet" for Matchmaker Scores. The third-highest score has been highlighted and its authors listed, flanking the goblet.

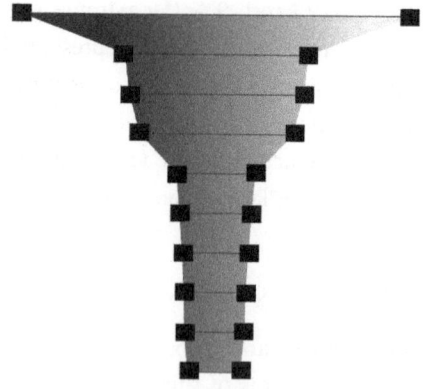

Figure 19.6. A slenderer goblet.

squares; higher scores are reflected by longer lines. These lines are placed in descending order over a field with a horizontal gradient so that color emphasizes the distance or nearness between two points. Two very distant points will cross from bright red to bright green (or, in grayscale representations, from dark to light gray). Figure 19.6, a slender vase (cast mostly in tepid ochre or, sans color, middling pewter), would represent the publications of a scholar who is not much of a matchmaker.

Integrated into a Google Scholar–like platform, these sorts of miniature visualizations could be useful for comparing scholars within a field and noticing the specific *charisms* of those who may not rank as highly according to traditional metrics (Figure 19.7).

Metrics very easily become goals, and so even subjective functions would lead to the pursuit of "subjective objectives." Unlike h-index and other impact-based metrics, subjective functions could also support multiple, conflicting behaviors:

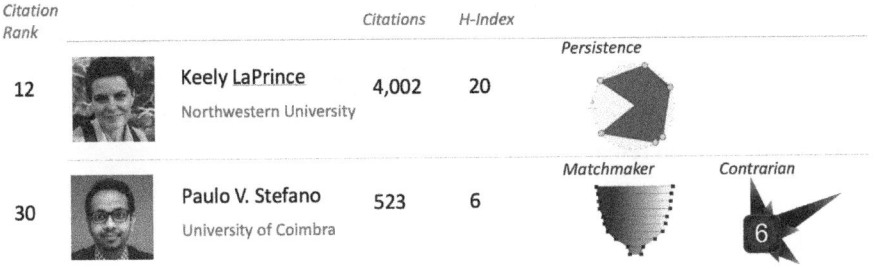

Figure 19.7. Comparing scholars beyond typical metrics. In the two rightmost columns, each scholar has selected one or two optional visualizations to represent themselves. (Image inspired by the way scholars are compared on Research.com. The two scholars here are fabricated, and their images are generated by artificial intelligence [AI].)

a self-styled contrarian might become even more contrary, while another person might experience their high Contrariness Score as a discomfiting glance into the mirror. Like any metric, subjective functions might still feel alienating and stifling—especially if used clumsily by institutions to evaluate faculty or to allocate resources (see Strathern).[8] Likewise, it would be important to attend to the aspects of humanistic thought that metrics—*any* metrics—would be powerless to capture.[9] Still, just as machine reading can reveal surprising facts about a corpus of novels without exhausting their meaning, personalized subjective metrics could serve as starting points for understanding one's own oeuvre. More strategically, subjective functions could come in handy should a researcher, a department, a field, or even a university need to push back against even more ill-fitting metrics.

WE NEED IT ALL

To compute these or other subjective functions, one would want to have access to nothing less than the full texts of as many terabytes of scholarship as possible, yet efforts to rethink humanistic metrics suffer from "a lack of machine-readable data with permissive licensing" (Konkiel, 5). In theory, a researcher with a sufficiently capacious hard drive (and a sufficiently relaxed approach to questions of copyright) could begin by downloading a cache of papers and books from Library Genesis (LibGen) or another of the "shadow libraries" discussed by Martin Paul Eve in chapter 6 of this book. But running calculations locally is one thing, building infrastructure quite another. Humanists would need to make these metrics part of a *public* system of information retrieval, an alternative Google Scholar where one could search and be searched, see and be seen, according to a toggleable menu of diverse metrics.

The subjective functions sketched in this chapter represent only four points in a large and underexplored design space. None of the four attends to questions of scholarly ethics, such as "citational justice" (Kwon). None directly attempts to measure anything like erudition. Scholars should be able to design their own functions to represent their own values, and others should be able to discuss and critique these functions. Imagining, arguing about, and building this extensible algorithmic infrastructure-for-thought should be a core metaproject of the humanities. The alternative is to remain "objectified."

NOTES

1. As a tooltip on any scholar's Google Scholar profile explains, this is "the number of publications with at least 10 citations."

2. This is another definition provided via tooltip on a scholar's profile; see Hirsch for the original definition of and rationale for this metric.

3. I am participating here in conversations about how metrics can be reimagined to better fit humanistic research. For instance, the HuMetricsHSS initiative has begun to imagine metrics that reflect the "shared values" of the humanities (Konkiel; see also Long). My goal is to consider how metrics could reflect values that may be largely unshared, if not totally idiosyncratic.

4. Zhang et al. found that LLMs (with some exceptions) performed competently on a variety of sentiment analysis tasks, though not always as well as smaller models trained specifically for these tasks.

5. This design is inspired by the Altmetric Attention Score "donut," which represents an individual published work's references on social media, the news, and elsewhere (Altmetric).

6. All badges were originally designed with color in mind. Where possible, these images have been adapted into grayscale for print.

7. In Figure 19.3, the first and last points representing references connect not to each other but to a central point; otherwise (if the first and last points were connected), the resulting shape would cover time prior to the first reference.

8. A funding body might bully a department or researcher into striving for a lower (or higher) Contrariness Score, for instance, regardless of whether this goal is at all meaningful from an emic/insider's perspective. On the other hand, subjective functions would be easier to hack than traditional metrics; a disingenuous scholar could larder their books with half-hearted ripostes.

9. As Eileen A. Joy (26) asked provocatively in a manifesto against metrics, "How do you measure the time spent not doing anything at all in order to open a space for thinking differently?"

BIBLIOGRAPHY

Altmetric. "The Donut and Altmetric Attention Score." Accessed September 20, 2023. https://www.altmetric.com/about-us/our-data/donut-and-altmetric-attention-score/.

Cottle, Richard W. "Objective Function." In *International Encyclopedia of the Social Sciences, 2nd Edition,* vol. 16, edited by William A. Darity Jr. (2008): 5–6.

Foltýnek, Tomáš, Norman Meuschke, and Bela Gipp. "Academic Plagiarism Detection: A Systematic Literature Review." *ACM Computing Surveys,* 52, no. 6 (2019): 1–42.

"Google Scholar Metrics." Google Scholar. Accessed August 20, 2022, https://scholar.google.com/intl/en/scholar/metrics.html#metrics.

Hirsch, Jorge E. "An Index to Quantify an Individual's Scientific Research Output." *Proceedings of the National Academy of Sciences* 102, no. 46 (2005): 16569–72.

Jenders, Maximilian, T. Lindhauer, Gjergji Kasneci, et al. "A Serendipity Model for News Recommendation." In *Joint German/Austrian Conference on Artificial Intelligence (Künstliche Intelligenz),* 111–23. Springer, 2015.

Joy, Eileen A. "'An Instrument for Adoration': A Mini-Manifesto *Against* Metrics for the Humanities (to Be Elaborated Upon at a Further Date)." In *Humane Metrics/Metrics Noir,* edited by meson press, 26–33. Post Office Press, Rope Press, and meson press.

Konkiel, Stacy. "Approaches to Creating 'Humane' Research Evaluation Metrics for the Humanities." *Insights* 31 (2018).

Kwon, Diana. "The Rise of Citational Justice: How Scholars Are Making References Fairer." *Nature.* March 22, 2022. https://www.nature.com/articles/d41586-022-00793-1.

Li, Jiwei, Michel Galley, Chris Brockett, et al. "A Diversity-Promoting Objective Function for Neural Conversation Models." In *Proceedings of North American Chapter of the Association for Computational Linguistics: Human Language Technologies,* 110–19. ACL, 2016.

Long, Christopher P. "Toxicity, Metrics, and Academic Life." In *Humane Metrics/Metrics Noir,* edited by meson press, 14–25. Post Office Press, Rope Press, and meson press.

Piwowar, Heather. "Introduction: Altmetrics: What, Why and Where?" *Bulletin of the American Society for Information Science and Technology* 39, no. 4 (2013): 8–9.

Strathern, Marilyn. "'Improving Ratings'; Audit in the British University System." *European Review,* 5, no. 3 (1997), 305–21.

Van Noorden, Richard, and Dalmeet Singh Chawla. "Hundreds of Extreme Self-Citing Scientists Revealed in New Database." *Nature.* August 19, 2019. https://www.nature.com/articles/d41586-019-02479-7.

Wood, Marshall K., and George B. Dantzig. "Programming of Interdependent Activities: I: General Discussion." *Econometrica, Journal of the Econometric Society* (1949): 193–99.

Yousif, Abdallah, Zhendong Niu, John K. Tarus, et al. "A Survey on Sentiment Analysis of Scientific Citations." *Artificial Intelligence Review,* 52, no. 3 (2019): 1805–38.

Zhang, Wenxuan, Yue Deng, Bing Liu, et al. "Sentiment Analysis in the Era of Language Models: A Reality Check." May 24, 2023. https://arxiv.org/abs/2305.15005v1 (preprint).

Appendix

Infrastructure Manifests

ALAN LIU, URSZULA PAWLICKA-DEGER,
AND JAMES SMITHIES, EDITORS

The infrastructure manifests included for the chapters in this volume are an experiment to explore what would be required for those working in the digital humanities (DH) or related areas to identify the principal infrastructures used for their professional work. (See what follows for a list of manifest files and their template file.) Each manifest represents the infrastructure used by one or more of the authors of each chapter in researching, preparing materials for, and producing their contributions to this book. However, two of the manifests—for chapters 2 and 8—represent additionally or instead the infrastructures of the platforms, services, or tools that they discuss as topics, thus showing how the manifest concept can be used generally to document infrastructure.

In the etymological sense discussed by Matthew Hockenberry at the opening of his "Manifesting Connection" (chapter 4), "manifests" are about making "something 'evident' or 'palpable.'" But unlike the Early Modern shipping manifests that Hockenberry begins by citing (e.g., those defined in the *Merchant's Magazine* in 1697 as "Transcript[s] of a Master of a Ship's Cargo, showing what is due to him for Freight from each person to whom the Goods in his Ship belong"), the infrastructure manifests in this book are intended to make evident both the payload of scholar*ship* (as it were) and the nature and cost of everything that from beginning to end does the loading, conveying, unloading, and anything else—that is, the infrastructures for extracting, shaping, mixing, connecting, supporting, framing, storing, transporting, labeling, and otherwise making, structuring, and communicating DH scholarship. After all, the premise of *Critical Infrastructure Studies and Digital Humanities* is that, however abstract scholarship may appear when it finally arrives on the receiving dock, it is always also a matter of actual materials, artifacts, organizations, and embodied intellectual and physical labor—everything and everyone needed, for instance, to create the physical copy of the present book and the online version of the same book.

In its present version (1.0), the template that we created for the infrastructure manifests in *Critical Infrastructure Studies and Digital Humanities* is limited in form and purpose. We designed it in consultation with authors in our volume just to start learning what such manifests should be. The template, which we asked at least one creator of each chapter to fill out as a spreadsheet, is prestructured with suggestions for some kinds of infrastructure to consider. There are rows for infrastructures under the following labels: "land," "materials," "energy," "transportation," "architectural," "civic, community, national, or regional," "institutional," "labor," "research-content," "tools," "networked platforms," and "high-performance computing." But we left the door open for individual authors to add other categories and to use free-form text in naming and describing infrastructure. We also sought primarily qualitative information and only some approximate quantitative data. Columns in the template ask for a description of the infrastructure that is the subject of each row, and they also inquire about attributes of that infrastructure by means of true/false check boxes or of simple quantification measures on a Likert scale. (True/false attributes include "control point for access to or use," "open source, open access, or public resource/utility," and "proprietary." Attributes measured on a 1 to 10 Likert scale include "how heavily used?," "reliability," and "satisfaction for user.") Other columns allow authors to register "ethical concerns" and add "comments."

In regard to their format, our infrastructure manifests are presently implemented as spreadsheets (originally Google Sheets) that for publication were transformed into downloadable, self-contained HTML files (and, for the blank template, also a Microsoft Excel file). In principle, however, manifests could—and should—be implemented in alternative input and output formats. After all, large spreadsheets are difficult to navigate, view synoptically, and reproduce in different page or screen sizes. Ideally, infrastructure manifests in the future will appear in multiple forms supported by a variety of portability, presentational, and archival data formats—including minimalist ones such as comma-separated values (CSV) or JavaScript Object Notation (JSON) files of the sort commonly used to transport lightly structured data between systems and formats. In the future, too, the information in infrastructure manifests could be implemented in markup languages and ontologies for linked data.

Loosely flexible in their categories and vocabularies, and provisional in their format, these version 1.0 infrastructure manifests essentially represent just a preliminary scoping exercise. They probe the authors of chapters—including the editors themselves as the authors of the introduction and this appendix—about what they feel is within scope to declare as their principal infrastructures and how best to describe such infrastructures.

Why should there be infrastructure manifests for DH? In the spirit of Susan Leigh Star's well-known dictum that infrastructure is normally "invisible" (382), our general goal is to make visible the infrastructure of knowledge production in a field where such infrastructure, as we argue in our introduction, is a constitutive "object of study." More specifically, we envision five discrete purposes for infrastructure manifests in DH, only the first of which is intended to be adequately addressed by the present version 1.0 manifests. We hope that the other four purposes will be addressed by others in the DH community in future iterations of manifests incorporating lessons from our present experiment. The five goals of the infrastructure manifests that we set as an agenda for DH are as follows:

1. To make visible the approximate scope, variety, and proportions of the kinds of infrastructure involved in knowledge work
2. To quantify some kinds of infrastructure
3. To reveal typical platform dependencies (and other dependencies), areas of scarcity, and areas subject to single (or dominant) control points of provision or regulation
4. To make visible geopolitical and sociocultural differences in infrastructure
5. To assess infrastructure comparatively on criteria of ethical, sociopolitical, and economic values, costs, and harm

We hope that publishing our infrastructural manifests will prompt others in the DH community to accompany their works with similar manifests; that our version 1.0 template will be evolved or "forked" for other research communities or purposes; and that DH might thus eventually move toward a shared understanding of, or productive debates about, the most important categories, attributes, quantities, and values (intellectual, sociopolitical, economic, environmental, cultural, ethical, and otherwise) of infrastructure to declare. Supplemented by metadata, quantitative data, and markup, future versions of manifests might also be machine-readable to allow for computationally assisted aggregation, analysis, and visualization. If DH is known in part for "distant reading," then the infrastructures that enable such methods should themselves be available for distant reading.

Being able to machine-process manifests would help overcome one limitation of the present infrastructure manifests that we acknowledge. Most of our manifests represent the infrastructure just of individual authors. In the case of chapters with plural coauthors, only some manifests (for chapters 5, 8, and 12) attempt to represent the collective infrastructure of those coauthors. This is because in practice, it can be difficult to combine authors' infrastructures in a single manifest without resulting in a hard-to-understand miscellany. Such combinations would be more meaningful if we could aggregate hundreds or thousands of authors to find statistically meaningful infrastructural patterns.

It is also important to note that the norm for the content of our infrastructure manifests is professional. Manifests are intended to declare the kinds of infrastructure that the DH scholarly community (and related scholarly and industry research and development communities) thinks can and should be made transparent under citable open-science, accessibility, accountability, provenance, reproducibility, collaboration, and other principles or protocols. Examples of such evolving norms include the following:

- Collaborators' Bill of Rights (Tanya E. Clement, Douglas Reside, Brian Croxall, et al.)
- Datasheets for Datasets (Gebru, Morgenstern, Vecchione, et al.)
- Data Provenance Standards (Data & Trust Alliance)
- Data Transparency Standard (IAB Tech Lab)
- FAIR—Findable, Accessible, Interoperable, Reusable (ALLEA)
- Fair Cite
- FAT—Fairness, Accountability, and Transparency in Machine Learning (FAT ML)
- Open Science Framework (Center for Open Science)

Excluded under the norms of the scholarly profession (and many other professions), therefore, is any expectation that special personal, medical, familial, or other private circumstances bearing on a researcher's needs for infrastructure should be declared, although authors are free to include such information if they wish. Privacy norms also bear on what should, or by regulation can, be made public about scholars' infrastructures and that of their collaborators or students. As Star notably observed, infrastructure is socially "relational" (380). Infrastructure manifests are not a one-size-fits-all lading list of goods. They are in part also social contracts that negotiate relationally between the goals of professionally shared knowledge and the needs, rights, and protections of individuals and organizations at different levels of advantage in the profession and society at large.

Access to Infrastructure Manifests (and Template)

Infrastructure manifests created by the authors and editors of this book are online as downloadable HTML files that can be opened locally in a browser (https://dhde bates.gc.cuny.edu/projects/critical-infrastructure-studies-and-digital-humanities /resource-collection/infrastructure-manifests). The template for these infrastructure manifests, which was originally a Google Sheet, is included for download as both a Microsoft Excel file and a PDF file. All the chapters in this book are represented, although chapters with multiple coauthors do not always have manifests from all the authors. The list of available infrastructure manifests is as follows, where author names identify who filled out a manifest and "et al." indicates that the manifest reports collectively on the infrastructure of multiple coauthors. An asterisk * indicates that the manifest documents the infrastructure of projects discussed in a chapter in addition to, or instead of, the infrastructure used by the author to write the chapter.

- Template for Infrastructure Manifest by v1.0
- Introduction (Infrastructure Manifest by Liu)
- Introduction (Infrastructure Manifest by Smithies)
- Chapter1 (Infrastructure Manifest by Beaulieu)
- Chapter2 (Infrastructure Manifest by Brown*)
- Chapter3 (Infrastructure Manifest by Montoya)
- Chapter4 (Infrastructure Manifest by Hockenberry)
- Chapter5 (Infrastructure Manifest by Pawlicka-Deger et al.)
- Chapter6 (Infrastructure Manifest by Eve)
- Chapter7 (Infrastructure Manifest by Cha)
- Chapter8 (Infrastructure Manifest by Gomez et al.*)
- Chapter9 (Infrastructure Manifest by Spence)
- Chapter10 (Infrastructure Manifest by Dodd)
- Chapter10 (Infrastructure Manifest by Parmer)
- Chapter11 (Infrastructure Manifest by Chen)
- Chapter12 (Infrastructure Manifest by Borda et al.)
- Chapter13 (Infrastructure Manifest by Wershler)
- Chapter14 (Infrastructure Manifest by Caranto Morford)
- Chapter14 (Infrastructure Manifest by Jacob)
- Chapter14 (Infrastructure Manifest by Patel)
- Chapter15 (Infrastructure Manifest by Kolb)
- Chapter16 (Infrastructure Manifest by Naidu)
- Chapter16 (Infrastructure Manifest by Sneha)
- Chapter17 (Infrastructure Manifest by Hannah)
- Chapter18 (Infrastructure Manifest by Miller Sylvia)
- Chapter19 (Infrastructure Manifest by Booten)

Some infrastructure manifests share information and sources. Examples include identical or similar descriptions and statistics about the materials or energy consumption of laptops, cars, or planes. In the future, the DH field could create a shared library of information, statistics, and sources as references for infrastructure manifests so that everyone need not reinvent the same wheel.

BIBLIOGRAPHY

ALLEA. "Sustainable and FAIR Data Sharing in the Humanities: Recommendations of the ALLEA Working Group E-Humanities." Edited by Natalie Harrower, Beat Immenhauser, Maciej Maryl, and Timea Biro, 2020. https://doi.org/10.7486/DRI.TQ582C863.

Center for Open Science. "Open Science Framework (OSF)," 2024. https://osf.io/.

Clement, Tanya E., Douglas Reside, Brian Croxall, et al. "Collaborators' Bill of Rights." Humanities Commons, 2021. http://dx.doi.org/10.17613/mvar-kj35.

Data & Trust Alliance. "Data Provenance Standards," 2023. https://dataandtrustalliance.org/.

Fair Cite. "Home page," 2012. https://faircite.wordpress.com/.

FAT ML. n.d. "Fairness, Accountability, and Transparency in Machine Learning." https://www.fatml.org/.

Gebru, Timnit, Jamie Morgenstern, Briana Vecchione, et al. "Datasheets for Datasets." arXiv:1803.09010 [Cs], 2019. http://arxiv.org/abs/1803.09010.

IAB Tech Lab. "Data Transparency Standard (v. 1.1)," 2020. https://iabtechlab.com/standards/data-transparency/.

Star, Susan Leigh. "The Ethnography of Infrastructure." *American Behavioral Scientist* 43, no. 3 (1999): 377–91. https://doi.org/10.1177/00027649921955326.

Contributors

ANNE BEAULIEU holds the Aletta Jacobs Chair of Knowledge Infrastructures at the University of Groningen, the Netherlands, and is the author of *Revealing Relations: Knowledge Infrastructures for Liveable Futures* (2026).

KYLE BOOTEN is assistant professor in the Department of English at the University of Connecticut, Storrs.

ANN BORDA is ethics fellow at the Alan Turing Institute, London, associate professor in the Melbourne School of Population and Global Health at the University of Melbourne, and honorary senior research associate in the Department of Information Studies, University College London.

SUSAN BROWN is Canada Research Chair in Collaborative Digital Scholarship and professor of English at the University of Guelph. She codirects the Orlando Project and directs the Canadian Writing Research Collaboratory and the Linked Infrastructure for Networked Cultural Scholarship.

TOBY BURROWS is senior honorary research fellow at the University of Western Australia. His recent projects include *DigiSpec: Scoping Future Born-Digital Data Services for the Arts and Humanities* and *Mobilising Dutch East India Company Collections for New Global Stories*.

ASHLEY CARANTO MORFORD is assistant professor of multiethnic American literature in the Department of English at Weber State University.

JAVIER CHA is assistant professor of digital humanities in the Department of History at the University of Hong Kong. He specializes in the integration of graph database technology in historical scholarship and engages in experimental projects that address the challenges posed by big data and artificial intelligence in the humanities.

JING CHEN is associate professor in the School of Art, Nanjing University.

ARIANNA CIULA is director and research software senior analyst at King's Digital Lab, King's College London. She collaborates in many digital humanities projects and is a coauthor of *Modelling Between Digital and Humanities: Thinking in Practice*.

Contributors

MAYA DODD is associate professor in the Department of Humanities at FLAME University, Pune, India. She is founding member of DHARTI, India, and director of the Milli Archives Foundation.

MARTIN PAUL EVE is professor of literature, technology, and publishing at Birkbeck, University of London.

ALLAN GOMEZ is the systems administrator for Philly Community Wireless. In his past life, Allan built low-power FM radio stations throughout Latin America and the United States.

MATTHEW N. HANNAH is associate professor of digital humanities in the School of Information Studies at Purdue University.

MATTHEW HOCKENBERRY is assistant professor of communication and media studies at Fordham University. His work focuses on the intersection of media, logistics, and global supply chains, past and present.

ARUN JACOB is doctoral candidate at the Faculty of Information, University of Toronto.

MIKE JONES is postdoctoral fellow of Indigenous and colonial histories at the University of Tasmania. He is the author of *Artefacts, Archives, and Documentation in the Relational Museum*.

LUCIE KOLB is professor of critical publishing and head of the MAKE/SENSE PhD program at the Basel Academy of Art and Design FHNW.

ALAN LIU is Distinguished Professor in the English Department at the University of California, Santa Barbara. Since he started the Voice of the Shuttle website in 1994 (https://liu.english.ucsb.edu/voice-of-the-shuttle-vos/), his books, essays, and projects have focused on the digital humanities and public humanities. His most recent book is *Friending the Past: The Sense of History in the Digital Age*, and recent projects are 4Humanities.org, WhatEvery1Says, and the Center for Humanities Communication.

IAN M. MILLER is associate professor in history at St. John's University, New York. His research interests are in the long-term interplay between changing ideas, changing institutions, and changing environments, especially in southern and central China, and in the use of digital texts and tools to explore new methods for writing history.

SYLVIA K. MILLER is director of scholarly publishing and research development at Duke University's John Hope Franklin Humanities Institute. She is the former director of the Mellon-funded Publishing the Long Civil Rights Movement project at the University of North Carolina at Chapel Hill, and she managed two international collaborative Mellon grants involving nineteen universities on five continents for the Consortium of Humanities Centers and Institutes. During twenty years in scholarly reference publishing in New York, she was executive editor at Scribner and publishing director at Routledge.

SARAH MONTOYA earned a PhD in gender studies at the University of California, Los Angeles. She is a Mellon Humanities Postdoctoral Fellow for the National Park Service.

SAUMYAA NAIDU is an independent researcher working on research that critically examines the role of design in digital technology, specifically in areas such as privacy, accessibility, and digital identities. Her areas of interest include design studies and digital cultures.

SHARIKA PARMAR is doctoral candidate in the Department of Humanities at FLAME University, Pune, India. Her research areas include modern South Asia, digital humanities, memory studies, and migration studies.

KUSH PATEL is associate professor of Contemporary Art Practice Master of Arts program at Srishti Manipal Institute of Art, Design, and Technology, Manipal Academy of Higher Education, Bengaluru, where they also lead and steward a digital humanities research and pedagogy space called the Just Futures Co-lab.

URSZULA PAWLICKA-DEGER is research manager in the Discovery Research program at Wellcome Trust. A former Marie Curie Fellow at King's College London, she is coauthor of *Digital Humanities and Laboratories: Perspectives on Knowledge, Infrastructure and Culture*.

MIRIAM POSNER is assistant professor in the Department of Information Studies at the University of California, Los Angeles.

JAMES SMITHIES is professor of digital humanities and director of the HASS Digital Research Hub at Australian National University, and author of *The Digital Humanities and the Digital Modern* (2017). He was previously professor of digital humanities and founding director of King's Digital Lab at King's College London. He has also worked at the University of Canterbury in Aotearoa/New Zealand, and in the government and commercial IT sectors in Aotearoa/New Zealand and the United Kingdom.

PUTHIYA PURAYIL SNEHA is research manager with the Open Knowledge Initiatives team at the International Institute of Information Technology, Hyderabad. Her areas of work and interest include digital media and cultures, open knowledge, arts and humanities scholarship and pedagogy, gender, and digital rights.

PAUL SPENCE is reader in digital humanities in the Department of Digital Humanities at King's College London. He is coeditor of *Multilingual Digital Humanities* and section coeditor for the digital modern languages section on Modern Languages Open.

Contributors

LIK HANG TSUI is associate professor in the Department of Chinese and History of the City University of Hong Kong.

DEB VERHOEVEN is professor and Canada 150 Research Chair in Gender and Cultural Informatics at the University of Alberta and founding director of the Humanities Networked Infrastructure (HuNI) Project.

MIGUEL VIEIRA is principal research software engineer at King's Digital Lab, King's College London.

DEVREN WASHINGTON is cofounder of Philly Community Wireless and senior policy organizer of the Movement Alliance Project, Philadelphia.

ALEX WERMER-COLAN is cofounder, long-term volunteer, and executive director of Philly Community Wireless. He also works as the academic and research director of the Loretta C. Duckworth Scholars Studio at Temple University Libraries, and serves as the managing editor of the Programming Historian in English.

DARREN WERSHLER is professor at Concordia University, founder and director of the Residual Media Depot, and associate director of the Milieux Institute. His most recent book, with Jussi Parikka and Lori Emerson, is *The Lab Book: Situated Practices in Media Studies* (Minnesota, 2021).

GRANT WYTHOFF directs the graduate program of the Center for Digital Humanities at Princeton University. His most recent book is *A User's Guide to the Age of Tech* (Minnesota, 2025). A cofounder of Philly Community Wireless, he now serves on its board of directors.